Praise fc

In this important book, Barney Wee and Agnes Lau have blended curren scientific information with ancient spiritual knowledge to guide us in ways to heal ourselves and our planet—now, rather than in some far-off future. I highly recommend that everyone read this work of profound insights into how the choices we make personally, can have a critical impact on the future of our planet.

-Dr. Irv S. Katz, Chancellor, International University of Professional Studies

If you want to know what's really going on at this moment in history, and what you can do to help create a better life and a more beautiful world, read this book. If enough of us follow its guidance, there is a real hope we can turn the tide and create a spiritually fulfilling and earth-friendly human presence on this planet.

-John Robbins, social activist and humanitarian,
Author of the bestselling The Food Revolution, *Recipient of the Albert Schweitzer Humanitarian & the Peace Abbey's Courage of Conscience Awards*

This book moved me. It moved me to be more aware, more conscious, more intentional. It so well captures the leading edge thinking in the morphic field of evolutionary consciousness. We must take heed of what these authors are saying. This is an important book.

-Dr. Margaret Smith (Harvard), former Lecturer of Claremont Colleges & Drucker Business School, Author of Nine Design Principles for Collective Intelligence & Prosperity

We live in a time of profound change and transformation. Living between paradigms, many of us are experiencing the world shifting under our feet—and we are struggling to maintain our balance. In this process, we are being called to imagine a new future for our collective wellbeing. In this important new book, Wee and Lau offer us a sensitive and insightful roadmap for navigating all aspects of our life and work. Taking a broad scope, they bridge science and spirituality in a way that offers each of us a glimpse into our hidden capacities. Theirs is a call to help us awaken to a more nourishing and sustainable world. Highly recommended.

-Marilyn Schlitz, PhD, Former CEO of IONS,
CEO of Worldview Enterprises, Co-Author of Living Deeply,
Senior Scientist of California Pacific Medical Centre

What are the Choices of Now? How do we create sustainable solutions? I want to congratulate you for choosing this book to reinvent the way you look at the challenges of our world. What could be a chain of lifetime discoveries is simplified here. A framework of noble values and knowledge is conveyed to make world-saving changes easy and doable, by anyone—right away!

-Dr. Aida N. Karazhanova, Economic Affairs Officer, Environmental & Development Division, United Nations Economic & Social Commission for Asia & the Pacific

Choices of Now **offers a thorough review of the world's vulnerable ecological and societal conditions. It's an impressive work displaying considerable scholarly research and offers well-considered and scientifically supported suggestions for what we can each do to improve humanity's condition. The authors clearly care about their message, delivering a rich and interesting read founded on deep underlying values about our world and our existence.**

-Jon Freeman, M.A. (Oxford), Organisational Mentor, Speaker & Author of The Science of Possibility & Future Money: Evolving Our Relationship with Finance

Barney Wee and Agnes Lau have written a very powerful book that needs to be read. They have taken a very thorough approach and make their case with clarity and intelligence. As our planet moves into an age that will likely be defined by environmental issues, we need a way forward that integrates all kinds of intelligence, including spiritual intelligence. The book offers a way through the greatest challenge humanity have ever faced.

-Dr. Jim Lockard, Spiritual Leader, Centre for Spiritual Living, Author of Sacred Thinking: Awakening to Your Inner Power

It's an encyclopedic review of the current human condition that proposes pragmatic solutions encompassing the physical, mental and spiritual realms. I heartily recommend it.

-Leonard Laskow, MD, Developer of Holoenergetics, Author of Healing with Love

This book inspires and enriches people's lives by helping them to live a healthier, more humane and meaningful life. This in turn hopefully will lead our communities, nations and world towards eco-harmony and eco-consciousness.

-Joyce Lye, Founder of Kampung Senang Charity & Education Foundation, an eco-conscious charity with IPC status

This book is a well-researched and comprehensive body of art that is both inspiring and thought-provoking. It reinforces how our daily choices make us irresponsible or responsible human beings. I created VeganBurg to inspire and excite the world to choose a plant-based diet as a means of restoring the planet and securing our children's future. ***Choices of Now*** **is undoubtedly the perfect gift for your loved ones and Mother Earth.**

-Alex Tan, Founder of VeganBurg - the world's first vegan burger fast-food joint

History has proven that we need to create and build on incomplete assumptions in order to evolve to a higher plane of existence. From the hunter-gatherer to the agricultural age, the industrial and now the information age, we have been developing our world based on "assumed absolute truths" that we now know were only presuppositions.

At different points in our history, Bacon (empirical scientific method), Descartes (mind-body dualism), Newton (universal laws of motion), and Einstein (general theory of relativity) have framed our views of our world in terms of what was correct, intelligent, or possible.

We now know, thanks to advances in science, that some of their theories were incorrect. Nevertheless, they served their purpose, as humanity has used them to evolve positively.

Curiously, the lesson that we draw from history is that we collectively need to embrace incomplete or "best-guess" assumptions in order to improve our world!

If you look at what is happening socially, politically, economically, and environmentally all around the globe, you can clearly see that holding on to our current assumptions, or for many of us (billions!), walking through life like a zombie can only lead us to a world of strife, conflict, shortage, and destruction.

What this implies is that in order for us to evolve toward a better world, we need to have the audacity to envision and embody new paradigm-breaking assumptions.

This book offers us new life-supportive presuppositions, integrating the best of science and spirituality, along with practical strategies to help us evolve toward a more harmonious, humane, and ecological world—individually and collectively.

-Barney Wee, co-author

Choices of Now

Choices of Now

Urgent Decisions for Co-creating the Future of Our World

Barney Wee & Agnes Lau

Wisdom Moon Publishing
2014

CHOICES OF NOW

URGENT DECISIONS FOR CO-CREATING THE FUTURE OF OUR WORLD

Published by Wisdom Moon Publishing LLC
San Diego, CA, USA

www.WisdomMoonPublishing.com

ISBN 978-1-938459-53-5 (softcover, alk. paper)
ISBN 978-1-938459-55-9 (eBook)
LCCN 2014905488

TABLE OF CONTENTS

Choices of Now

GRATITUDE

There are many people without whom this book would not have come into reality. When I decided to write it, I didn't know what I had signed up for: I thought it would take six months to finish; it turned out to be a four-and-a-half-year research and self-purification project.

First of all, I would like to thank Derek Tang for casually passing me Ervin László's book *Chaos Point*; it planted a splinter in my brain to do something good for our world.

When I started to write I realised the vastness of topics I needed to cover and I enrolled my partner, Agnes Lau, to co-write the book with me. The amount of hours and effort she poured into the book in the midst of her hectic life is—beyond words. The writing wouldn't have been possible without her help.

The time and dedication required to write this book demanded that I be left undisturbed by office matters. So my business operation team, J.T. Kua, Alfred Chung, and Joseph Chng, ran the heavy operational demands of the business by themselves. At home, my able and devoted housekeeper, Mrs. Marina Comilang, took good care of my autistic son so that I could focus on the writing. I'm deeply grateful for their understanding and undying support.

Along the way, when I stumbled into difficulties, there were several generous people who synchronically showed up to assist me with their knowledge and insights. I'm very grateful to Cheryl Pearson (USA), C. Johan Masreliez (Sweden), David An Wei (China), Michael Yu (China), Aruna Ashok (India), Marie-Ann Yong (Malaysia), Sathya Seelan (Malaysia), Yogas Lam (Philippines), Cynthia Wihardja (Indonesia), Gunawan Suryanegara (Indonesia), Teerasak Wongpiya (Thailand), Michael Phao (Thailand), Fabian Lim (Singapore), Lee Peh Gee (Singapore), and Tan Seng Hong (Singapore). They all offered their precious time to answer my questions about disparate topics, from time acceleration, expansion of the Universe, to cultural and political issues around Southeast Asia.

I'm immensely thankful to Sangeet Duchane (USA) and Flavia Pal (Romania) for their editing work. Sangeet gave the book a significant restructuring so that it would flow more smoothly. Flavia ploughed through the book for consistency and integrity and her dedication added new perspectives and richness to it. I feel so abundantly blessed with, and deeply grateful for Flavia's contribution and devotion, not just to the editing process, but to the mission of the book as well.

The message we share in this book incorporates the wisdom and brilliant work of many researchers and pioneers across many fields of science and spirituality. Their quest for answers to explain our complex world, their audacity to challenge the unknown, their relentless pursuit of excellence, and their tireless dedication to bring light into darkness and reduce human suffering makes them co-authors of this book. We give our heartfelt appreciation to Ervin László, Ph.D., Dr. David R. Hawkins, Gregg Braden, Bruce Lipton, Ph.D., Marilyn M. Schlitz and the scientists at the IONS, Daniel Goleman, Ph.D., Dacher Keltner, Ph.D., Dean Radin, Ph.D., Lynne McTaggart, Dr. Leonard Laskow, Peter Russell, Robert Dilts, T. Collin Campbell, Ph.D., Paul Hawken, and John Robbins.

INTRODUCTION

Dear Reader,

As the editor of the book, I have created in my mind a picture of how I would like the future of our common home—our planet—to look. I have then made a commitment to myself and Mother Earth that day after day the smallest of my actions and my thoughts will mirror this image of harmony and abundance. I have made a choice to support life on earth instead of destroying it and it is my wish that this book will do for you what it did for me: draw you nearer to the realisation that there is still hope for the future of mankind as long as we—you and I—change our ways.

It is not by some random accident that you are now holding this book in your hands; you were drawn to it because you too are probably aware of the dire straits humanity is currently in; like me, you are probably hearing the planet's cry for help and are wondering how much further we can push the limits; and the truth is, not much further. If we continue pushing, we will very soon hit the fall-off point and there will be no turning back, in spite of our deepest regrets.

This book has done for me what I never would have had the audacity to do: it has distilled years of research and extensive reading on various topics, including science and spirituality, into this easy to understand, coherent text; it has helped me turn the change I was only fantasising about into an executable possible reality, it has brought it to my doorstep, and has spared me years of being out there in search for meaning, explanations, and connections between the tons and tons of information available.

The intention behind writing this book is to open our eyes and present us the reality we—planetary citizens—are facing today. Also, it is intended to help us draw conclusions and show us that we, as individuals, still have the power to do something about the biggest global problems that we have unwittingly created with our daily lifestyles. And for this, it provides us with a step-by-step action plan with immediately applicable daily actions we can take to start creating a sustainable future for our dying planet. By reading the book you will notice how these intentions are manifested through words and how understandably they are woven together.

The convergence of so many large scale issues highlights that we can no longer turn a blind eye to what is happening around us: the beautiful world that once was is rapidly fading to grey.

The question today is: given the chance to improve our future, are we willing to take it? Hence, the title of the book—*Choices of Now*.

This book is centred around us, homo sapiens, and the ways in which we can harmoniously evolve with the planet's ecosystems toward higher planes of consciousness. The information presented here is complex and often hard to digest; we are not denying that. At the same time, it is the authors' intention to make the reader aware of the simple fact that in order for change to occur, individuals—you and I—must take full responsibility to live in a way that supports rather than depletes life.

It is beyond the power, ability, and resources of any single government, corporation, or scientist to resolve the global problems, the multi-systems collapse we are encountering today. The biological, social, economical, political, and environmental issues we have call for the paradigm shift of every individual.

But let us face the truth here, leaving all masks aside: how many people really care about the planet? The vast majority care about themselves. So unless people can see how their livelihood and the livelihood of the ones they love are being threatened, they will see our planetary problems dissociatively.

People don't care about the world's problems when they don't think it is their responsibility to do something about them.

People won't make an investment of time and energy, unless they understand how things work, and how they can make a difference.

People won't change until they see that they are the ones causing their own pain.

People won't attempt to change when they don't see that there's hope and that there are others who are successfully making positive changes.

People need to see that there *are* intelligent choices, that there *are* groups that will support their change, that there *are* simple and powerful ways to make a positive change impacting the world at large.

So, the ultimate message of the book is *not* to give readers a comprehensive knowledge of climate change or environmental news; it is also *not* to tell people about the latest scientific breakthroughs and dazzle them with leading-edge technologies. These two features of the book are just means to an end. The ultimate message of the book is to tell readers that the way we are living our lives (the way we work, recreate, shop, eat, travel, the way we define success and happiness) is causing strife, chaos, decay, war, and unhappiness. We need a new vision of the kind of world we want to live in.

I am hoping that, after reading this book, you will join the growing life-supportive community in our attempt to start co-creating a new world that our children and future generations can have the privilege to enjoy, a world with food and shelter for everyone, with peace and cooperation between nations. This is our chance to make it possible for future generations to say that at the turn of the 21st century, individuals and nations worked together for the first time to revive our planet and ensure our upward evolution.

The effectiveness of the message of this book depends on our willingness to be part of the solution to our common problems, on our genuine desire to make changes in our current lives. As you progress through the chapters I will ask you to hold in mind that you are presented with real facts and if you happen to feel deeply moved by the information exposed, this is due to your sense of humanity; no external thing or person can "make" you feel the way you do; it comes from within you, from a space of deep understanding and compassion for all living things and from a desire to see happiness and beauty around you, instead of decay and pain.

I invite you on a journey of truth, scientific evidence, and spiritual wisdom at the end of which you are given the key to a door that can open up a whole new beginning: a new good life on earth for everyone.

Flavia Pal
Editor

PROLOGUE

Looking back at my personal voyage with this book, I remember that one of the most shocking environmental news that hit me viscerally was the 2004 Indonesian tsunami. I was shocked and felt overwhelmingly helpless about the millions affected. Soon after, an unshakable feeling of dread took hold of me and had me captive for years. From 2004 on, a slew of disasters raged around the planet: in 2005—hurricane Katrina, in 2006—the Java devastating earthquakes, in 2007—the South Asian floods, in 2008—the Sichuan earthquake. As these catastrophic events were unfolding, I was asking myself, "What is our world becoming?" "Is nature or God trying to tell us something?" These were the questions that spurred me to find answers.

In September 2008, I received a book that literally shook my world: *Chaos Point*, by Ervin László. I remember reading it in a shopping mall, looking at the people around me and wondering how many of them were aware of the fragile state of our planet and that our way of life could be adversely changed in an instant.

The clear sign that sparked my desire to write this book came in February 2009, when I serendipitously found something truly amazing: as I was clearing my storeroom, I dropped a stack of papers on the floor. I couldn't believe my eyes: I stood there, in utter wonderment, looking at them: what caught my eye was my college English paper that I had written about man's choice to ignore the balance of nature, creating a host of ecological disasters; I had written this paper 27 years before! It was as if I knew at the age of 18 that I needed to write an article to encourage my 45-year-old self to commit to a major project in life!

However, I knew that if I didn't change my life routine, I wouldn't be able to write the book. Between running the office, training every weekend, consulting with corporations on weekdays, and coaching individuals, there was simply no space for the book. Furthermore, back in 2009, I knew I was part of our global problems: my family and I were using the air-conditioning, water, electricity, and throwing away food with no consciousness of conserving natural resources. Shopping for things we didn't really need was for us a weekly form of recreation. In these conditions, I knew I was not aligned to the message of the book. So, after months of contemplation and discussions with my partner, we made a bold and scary decision: we were going to sell our office and work from home. We cut down the volume of our business to make time for writing and to actively simplify our lifestyle.

But getting into the new way of living was not easy; we had to deal with our old habits and secondly, and then to pace and lead loved ones into the new lifestyle. Simple things like reducing water flow during dishwashing and bathing, or turning off power points in the home, were a constant struggle. We also had to reinvent our business totally in order to make time for writing; on the one hand it was exciting, but on the other, very painful, as it was against my grain to turn down business coming to me.

And so I began the most challenging project I had ever undertaken; I would spend 5–6 days a week, 10–14 hours a day, glued at my desk, frequently overwhelmed by the information discovered and the laborious task of conveying it in few words. With time, I realised that the problems, revelations, and solutions I was uncovering in the book were mirrored in my life, as they became frameworks and principles guiding my daily actions. This book truly became my teacher.

In my research, I came across many mentors whom I've never met; Ervin László, Bruce Lipton, Gregg Braden, David Hawkins, Lynne McTaggart, Collin Campbell, Duane Elgin, Marilyn Schlitz, and Michio Kaku are some of the luminaries whose wisdom helped me to make sense of the rapid and vast changes in our world and to re-create my life.

Writing this book came with some painful lessons; my serious intention to walk the talk caused severe friction in my close relationships, especially with my partner. Even though she was helping me write certain chapters, I became frustrated with the progress we were making. Moreover, our change of lifestyle caused great discomfort in her, but I was expecting her to follow my lead and adapt quickly. There were many others who did not live up to my expectations of how one should be part of the solution to planetary problems, which naturally triggered conflict and resistance in them. I had come to a point where I felt I was alone and the load of completing the book was entirely on my shoulders. On top of that, my body reacted to these problems through a severe skin condition; it was very hard to bear it all. My dreadful long hours of silence, loneliness, doubt, and unworthiness culminated with a void in my spiritual practice; I felt I had lost my connection with God.

I later discovered that these moments were for me the Dark Night of the Soul: I had to rely on God to help me continue writing the book, and most of all to help me surrender my grief, despair, and remorse. In my passion to do things right, I had ostracised the people that mattered most in my life; and although I had sworn to support life, I was belligerent and unforgiving. (*My story continues in the epilogue.*)

Overview of the Three Main Parts of the Book

***Part I*, Time of Great Change Has Come** explores the elusive concept of *time* and our intimate connection to it through the lenses of science and ancient traditions, delving into the enticing fields of astrology, astronomy, archaeology, and many more. "What is going to happen to our planet and our species? What extraordinary changes will the near future bring?" We discover that we are now at an evolutionary crossroads and we can vote for our future by consciously choosing either our current path of devastation, or a new one leading to a better world for everyone.

***Part II*, The First Path—Strife and Destruction** thoroughly examines the present state of our world by exposing the truth about the condition of our planet, our global society, neighbouring communities, and individual lives, closely examining vital aspects that directly affect the survival and evolution of our species. This second part daringly elucidates the answers to questions like, "How have we come to live the planetary crises of today?" or "Where are we heading if we continue going down this same road?" based on scientific evidence and the wisdom of great spiritual traditions. We find out here that being aware of what is happening around us is critical if we want to make an informed decision about our future.

***Part III*, The Second Path—Transformations and Growth,** examines individual and collective worldviews, beliefs, and values that have supported our existence in the past, but are obsolete today. It gives us hope about the future by guiding us on how to evolve into an ecologically wise species, by introducing the revolutionary concept of spiritual-science. This part of the book puts forth specific, easy-to-implement strategies that anyone can apply so that we can usher in a new world of compassion, cooperation, and abundance. We are shown how to go beyond our current limitations and how to use the tremendous power within each of us to tap into the collective field of consciousness and co-create a better world with Mother Earth.

On a global scale when **I** is
transformed into **We**, our
Illness will be turned into ***Wellness***.

PART I
TIME OF GREAT CHANGE HAS COME

Almost unanimously, people around the world are feeling that lately time has sped up. It does not matter if you are talking to Aborigines in Australia, an 84-year-old farmer in Kyoto, an African grocery shop owner, a British teenage musician, or a Chinese business executive—they are all saying the same thing. Regardless of our age or what we do personally or professionally, we are all feeling that time is like fine sand, seeping through our fingers, and we are desperately trying to hold on to it.

This phenomenon is disconcerting to many; we sense it, but we can't explain it. Many people think that it is all a matter of time management and they are trying to get more things done, they're packing their schedules, sleeping less, and moving around more quickly. The same group of people is saying that they don't have time for themselves, often feeling drained, or having a sense of emptiness inside; and that is if they do get a breather in their day to reflect on what is happening to their lives.

Time is speeding up. The hours, minutes, and seconds are the same, but our experience of time is accelerating. What is tampering with our perception of time? If time is racing through our lives, what is it racing up to? And more importantly, what is time trying to tell us?

These were the questions that initiated my search for reasons, for a way to explain what is going on with me, my family, my work, my country, and my world. Like a fish, I was asking, "Where's the water?" Well, the only way I could see the water was by being out of it. The water I am referring to symbolises the things that I am busy with in my life.

So, I started making time to do research: I read books, I did Internet searches, watched scientific and esoteric programmes, and I talked to people. But most of all, I started experimenting. I began to introduce new ideas and activities in my life, to have more moments of stillness and I further simplified my daily routine. Gradually, moments of synchronicity and serendipity started to link the dots for me. This is how I knew what I had to do and why I wanted to do it. In the following chapters, my mission is to share with you my discoveries and realisations.

In this discovery process I have found important aspects of scientific studies, ancient wisdom, and several ancestral prophecies that help to explain why our perception of time is changing. They also help us understand what is happening in our world and they assist us in finding empowering ways to move forward. Both scientific and spiritual knowledge seem to be in agreement that we are in a time of great chaos, and also a time of choice: a choice either to awaken to a new consciousness—a new awareness or perception—in order to co-create a new life-supportive world, or to flow along our unconscious path toward destruction. Whether we decide to stick to our old familiar ways or to move on to new unknown territory, the chapters of this book emphasise that it is now that we need to make these world-changing choices.

I share similar understandings with possibly hundreds of millions of people around the world about the fact that we are facing crises in many areas of our lives; and this is because we are out of balance with each other and with our world. In order for us to change, we must first take a conscious step back from our daily life and look at it from a different angle, so that we can unlearn and relearn to create new possibilities.

In this book, the step back is symbolised by the *Pause* section in each chapter that helps us to stop and consider our situation from a different perspective, and to see how we can begin to heal certain areas in our lives by embracing new ideas and making certain important lifestyle or attitudinal changes.

CHAPTER 1
WHAT'S HAPPENING TO OUR TIME?

TIME CONSCIOUS

"What time is it?" we ask ourselves and look at our watches at least a dozen times a day; time is a crucial aspect of our human experience. We structure and measure almost everything with it. We say things like, "Time and tide wait for no man," "Time flies by so quickly," "Our time is up," "We need to buy more time," "We are running out of time," and probably the most famous phrase of all, "Sorry, no time."

DEFINING TIME

Time surrounds us; it permeates all aspects of our experience and we are deeply affected by it. Though it is a familiar and intimate part of our waking experience, we cannot see, hear, smell, taste, or touch it. Yet we do feel it. We have a deep acceptance that the future is open until it becomes the present, and that the past is fixed. As time flows, the structure of fixed past, immediate present, and open future gets carried forward in time. This structure is built into our language, thoughts, and behaviour. How we live our lives depends on it.

Over the centuries, poets, philosophers, and scientists have been trying to define time and none of them, so far, has been able to come up with a satisfactory description. Saint Augustine of Hippo, a prominent philosopher of the 5th century, said, "What, then, is time? If no one asks of me, I know... But if I wish to explain to him who asks, I know not." In the 17th century, Isaac Newton battled with Gottfried Leibniz, a renowned German mathematician, on whether time is absolute or relational. Even Einstein, the man who redefined time and space in 1905, said, "People like us, who believe in physics, know that the distinction between past, present, and future is only a stubbornly persistent illusion."

More recently, in 2007, the BBC featured *Cosmic Time*, a show presented by Michio Kaku, a celebrated American physicist. He brought the audience through the mechanics of quantum physics, astronomy, and cosmology and concluded by saying that "the time that we feel passing, the time that we know and trust may be something of an illusion; an illusion that allows us to make sense of our place in this tiny corner of the cosmos."

So, what is time? The truth is that scientists are still searching for a good definition, because time can speed up, slow down, or even come to a halt. Many experts, after multiple elaborate experiments and calculations, have concluded that time doesn't seem to exist! At this point in our evolution we can safely say that we still do not have a full understanding of what time is. And since this is so, we owe it to ourselves to explore other ways of looking at time.

COMMODITISATION OF AN ILLUSION

Despite its total intangibility, we perceive time as a commodity: we try to save time, we hate to waste it, we declare that "time is money," we say we will make time for important activities, and when we need to catch our breath, we call for a "time-out." In our quest to understand this global phenomenon we have mentioned earlier whereby people are feeling that time is accelerating but their watches mysteriously seem to disagree, we can't help but wonder if it is not the way we think about time that is speeding it up. After all, our thoughts do translate into actions and activities. To begin solving the puzzle, let us take a closer look into our lives.

PACE OF LIFE

It used to take a much longer time for us to do things. Say, for example, to travel from point A to point B. We've gone from horses to locomotives and automobiles, and now airplanes and jets. What used to take weeks or days has been reduced to hours. In fact, time spent travelling is largely considered to be a waste, so we try our best to shorten it.

Communication and the exchange of information used to take months or weeks, but now—a few seconds. In the past, we went to universities, libraries, and/or bookshops to obtain information. In the 1960s and 1970s, we relied on personal meetings, mail correspondences, and telex to communicate. Today we have the Internet, electronic mail, mobile phones, and television. In fact, we are struggling with information overload. The irony is, in order to save time we are increasingly spending more time and money to chase after the newest or latest software and gadgets to speed up the management of our information.

Without our realisation, such constant bombardment of information is turning us into a civilisation of people with some form of obsessive compulsive disorder. We are frequently and ceaselessly clicking our mobile-phones and television remotes, and checking our email. These multimedia-related repetitive behaviours and distractions combined with limited amounts of outdoor activities and engagement with people have become our daily routine. This, perhaps, has somehow contributed to our perception that time is moving faster.

Time has been accelerating with such velocity that people's lives are increasingly overwhelmed by stress, exhaustion, and the futility of satisfying the multiple demands made on them. In a struggle to prevent their lives from spinning out of control, people are collectively racing against clock-time, which is out of sync with nature.

Pace of Evolution

The effect of the rapid acceleration of growth and development is not restricted to humans, but also affects all life on earth. Change that used to happen very slowly is now happening at breakneck speed. Early evolution on the planet was very slow. About four billion years ago, simple life forms emerged. Then, a few hundred million years ago, came the vertebrates with central nervous systems. A couple of million years back, hominids came into existence, followed by *Homo sapiens sapiens* about two hundred thousand years ago. A system for communicating ideas and feelings using sounds and gestures arose tens of thousands of years ago. Then, things really began to speed up.

The movement into towns and cities happened only a few thousand years back. Just three centuries ago, the industrial revolution started. And most recently, the information revolution has begun, so it is only a few decades old. Technological breakthroughs proliferate in society in years and sometimes months, rather than centuries. Every evolutionary level or phase happens faster than the previous one, and this pattern is set to continue in the future.

Coming to Zero Time

Ray Kurzweil, author of *The Singularity Is Near* (2005), has asserted that if computing power continues to double every 18 months (which it has for the past 50 years), then, sometime in the 2020s, we might have computers that equal the capabilities of the human brain. From there on, it is a small step to a computer that exceeds the human brain. At that point there would be no need to make computers; super intelligent machines would be able to make better ones and they would do so faster than humans. What would happen then? Some say that humans could become obsolete; machines would take over the world. Others believe that there would be an amalgamation of human and machine intelligence—easy transference of information between the Internet and the human mind. If this happens, we would have a radical new era of evolution on planet Earth.

The idea that there might be a singularity in human development was presented by Terrence McKenna in his book *The Invisible Landscape*

(1976). By incorporating the ancient wisdom of the *I-Ching* of China (also known as *The Book of Changes*), which dates back to 2800 BCE, McKenna developed a mathematical fractal function called the "timewave." The timewave is represented as a graph with peaks and troughs, which, when mapped onto the timeline of human history, correlates with four thousand years of significant historical events in an amazing way. The peaks and troughs coincide with major events of the past: ca. 500 BCE, when Lao Tzu, Plato, Zoroaster, and the Buddha exerted a major influence on human consciousness, ca. 1500, when the Spanish explored the New World, the 1900s, when the world wars occurred, and many more events.

The most striking characteristic of McKenna's timewave is that its shape repeats itself, but over shorter and shorter intervals of time. The recurring pattern of the timewave that took place in the 1960s happened 64 times faster than the previous interval. In 2010, the pattern was reiterated 64 times faster still. The timescale is reduced and compacted from months to weeks, days, and eventually moves toward zero. Based on McKenna's research and calculations, what he calls the "timewave zero" ends on a specific date, 22 December 2012. According to McKenna, the Universe has a powerful attractor at the end of time that dramatically increases interconnectedness of all things.

Evidence of Time Speeding Up

Atomic Clocks

Originally, a second was defined as 1/86400 of a mean solar day, as determined by the rotation of Earth around its axis. By the middle of the 20th century, it was apparent that Earth's rotation did not provide a sufficiently uniform time standard, so, in 1956 the second was redefined in terms of the annual orbital revolution of Earth around the Sun. In 1967, the second was redefined in terms of a physical property: the oscillations of an atom of caesium as measured by an atomic clock, which has an accuracy of one ten-billionth of a second per day.

Since Earth turns a little more slowly each year, if a second or two were not added occasionally to the world's atomic clocks, timekeepers would eventually be saying noon, even though the Sun would be just rising over the eastern horizon. More importantly, many electronic navigation and communication systems, including the global positioning system (GPS) which can tell people exactly where they are on the planet, depend on extremely precise measurements of time intervals.

Explained simply, Earth takes more time to complete one rotation around its axis. It takes 86,400.003 seconds to complete one rotation, and, as this increase in time adds up to one second, then a second has to be added to the atomic clocks around the world. As Earth is slowing down, time, which was held constant by the atomic clock, is passing at a faster rate. Since 1972, at an average of one to two seconds per year, 24 seconds have been added to the atomic clocks around the world!

Universe Expansion and Dark Matter

In 1998, NASA's Hubble Space Telescope captured distant supernovae—big stellar explosions—which showed that a long time ago the Universe was actually expanding more slowly than it is today. In other words, the Universe is not contracting, but is expanding at an accelerating rate. No one knows how to explain this expansion, but something is causing it.

One of the theories speculates on the existence of mysterious *dark matter* and *dark energy*. Einstein's cosmological constant states that empty space possesses its own energy, and since this energy is a property of space itself, it would not be diluted as space expands. As more space comes into existence, more of this energy-of-space would appear. As a result, this form of energy would cause the Universe to expand faster and faster.

Another explanation for dark energy is that it is a new kind of dynamical energy field, something that fills all of space, but whose effect on the expansion of the Universe is the opposite of that of matter and normal energy. According to emerging theories in the field of quantum cosmology, the passage of time is directly related to the expansion of the Universe. If the expansion of the Universe is accelerating, the course of time should also be accelerating.

Up until recently, dark matter had no definitive measure. On 6 January 2010, BBC news reported that a team of US astronomers found a giant halo of dark matter surrounding our galaxy, in the shape of a flattened beach ball. This is the first definitive measure of the scope of the dark matter that makes up the majority of mass in galaxies.

Ukrainian Scientists Discovered Time Acceleration

In 2004, Ukrainian physicists made a remarkable discovery that changed our perception of time. Two scientists of the Institute of Quantum Physics in Kiev, Dmitro Stary and Irina Soldatenko, began an experiment in the early 1970s and continued it for 30 years. They measured with precision ordinary distances, such as the length of the iridium standard meter rod, which can be determined using modern high-precision quantum devices. After all, if the Universe is expanding, the process will affect not only the edges of the Universe, but the Universe in its entirety. This means that the iridium standard meter rod will also have to become longer after some time.

Stary developed a unique computer program that enabled the scientists to compensate for and correct the mistakes of their electronic devices. In the end, after tedious and lengthy research work, the Ukrainian scientists managed to use their data to calculate more accurately the age of our Universe, as well as to acquire other significant data.

The major outcome, however, which was never intended by the scientists, became obvious only during the final stages of the experiment. After conducting a thorough analysis of the data obtained, they discovered that not only does the Universe expand, but time does indeed tend to accelerate as well!

The results of Stary and Soldatenko's research turn our preconceived notions of the Universe upside down. Their findings have not been totally accepted by the scientific world. In the meantime, the results are being tested in many laboratories worldwide. It is likely that many researchers today agree that this is a fundamentally new physical phenomenon, which might trigger new breakthroughs in the world of science.

So, what does all this mean? How does it affect us that time on Earth is speeding up because of the slowing down of Earth's rotation and the acceleration of the expansion of the Universe? What would it mean for us to reach timewave zero? Science can only take us this far; now it's time to take a look at ancient prophecies in order to get a more comprehensive picture of what time is trying to tell us.

Prophecies About Time

The study of the beginning and ending of time has been included in ancient teachings in many cultures around the world. Over the last 6,000 to 8,000 years, group after group has been predicting a significant time when the world as we know it would come to an end. Does that refer to the end of the world, the destruction of the planet, of all life on Earth? Though some interpret the predictions that way, many believe the predictions indicate major shifts in the way we live our lives on this planet. The nature of the shift comes in several forms, but many say it has yet to be determined.

Are we living in the predicted time and will the choices we make now determine how the shift will unfold?

Multi-Cultural Obsession: Calendars

We often find predictions about the end of an age and the beginning of a new one in the calendars created by ancient people. Interestingly, the word *calendar* itself is derived from the Latin word *kalendae*, meaning "accounts book"—the first day of every month being *the calends*, or the payment date for debts. This helps to explain the pervasiveness of the societal programming that "time is money," which is an artificial or inorganic way of living with time.

Almost every culture has its own system of timekeeping that dates back several centuries or millennia. Early civilisations that had complex agriculture-based urban settlements, with thorough writing systems have left evidence that people were keeping track of days, months and years. The Chinese, Indian, Hebrew, Islamic, Egyptian and many other cultures have produced very different calendars. Each culture has different numbers and lengths of months, and numbers of days per year. Some have based their calendars primarily on lunar, solar, or on both cycles. Others have used stars and other celestial bodies to plot them. Still others have incorporated geographical and climatological observations into their time-tables.

The 12-month calendar, which currently serves as the world's standard of time is called the *Gregorian calendar*, named after Pope Gregory XIII, who revised the previous Julian (introduced by Julius Caesar) calendar in 1582. Both the Gregorian calendar and the clock are based on the Babylonian model, which substituted a measurement of space for a measurement of time. However, as we will see, time is not space; time is of the mind.

Ancient Mystery Date: December 2012

Mayan Timekeeping

In any discussion of calendars and ancient civilisations, one must pay special attention to the extraordinary accomplishments of the Maya of Central America (AD 250–900). Mayans were obsessed with timekeeping and spent centuries observing heavenly bodies. They strove to find certain repeated patterns in their observations and recorded data, which could then be used as a guide for predicting the future. Unlike most Western cultures, the Mayans believed that patterns of events repeated themselves every 260 years, and that this period of time was part of a larger cycle of 5,125 years. They believed that the future was present in the past.

What is also astonishing about the Mayan calendar is that it details immense passages of time and includes such accurate mathematical calculations that astronomers today cannot explain how ancient people made them. To this day, modern Maya keep track of great cycles, as well as local time, by using this centuries-old system. Experts like Michael D. Coe, an eminent archaeologist of Yale University, tell us that the Maya have "not slipped one day in over twenty-five centuries." In fact, the

Mayans have pinpointed solar and lunar eclipses thousands of years ahead of time!

The Mayan timekeeping system is not only astronomically accurate but also prophetic in nature. Its prophecies indicated the arrival of the Spanish conquistadors in the Yucatan in 1520, the 1776–1789 American and French revolutions, the American Civil War and the assassination of Abraham Lincoln in the 1860s, the Second World War, the political assassination of John F. Kennedy in the 1960s, the Vietnam War, and many other events.

Using this advanced calendar system based on astronomy and cosmology, the Maya did something that is unthinkable to us today. Without high-speed computers, powerful telescopes, and satellites, they calculated the movement of Earth and the entire solar system in connection to the centre of the Milky Way galaxy. John Major Jenkins, author of *Maya Cosmogenesis 2012*, calls the alignment of Earth with this centre, "the galactic alignment." This rare astronomical event happens only once in every 26,000 years. These vast time-cycle calculations include the so-called "Mayan long count," which spans 5,125 solar years and is known as one "world age." The recent end of a long count marked the completion of the Fifth World Age. This was scheduled to happen on 21 December 2012. This date also marks the gateway to a new great cosmic cycle of approximately 25,826 years. Coincidentally, it is only one day earlier than the date Terrence McKenna arrived at in his calculation using the *I-Ching* (McKenna did not know anything about the Mayan calendar at the time he made his calculation.).

There have been many speculations and conjectures about what 21 December 2012 meant and what was going to happen. According to Mayan scholars, the past four world ages ended in cataclysms and the planet has lived through fire, wind, earth, and water. The current 5th cycle, from 3114 BCE to 2012 CE, is the synthesis of the previous four world ages and it is called "the Age of the Caban." In the Mayan language, *caban* means sudden movements (earthquakes) and abrupt collapses of structural systems. In Mayan astrology, in this period Earth is rigidly holding on to the past; it needs to be broken loose from the quicksand of time to be catapulted forward. This period is also known as "the triumph of materialism," "the transformation of matter," and "a time of great forgetting" (that we are one with nature). Each world age ends in a cataclysm, which may not just be geophysical in nature, but also biological and social. And each new cycle brings about an improved form of life.

Astrological Timekeeping

The 2012 winter solstice marked the end of the Mayan long count cycle (5,125 years that started in 3114 BCE) on 21 December; it also signalled the completion of an even larger scale: the great precessional cycle that started 25,920 years ago. It was then that Earth began its journey through the celestial path of the 12 zodiac signs of the astrological system. Satellites today reveal that it takes Earth about 72 years to pass through one degree of the Zodiac orbit. Thus, Earth needs about 2,160 years to pass through one zodiac sign of 30 degrees. The Piscean Age, which began about 2,000 years ago, coincided with the arrival of Jesus. We are currently transitioning into the constellation of Aquarius, which heralds the age of:

- embracing more mystical/spiritual methods in self-help;
- looking inwards for answers, instead of fixating on outward possessions, money, and adoration;
- breaking free from centuries of indoctrination and collectively moving against corrupt and unethical societies/governments/organisations;
- the unification and synthesis of different beliefs, culminating in a unified awakened consciousness;
- a non-possessive form of love, friendship, and collaboration;
- exploring outside the confines of Earth—expanding; and establishing the new frontiers of space.

In astrology, a *precession* is the motion of a spinning object in which the body wobbles so that the rotation axis forms the shape of a cone. One complete wobble of Earth's axis takes approximately 25,920 years. As the planet crosses the threshold of the equator of the Milky Way, it not only begins a new 5,125-years world age, but every fifth time, it also completes the precessional cycle of having moved through all the twelve zodiac constellations, thus beginning the next great precessional cycle.

The ancients knew that endings of cycles were always filled with upheaval and turmoil. The ending of one of Earth's precessional cycles is a catalyst for tremendous change on Earth. Our understanding is that as Earth and its inhabitants complete this precessional cycle, everyone will go into a "healing crisis" before relaxing into a new and higher state of order.

As we will discuss in more depth in Part III, to facilitate this healing process, we are to accept unconditionally, without judgement, all that we experience in our lives and all that we feel emerging within ourselves in every moment. In other words, we are encouraged to embody the principle of compassion—the acceptance of all conditions. At the same time, we are signalled to stop blaming others and the external events in our lives for our misfortune, and stop waiting for something from outside of us to come and save us; we must realise that we have the power to co-create our reality from the inside out, moment by moment.

Hindu Vast Periods of Timekeeping

Hindu culture was one of the earliest on earth to believe in repeating cycles of world ages. Their ancient texts talk about a cycle of four world ages:

The Golden Age	Satya Yuga	Duration: 1,728,800 years; age of extreme splendour; death occurs only when willed; there are no wars, famines, strife or evil —peace prevails.
The Silver Age	Treta Yuga	Duration: 1,296,000 years; age of the beginning of entropy; beings begin to be corrupted and evil; beings live a shorter time.
The Bronze Age	Dvapara Yuga	Duration: 864,000 years; age of the "fall" of humanity; beings are divided between virtue and sin; evil and corruption spread; beings live a much shorter time.
The Iron Age	Kali Yuga	Duration: 432,000 years; age of darkness; corruption and evil become the driving forces; greed, wars, famine, and disease spread across the planet.

. It has been foretold in ancient Hindu texts, such as the Mahabharata, that the Kali Yuga would display these signs:

- anger and rage: "And when end of the Yuga comes, sons will slay fathers and mothers. And women [...] will slay husbands and sons." "[...] men will begin to slay one another, and become wicked and fierce and without any respect for animal life [...]";
- deceit and avarice: "And friends and relatives and kinsmen will perform friendly offices for the sake of the wealth only that is possessed by a person." "And when the end of the Yuga comes, everybody will be in want."
- lust and aversion: "[...] and the highways will be filled with lustful men and women of evil repute." "And girls of five or six years of age will bring forth children and boys of seven or eight years of age will become fathers."
- environmental collapse: "The inhabited regions of earth will be afflicted with dearth and famine [...]" "And the deity of a

thousand eyes will shower rain unseasonably. And when the end of the Yuga comes, crops will not grow in abundance."

- cosmological disasters: "And all the points of the horizon will be ablaze, and the stars and stellar groups will be destitute of brilliancy, and the planetary conjunctions will be inauspicious." "[...] and innumerable meteors will flash through the sky, foreboding evil."

Many of the signs above are familiar to our world today.

According to experts—cosmologists, theologians, philosophers, and anthropologists—the end point of the Kali Yuga was December 2012. This was also the starting point of the next world age in the Hindu system: the Satya Yuga—the Golden Age of Man.

The Yugas: The Acceleration and Hardening of Time

According to Jay Weidner, hermetic scholar and author of *The Mysteries of the Great Cross of Hendaye,* the times listed in the Hindu world ages are not literal, but may be metaphorical. To gain a new understanding, he suggests that we look at yugas from a hyper-dimensional perspective. In modern physics, a four-dimensional space (of which time is one aspect) is often represented in the shape of a hypersphere (a hypercube—a cube within a cube). According to alchemists, each object in the three-dimensional world is an affectation of the four-dimensional space. Humans, animals, plants, planets, and stars are the solids inside the four-dimensional energy flow.

As energy flows inwards, into the centre of the hypercube, it works like a vortex, similarly to a tornado. Tornados are made up of the same air that surrounds us but, as this air spins, it takes on a strange solidity. The spinning of the air into the vortex tip of the tornado, near the ground, becomes iron like, capable of ripping and hurling houses, trees, and lorries. It could be said that this hardening of vortical forces from the fourth dimension is what makes up the solidity of our three-dimensional space. Alchemists from all ancient traditions knew this; they called the four-dimensional space "spirit," and the three-dimensional space "matter."

Using the tornado analogy, as the vortex spins inwards, into the hypersphere, it becomes fast, violent, and dense, which in the Hindu four world ages represents the phase of the Iron Age. As it spins faster and faster into a narrower and narrower point, it would eventually compress to the point where it has nowhere to go but outwards again; this is the zero point. As it expands outwards, time slows down. When we finally reach the zero point, the Golden Age begins—this is when time will slow down. Without getting stuck in details, the spaces between the zero point and the outer vortex are also occupied by the Silver and Bronze Ages. As the energy of the vortex moves inwards, into the hypercube, it spins faster and becomes harder. This brings in the Kali Yuga, or Iron Age.

This is why in the age of the Kali Yuga each second feels shorter than the second before it. As we spin toward the tip of the tornado, each day, each month, and each year appears to be going faster than the preceding one. The application of modern physics and ancient alchemy allows us to look at the years stated in the yuga system differently; they are actually symbolical representations of our temporal experience, rather than actual years we live through in time. So the Golden Era feels much longer than the Iron Age.

Modern Discoveries Pointing to 2012

Astronomy: Black Hole in the Milky Way

Our experiences of time and ancient prophecies are not the only indications that something momentous is happening in our time. There have also been several scientific discoveries and calculations that pointed toward possible unusual events around 2012.

As mentioned earlier, John Major Jenkins discovered that the ancient Mayans calculated the movement of Earth and the entire solar system in relation to the alignment of the Sun with the centre of the Milky Way galaxy (the galactic alignment happening once in every 26,000 years). The Mayans refer to the centre of the Milky Way as the "Dark Rift" or "Xibalba-Be," which means "the dark road to the underworld." What could this rare astronomical alignment mean? Could it be that our solar system is passing through a region of space whose conditions differ vastly from the ones of our current region?

The Mayan mythological explanation of this rare alignment is very complex. In essence, its significance relates to the transcendence of opposites. It goes beyond the union of male and female opposites and involves the non-dual relationship between infinity and finitude, and the union of our higher and lower natures. Likewise, time is restored to its relationship with timelessness, whereby human consciousness reclaims the timeless eternal perspective. Therefore, the manifested world of appearances is restored to limitless possibilities.

Astrophysics: Solar Maximum

Another astronomical prediction coincides with the date observed by esoteric traditions. Astronomers have been studying the Sun for more than 150 years, and particularly since the year 2003, the Sun has become very turbulent. In 2006, scientists from NASA (US National Aeronautics and

Space Administration) and NCAR (National Center for Atmospheric Research) predicted massive solar activity in 2012-2014, with storms stronger than the Carrington event in 1859. In September 1859, within hours of a colossal solar flare—a sudden, large release of energy on the surface of the Sun—billions of tonnes of solar plasma penetrated Earth's magnetosphere, inactivating the Victorian-era electrical systems, magnetometers, telegraph networks, and causing numerous fires.

Whenever the events take place, one inevitable fact warrants our concern—our civilisation is heavily dependent on the globally connected energy network. A much smaller solar storm in 1994 caused major malfunctions to two communication satellites, network televisions, and nation-wide radio services throughout Canada. Other storms affected systems ranging from mobile phone services, GPS, hospital life-support systems, and electrical power grids. According to the US National Academy of Sciences, if a Carrington event does occur, it could disrupt or destroy at least thousands of major transformers in just two minutes and would cut off power for more than hundreds of millions of people. Its impact could cost trillions of dollars and recovery could take more than three to four years.

Geophysics: Weakening Geomagnetic Field

Earth possesses an electrically charged magnetic field, consisting of two oppositely charged separated poles, called *dipoles*. This magnetic field is generated by spinning molten fluid at the centre of Earth. This field extends to about 36,000 miles into space and acts as a magnetic shield to deflect harmful solar winds and other galactic waves away from Earth.

A discovery at the end of 2008 is now worrying scientists. NASA's Themis Spacecraft measured an opening in Earth's magnetosphere which was 10 times larger than previously estimated, or 4 times the size of Earth. The hole allowed up to 20 times the normal amount of solar particles to enter Earth's atmosphere. According to scientists, this opening in the magnetosphere is a strong indication that Earth's magnetic field has weakened.

The idea that Earth goes through drastic magnetic changes in geological time was confirmed in 1999 when two teams of scientists completed a drilling project by going deeper into the ice of Antarctica than anyone had ever gone before. The ice cores revealed 420,000 years of layered frozen air bubbles containing yearly records of airborne elements and compounds, rain and snow, and microscopic life and dust. Using new scientific methods, the ice cores have also revealed the strength of Earth's magnetic field in the past. With that information, scientists postulated that

Earth has gone through 14 magnetic reversals (north and south poles traded places) in the past 4.5 million years.

More support for the possibility of a magnetic reversal is found in a paper published in *Nature Geoscience* in May 2008, by geophysicist Mioara Mandea, from the GFZ German Research Centre for Geosciences, and her Danish colleague, Nils Olsen, from the National Space Institute of Copenhagen. They discovered that the movements in Earth's molten fluid are changing surprisingly fast, and this change is affecting the magnetic field of the planet.

The very precise measurements of Earth's magnetic field delivered by satellite data and ground observations over the past 9 years have made it possible to reveal what is happening 3,000 kilometres beneath our feet. "The rapid changes in the molten fluid may suggest the possibility of an upcoming reversal of the geomagnetic field," said Mioara Mandea. According to both scientists, a flip in the magnetic poles typically involves a weakening in the magnetic field, followed by a period of rapid recovery and reorganisation of the magnetic polarities.

Not only would a magnetic reversal cause a geophysical and astrophysical disaster, but it would dramatically impact the human nervous system. Earth's magnetic field has a pervasive impact on human biological processes and the perception of time and space. In 1993, Weiser, Fuller, and Dobson, a team of multi-disciplinary scientists, discovered that the human brain contains biogenic magnetite, which responds to subtle magnetic energy, such as Earth's geomagnetic field. This discovery, along with many other scientific principles, suggests that a magnetic reversal could trigger a collective shift in human consciousness.

Based on recent scientific studies, the following scenarios could upset the delicate balance of Earth's magnetic dynamo and trigger a reversal:

- a major asteroid or comet impacting Earth;
- a volcanic eruption or earthquake from within Earth;
- a large orbiting asteroid or planet moving close to Earth;
- gravitational or gamma rays affecting Earth when crossing the galaxy's equator (galactic alignment); or
- huge solar flares totally disturbing Earth's magnetic field.

Many of the scenarios outlined above have been predicted for 2012–2017.

Climatology: Multi-System Chain Reaction

Latest updates from climate, oceanic, atmospheric, and environmental studies in 2008 and 2009 have sharply increased the trepidation of many scientific, governmental, and humanistic bodies around the world. Initial estimates by climatological and environmental experts have been drastically revised. The points of no return have been shifted from the end of the 21st century to mid-century, then to the next 25 years, and now, for some areas, to the next 5 to 10 years.

Three critical areas of concern that deserve our full attention are global sea levels, atmospheric carbon dioxide, and atmospheric warming. From the Intergovernmental Panel on Climate Change (IPCC):

- its 2008 Report confirmed that sea levels around the world have risen 150% faster than previously predicted. Research in the Antarctic in February 2009 reported that two large glaciers were sliding into the sea. In March 2009, 2,500 researchers in Copenhagen predicted that global sea levels could rise by over 1 meter in approximately 20 years' time. This would inundate 30% of the land mass on Earth, displacing hundreds of millions of people around the world;
- air quality around the planet is severely deteriorating, and its negative impact on all life forms is increasing. Carbon dioxide emissions rose from 1.1% between 1990 and 1999, to over 3% (more than double) between 2000 and 2004;
- forecasts in the 1990s said that there would be a three degrees Celsius increase by the end of the 21st century; present experts predict that this could happen within a decade (by 2020). A three degree global warming would cause serious disruption in human life and economic activity; a six degree warming would make most of the planet unsuitable for food production and human habitation.

In June 2012, the US National Oceanic and Atmospheric Administration released a report (involving 22 scientists from 5 countries) noting that April 2012 was the 326th consecutive month in which global temperatures exceeded the 20th century average. And this is due to long range climate change cycles, rampant population growth, and changes to the environment caused by humans: the burning of fossil fuels and the conversion of 43% of the planet's land to farms or cities; this has an incalculable impact on global ecosystems. Such shifts in weather patterns can cause the collapse of biodiversity and food supplies, the spread of deadly diseases, social unrest, economic instability, and massive loss of human life.

According to Anthony D. Barnosky, Professor of Integrative Biology at the University of California, "people have become a geological force in their own right, and we are changing the planet in ways every bit as dramatic as major geological events." "We are becoming much more dominant in Earth by our sheer numbers and the way we use natural resources."

Post-2012 Interpretations

While there was no global devastation on 21 December 2012 (as depicted by the Hollywood movie,*2012*), there are still ongoing studies and intellectual exchanges between Mayan scholars, evolutionary and consciousness experts regarding the true meaning of the year 2012. On one hand, the indigenous Mayan elders and experts in Guatemala are upset with people around the world, including their own governments and tour-related businesses for twisting the truth over the 2012 predictions. Felipe Gomez, leader of the Maya Alliance Oxlaljuj Ajpop, said that it's an end of one of their major calendar cycles, and not of the world. He also said to Agence France-Presse that the new time cycle—simply put— "means there will be big changes on the personal, family and community level, so that there is harmony and balance between mankind and nature." Carlos Barrios, a Mayan priest, anthropologist, and Mayan Calendar expert, interviewed 600 other Mayan elders to provide the world a comprehensive understanding of 2012. Stated simply, he said that in the midst of global environmental destruction, social chaos, war, and other Earth changes, "Humanity will continue, but in a very different way. Material structures (old political, economic, and social systems) will change. From these changes we will have the opportunity to be more human."

The subject of Mayan astronomy, astrology, and metaphysics is just too vast to be discussed in detail. However by synthesizing the post-2012 perspectives of leading Mayan Calendar researchers like Carl Calleman, John M. Jenkins, and other Mayan, we would see that they revolve around the following important messages:

- It is now possible for human beings to "download" unity consciousness from the metaphysical realm to promote ecological change (more information will be shared in Parts II and III).
- The number of people spiritually awakened in the world (according to Oneness University in India) has increased markedly since Oct 2011 and this awakening will continue to intensify in the years to come. The "veil" of materialism has been lifted.
- More people around the world will overthrow egotistical and oppressive regimes characterized by exploitation and deception and replace them with enlightened and compassionate ones. In the process, people have to surrender their egos and endure material and environmental hardships.

- There is no instant-enlightenment, nor fated doomsday. However it is a time of deep renewal. Positive change is facilitated through intentional and consciously-directed efforts of human beings.

To add on to the Mayan Calendar messages, John Perkins, a former chief economist of a major international consulting firm and the author of New York Times Bestseller *Confession of an Economic Hit Man,* said that many of the economic, environmental, and social problems are connected to predatory capitalism. This unecological system promotes planetary-wide exploitation to benefit a small number of already very wealthy people. So much so that CEOs of big corporations—together with politicians and the media—control human and natural resources around the world, rather than governments. Mr. Perkins further added that in their relentless pursuit to increase greater fortunes, they have polluted air, water, and earth, relegated countless numbers to the status of unemployed, and doubled the gap between the few who live luxuriously and the billions who are malnourished or starving. They exemplify that egotistical regime described in the Mayan creation myth—the Popul Vuh.

With lightning speed information sharing across all corners of the Earth through internet and mobile phones, most people are seeing that we are standing on a shore that is threatened by escalating waves of economic, social, and environmental disasters. In short, we must abandon our separative, gluttonous, and exploitive ways, in favour of lifestyles and systems that will pass on to our children and grandchildren a world they can inhabit and continue to help evolve. Simply stated, we must become sustainable.

With regards to the 2012 prophecy, it is unwise to have just one point of view. One possible interpretation is that it is one of the most elaborate hoaxes in history. But the truth is, many evolutionary theorists, scientists, and spiritualists around the world do not believe in this interpretation. More information about the possible interpretations of the 2012 prophecy continues in Chapter 2, page 45. You will see that the authors, after intense research across several disciplines, have accurately postulated many of the post-2012 interpretations that are currently held by many scientists and spiritualists.

Pause: If Ancient Rocks Could Speak

Through science and spirituality we are being given startling information about what is happening to our time. We are given many possibilities, yet we do not have a unified framework to understand what these different sources of information are trying to tell us. By tapping into another source of ancient knowledge, we can have a clearer picture of what is going on and where we are heading.

Scholar Robert Bossière spent years with the Hopis—Native Americans living in northeast Arizona. He wrote a book called *Meditations with the Hopi* (1986), and according to their tradition we are currently in the Fourth World Age. The first world was destroyed by fire, the second one by ice, and the third one was cleansed by water. The Hopi descriptions of the end of time are remarkably similar to Earth's geological record. Recent scientific discoveries have confirmed that Earth went through a turbulent period of earthquakes and volcanic activity about 20,000 years ago. This was followed by an ice age around 11,000 years ago; then, roughly 4,000–5,000 years ago, there was a deluge, which was believed to have been the biblical flood—and all this is consistent with Hopi legends.

We have seen that the Indian and Mayan systems of time cycles describe a meaningful relationship between celestial, planetary, and human events. The same is true of the Aztec and Egyptian systems. This macrocosmic/microcosmic relationship parallels the ancient axiom "as above, so below." The Hopi tradition also provides an important prophetic message from our past.

In the Hopi village of Oraibi, a prophecy was etched into a stone, known today as the Hopi Prophecy Rock. Nobody knows how long the images (petroglyphs) have been there or who created them. Through scientific testing, the rock is estimated to be about 10,000 years old. The message on this rock not only foretells events to come but also points toward a way of being that would help humanity survive the great changes that come with the end of the Fourth World Age.

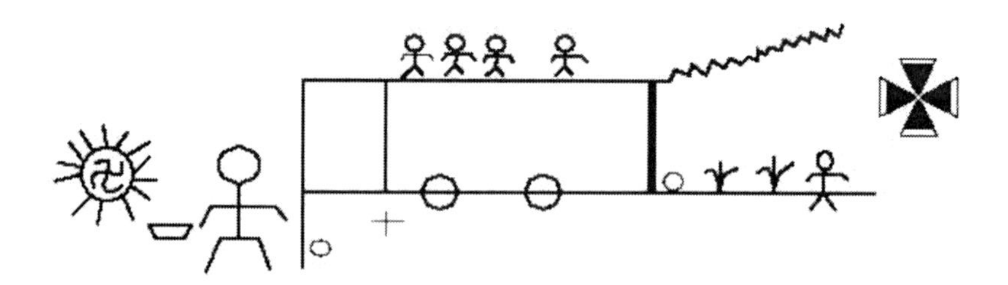

A replication of the petroglyphs in Oraibi, Northeast Arizona, U.S. Used with permission from http://www.crystalinks.com/hopi2.html

In his book, *The Rocks Begin to Speak* (1973), Lavan Martineau deciphered the hidden meanings of the images. Martineau was raised by Native Americans and was fluent in the sign language used by many tribes.

During the Second World War he was exposed to the science of cryptology, which helped him formulate his work in understanding the Native American petroglyphs. While there are many levels of meaning, the crux of this petroglyph's message revolves around the parallel lines, symbolising two possible paths that mankind may choose. The lower path leads to abundant crops, health, and peace; the upper path leads to strife, shortage, and suffering.

A third path was also given—a vertical line that connects the jagged and the smooth ones. It represents the path of choice. According to the Hopis, as we come close to the end of the Fourth World Age, many people will be confused or awakened by the chaos on earth. Those who move beyond the third vertical path and onto the upper jagged path will be heading toward the path of destruction. Therefore, the third vertical path is the point of no return. As we will see in Part II, there are reasons to believe that humanity is currently at that point.

The Hopi prophecy offers us choice. In crisis, we are presented with both destruction and peace, breakdown and breakthrough. Many other prophecies, like the Mayan cosmology and the visions of Nostradamus (French seer, 1503–1566), also tell us that we have a choice. All these prophecies—which contain some predictions that have already come true—state that we are the ones determining how we want our story to unfold. According to Robert Bossière, "The prophecy does not say what the third upheaval of earth will be. For it depends upon which path humankind will walk: the greed, the comfort, and the profit, or the path of love, strength, and balance."

The Hopi message for humanity is to live simply, humbly, and in a balanced way, for the sake of all that lives—man, animal, plant, earth, and the sacred water. This message can be found in their oral traditions, handed down from one generation to another. The Hopi elders told their young the story of their origin: the arrival of the first people on earth, the meeting with the Creator, and what "He" said to the people of earth:

My life is simple.
All I have is my planting stick and my corn.
If you are willing to live as I do and follow my instructions,
the life plan which I shall give you, you may live here with me
and take care of the land.
Then you shall have a long, happy, and fruitful life.

—Hopi oral tradition

Redefinition of Crisis

In Greek, the word *crisis* (κρίσις) can mean a turning point, where a choice has to be made based on one's ability to discern. In the medical dictionary, a crisis is the turning point in an acute disease—it is the point of change, rather than the status quo. In one teaching in Jewish Kabbalah, a crisis is a sign of the end and also the threshold for a new beginning. In Sanskrit, we have vimarsha (that is, vimarśa): a critical juncture where one faces a test. From these various descriptions, it is obvious that in a crisis, two different scenarios will be simultaneously presented and one has to make a discerning choice.

These prophecies urge us to seize the 50% chance of creating a new life, rather than keeping to the status quo. This process of choice can also be seen as a test where we choose to end our old ways, while we reinvent ourselves to create a new world. It is clear for us now that staying on our current path of business-as-usual, isolating ourselves, disconnecting from Mother Nature, and increasing our consumption of non-essential wants, is leading us to destruction. The path to a new world, where all of us truly want to belong, revolves around these simple memes: live simply and organically, actively practise connecting to the collective consciousness, and increasingly act in ways that support life and nature.

Biblical Prophecies

There is no question that the endings of great ages were seen as times of crisis; a deeper meaning of crisis was brought to light in a landmark archaeological find—the Great Isaiah Scroll. Isaiah was a prophet in the Hebrew Bible, which was later adopted as the Old Testament of the Christian Bible. This scroll was found in a cave near the Dead Sea in 1946; today it is one of the documents known as the Dead Sea Scrolls. The scroll has been dated to 150 BCE (2,160 years old), and it is the most complete scroll of the hundreds that have been found. It is so precious that it is heavily protected in a vertical steel cylinder that retracts into the floor, inside a steel and concrete vault in the Israel Museum.

In this scroll, Isaiah talks about his visions, as he has seen the possibilities in our future. Isaiah implores us to live our lives in a particular way, so that we will not have to experience the unimaginable things that he has seen. He views a time in our future when the foundations of earth will shake, great cities will be blackened, and only a few people will be alive. However, at the same time, he also describes a vision of hope, where a new earth appears and all the things of the past are not remembered, and the sound of weeping is no longer heard. He goes on to explain how water will spring forth from the deserts of earth, the deaf will hear, and the blind will see. From our current perspective, such descriptions are confusing. We ask, "What is the actual scenario?"

From a linear perspective, one might conclude that the suffering must come first, before the peace and healing can occur. From a non-linear perspective, though, Isaiah could be sharing with us possible future scenarios, which depend on our choices. Seers of our past have recognised that their visions merely portrayed possibilities for a given moment in time, rather than events that would occur with certainty. Each possibility has been based upon the conditions at the time of the prophecy. As conditions changed, the outcome of each prophecy reflected the change. The same line of reasoning reminds us that by changing our course of action in the moment, we can redirect the outcome of our future.

Later descriptions within the Great Isaiah Scroll give us an important clue: Isaiah is imploring us to focus on something important in order to avoid a cataclysm. Isaiah said, "From within a mountain we will find a refuge for the distressed, shelter from the rain and heat [...]. From this mountain, the web that is woven onto all nations will be destroyed." For hundreds of years scholars have taken these words literally and have tried to look for a physical mountain throughout the Middle East, and they have not found it.

Gregg Braden, a computer scientist who has spent more than 20 years studying remote spiritual sites in Egypt, Peru, Tibet, and America, has found hidden meanings encoded in the mysteries of our most cherished traditions. A *New York Times* best-selling author, Braden stated in his book *The Isaiah Effect* that the mountain Isaiah was referring to must be understood in the context of the language it was written in at that time, which was primarily Hebrew and Aramaic. Braden discovered that the mountain Isaiah speaks about is not a physical mountain, but a symbolical one. He discovered that in Hebrew the word *mountain* represents "heavenly Jerusalem," which in further research into biblical Hebrew translates into "vision of peace." So, if we insert this translation into what Isaiah said, we obtain, "From within a vision of peace we will find a refuge for the distressed, shelter from the rain and heat [...]. From this vision of peace, the web that is woven onto all nations will be destroyed"; and this brings about a very different message.

This *vision of peace* is more than just a phrase to Gregg Braden; he interprets it as an inner technology that allows us to focus from within our hearts. In doing so, we speak the language of God. It is from this heart-based technology that we manifest the experiences of healing and peace in our world.

With the discovery of the Dead Sea Scrolls, a small group of spiritual practitioners and healers was made known to the world: the Essenes (200 BCE–100 CE). This group studied many different forms of ancient knowledge (coming from Zoroaster, Hermes Trismegistus, the Laws of Moses, etc.) and turned this knowledge into a powerful philosophy for living. According to the Essenes, the world around us resonates with, and is transformed by, what we become in our hearts. In other words, if we want to experience peace, joy, love, and compassion in our world, we must vibrate those emotions in our hearts and act in accordance with them. This literally means that we must be—in our hearts—the change we want to see in our world.

CHAPTER 2
SCENARIOS AND TIPPING POINTS

The idea that we are going to experience some kind of planetary-scale major upheaval involving geophysical, environmental, biological, and social change is alien and unwelcoming to the majority of people. However, burying our heads in the sand like ostriches is not going to help us at all. In fact, it may be disastrous. By not being fully aware of how our way of living is impacting our future world, we might just bring ourselves unconsciously closer to the catastrophic end we fear might happen. Gaining awareness of the future scenarios also enables us to anticipate and utilise our resources optimally in a given situation. In order for us to make the right choice, we must first be fully aware of what is happening in our world, and then anticipate the likely scenarios.

There is an ongoing extensive debate about the likelihood of certain "extreme" scenarios occurring in the next few years. Most predictions take only one field of data into consideration. They fail to evaluate the possibility that when a critical point is reached in one field, this could push the other fields toward a point of no return. For example:

- a rise in carbon dioxide in the atmosphere leads to global warming; this produces drought in some areas and floods in others, thereby affecting the vegetation in many regions. That, in turn, limits the biosphere's capacity to absorb carbon, which leads to a further increase in temperature;
- a change in a country's economic policies can drive certain economic sectors (e.g. agriculture and mining) toward unecological expansion, leading to deforestation, the intensification of land degradation, and environmental pollution. This, in turn, contributes to global warming and biodiversity destruction;
- a decrease in the quality of air in urban and industrial areas leads to a deterioration of public health, with serious social and economic impact;
- a rapid rise in global temperatures affects ocean current circulation. These new currents drive warm water into the deep sea, which causes rapid decomposition of frozen methane hydrate deposits, this being another factor that exacerbates global warming.

As the information in Part II will demonstrate, with the unexpected acceleration linked to incalculable precipitating factors that are reinforcing each other, there is now a clear possibility that any of the vital trends we are experiencing at this time could reach a threshold point between 2012 and 2020. Based on Edward Lorenz's famous butterfly effect, where multi-systems amalgamate to form a larger complex system, a shift in one of the sub-systems at a critical juncture can result in a total shift in the entire system. This is commonly referred to as the *tipping point*. The chain-reaction sparked by it will impact not only the immediately affected areas but whole continents, and very likely, the entire planet.

What Are Likely Scenarios in the Near Future?

There are many legitimate websites and books that expound in great detail on this subject. After 24 months of research, we have decided to focus on some of the most probable global-shift scenarios in the near future that would be potentially devastating. It is not our intention to write exhaustively on these subjects, as many of these scenarios would require years of study. Instead, we want to provide a broad understanding of the severity of the upcoming global changes. The key takeaway from this section is to be informed and mentally, emotionally, and spiritually prepared.

Financial Doomsday

At the time of this writing, analysts and experts are saying that developing nations around the world would prevent the US dollar from collapsing, since two-thirds of the global reserves are held in US dollars. However, many experts strongly advise people to be prepared for the worst, as they successfully predicted the US-triggered global financial crisis of 2007–2009, which resulted in a $4,100,000,000,000 debt of the top five US investment banks in 2007. Authors like Michael J. Panzner (*Financial Armageddon*), Michael Maloney (*Rich Dad's Advisors: Guide to Investing in Gold and Silver: Protect Your Financial Future*), David B. Wiencke (*How to Survive the Dollar Collapse*), and many others are certain that a total financial collapse is highly likely to happen.

What would this scenario be like? There would be overnight inflation in the price of oil, and transportation would come to a halt. Electricity that powers cities and homes would soon be disrupted. The cost of all imports would rapidly inflate and people would be immediately affected by the high price of food. Countries that are heavily reliant on imports to survive would be severely impacted. Unemployment would be prevalent and banks would have no real money for businesses and individuals.

Civil unrest would break out in many parts of the world and martial law would have to be declared. Soldiers would be posted at gas stations, supermarkets, and warehouses to ration food and fuel. In many parts of the world gangs would proliferate, looting and killing people who get in their way. Hundreds of millions would die from starvation, crime, and/or disease.

A light version of this modern day catastrophe can be extrapolated from Argentina's economic collapse in 2001. Argentinean author Fernando F. Aguirre gave a "raw and shocking" account of his ordeal during the socio-economic collapse of his country in his bestselling book, *The Modern Survival Manual: Surviving the Economic Collapse* (2009). Readers say that it is a very important book on modern survivalism.

According to many world-shift experts, from both the scientific and spiritual communities, this impending catastrophe is necessary to bring about the global shift in human civilisation and consciousness. This event would trigger a major reprioritisation of human lives, ungluing people from money.

Viral Wipe-out

The largest killers in human history have not been wars, as we may think, but pandemic diseases. The bubonic plague has killed 250 million people since the 14th century. Smallpox, which was successfully controlled in the 1970s, has claimed 500 million lives. The Spanish flu of 1918, an H1N1 subtype Influenza-A Virus, similar to the recent Swine Flu pandemic, took 50 million lives in just two years.

With the increasing encroachment of human habitation and activities into wildlife and nature, the opportunity for transference of animal viruses to humans is greater than ever. The exponential increase in the demand for meat, eggs, and milk has produced the expansion and intensification of factory farming of domestic animals. "Factory farms are super-incubators for viruses," says Bob Martin, former executive director of the Pew Commission on Industrial Animal Farm Production. In densely packed factory farms, flu viruses in pigs, for example, are often exposed to bird, bat, and human flu viruses, leading to the rapid recombination and replication of the viruses. When these animals are shipped to and from different locations, these recombined viruses are enabled to reinvade and bypass the animals' immune systems.

Scientists need six months to develop a vaccine for a new strain of flu, and there is no way of telling what the original virus will have evolved into in that period of time. Scientists are concerned that rapid viral evolution is making it more and more difficult to trace the source of any given virus. Each time a virus changes, it typically becomes more resistant to pharmaceuticals and renders vaccines useless. Moreover, animals in factory farms are injected with growth hormones and antibiotics. When these synthetic concoctions are given to animals that are highly stressed with neurotic hormones, the animals may become breeding grounds for what scientists call "killer viruses."

Health authorities around the world are on high alert for the possibility of a virulent killer. Scientists are intensely studying all available data on the nature of flu viruses, tracking back to the Spanish flu in 1918. WHO (World Health Organisation) Director-General Dr. Margaret Chan has urged greater worldwide surveillance for any unusual outbreaks of influenza-like illness. With increased frequency and ease of travelling in the world today, the threat of a viral wipe-out is very real.

Global Fire-Cracker End

In the exact moment of writing this segment, we discovered a breaking top story in Singapore's daily newspaper *The Straits Times*: "US and Russia Plan Radical Arms Cuts." The two nuclear superpowers are planning unprecedented cuts to their Cold War arsenals of nuclear weapons under a new arms reduction deal. Both countries have been trying to find a replacement for the 1991 Strategic Arms Reduction Treaty (Start I), which led to the biggest reduction in nuclear weapons in history, but have so far failed to reach a new deal.

At the same time this was happening, Iran's President, Mahmoud Ahmadinejad, dismissed a year-end deadline set by the USA's Obama Administration and their allies, for Tehran to accept a UN drafted deal to swap enriched uranium for nuclear fuel. President Ahmadinejad said, "If Iran wanted to make a [nuclear] bomb, we would be brave enough to tell you." The Iranian leader lashed out at the United States, vowing Iran will stand up against the US attempts to "dominate the Middle East."

There are also escalating tensions in other parts of the world: between India and Pakistan, Taiwan/Japan and China, Iran and Israel, and Turkey and Syria. All of these are potential hot spots that could quickly escalate to a worldwide nuclear war because of the alliances between these countries and others for political, religious, or historical reasons. Other countries with nuclear capability include the UK, France, and North Korea.

A nuclear war could be initiated accidentally, aggressively, pre-emptively, and/or in retaliation. While some speak of a limited nuclear war, that scenario is unlikely. Any nuclear war could easily escalate and spiral out of control because of the use-them-or-lose-them strategy. If a country does not use their nuclear weapons quickly enough, they might end up being completely destroyed by the enemy's attack.

Another possible trigger of global nuclear war could be the activities of terrorist organisations that operate in Asia, Europe, South America, and Africa. Some are ideological, others political, and many others religious. In order for them to get their messages heard and remembered, non-combatants must be targeted. They use unconventional warfare on innocent civilians to pressure those in power to change to meet the demands of their agendas. The ease with which terrorists can get their hands on nuclear weapons is one of today's greatest security concerns. Several studies conclude that improvised nuclear devices (IND), built from stolen or diverted fissile materials (either plutonium or highly enriched uranium) pose a real and alarming threat. Terrorist nuclear activities could be mistaken by nuclear-equipped countries to be attacks by another sovereign nation and an erroneous nuclear chain reaction might ensue.

The atomic bombs used in Hiroshima and Nagasaki during the Second World War were estimated to be in the yield range of 15–20 kilotons. They killed approximately 100,000 people and injured 100,000 more. Today, the nuclear bombs in the United States and Russia are in the 15,000–20,000 kiloton range, a thousand times stronger. It is hard to imagine the devastation these bombs would create and the long-term radioactive effects on all forms of life and the environment.

A big threat to countries with nuclear power plants and eventually to the whole world is the force of nature; before it, our technology has been proved powerless many times. One such example is the natural disaster that hit Japan's Fukushima Daiichi power plant on 11 March 2011, taking down the myth of the safe nuclear power plant. A 9.0-scale earthquake and a subsequent tsunami disabled the cooling systems of the plant's reactor, leading to nuclear radiation leaks and triggering a 30km evacuation zone surrounding the plant. According to the BBC, in December 2011, engineers brought the plant to a "cold shutdown condition," but decades will be needed to decontaminate the area and to decommission the plant altogether.

The consequences of such a catastrophe are many. Aside from the obvious but very serious safety and health hazards to the neighbouring communities, or the socio-political and economic disruptions, Hans-Peter Durr, former executive director at the Max Planck Institute for Astrophysics, Germany, says that nuclear power plants can contribute to the increasing number of nations with nuclear weapons, terrorist attacks and "dirty bombs," or radiation disasters due to mishandling or natural calamities. To these, in the light of what has been discussed previously, we can add massive solar flares—that would create an instant collapse, as nuclear power plants are controlled by computers and use electricity to operate, and a shift in magnetic poles—that would greatly affect the function and accuracy of instruments used in nuclear power plant control rooms. With 438 nuclear power plants in the world (as of 2012), the probability of a global catastrophe is close to certain.

The Chinese celebrate festive occasions like marriages and the New Year with fire-crackers; it is a symbolic gesture of chasing away evil spirits and welcoming the new. Perhaps the world will have a fire-cracker end—not to chase away evil spirits, but to turn billions of people into spirits. It is said that the world has enough nuclear fire-crackers to kill the entire planet 30 times over.

Informational Revolution

There is another situation that could lead to an outbreak of war, economic recession, and mass social upheaval, and that is the increasing lack of security of sensitive information over the Internet. With the advent of computers and the Internet, the world has been forever changed. These technological advances have modified almost every aspect of human existence, from the way we do business to the way we socialise; some evolutionary scientists are saying that the course of human evolution has also been changed. How we store and retrieve vital information and the way we manage control systems in electrical power-plants, aviation, military, hospitals, banks, and so on has been affected as well. It also shapes the way we learn and spend our recreational time, how we process complex data from our environment and outer-space through computers, and how we update ourselves on what is happening around the world.

Since most of the vital data on human activities and life support systems are currently stored electronically and digitally and can be accessed anywhere in the world fibre-optically and wirelessly, information security has become the number one national security priority in most countries. Despite all of the efforts put into security, costing billions of dollars, no one can fully guarantee that critical information is absolutely safe.

On 30 November 2010, the *Jakarta Globe* (an Indonesian newspaper) headlines read "US Scrambles to Contain Global WikiLeaks Fallout." On the same day, the *New York Times* declared, "The release of more than 250,000 classified US State department documents has sent shockwaves through the diplomatic establishment, and could strain relations with some countries, influencing international affairs in ways that are impossible to predict." Reuters said, "What WikiLeaks has shown is how much sensitive data can now be stolen at one go and how widely it can be disseminated." Less than a week later, the BBC reported that a "list of facilities vital to US security [was] leaked." Former UK Foreign Secretary, Malcolm Rifkin, said, "This is further evidence that they [WikiLeaks] have been generally irresponsible, bordering on criminal [...]. This is the kind of information terrorists are interested in knowing."

These articles were talking about the November 2010 release of sensitive diplomatic cables published by a site—WikiLeaks—which has gained a reputation for publishing sensitive material from governments and other high-profile organisations. Before this incident, in October 2010, the site released almost 400,000 secret US military logs detailing operations in Iraq. Much of this highly sensitive and classified information was uploaded to the website anonymously, and the team of volunteers at WikiLeaks decided what to publish. Despite warnings and attempts from governments and corporations to shut down the website, WikiLeaks has boldly remained online. It has moved its operations between several companies and countries and has persuaded volunteers to set up mirrors of the website, hosted on different servers all over the world.

Recent sensitive, embarrassing, and/or dangerous information released by WikiLeaks includes:

- Pakistan: the United States has for years led secret efforts to remove highly enriched uranium from Pakistan;
- China: Cyber-attacks against Google "were part of a co-ordinated campaign of computer sabotage carried out by government operatives [...]";
- Saudi Arabia: King Abdullah urged the United States to attack Iran, to destroy its nuclear programme;
- Yemen: On US air strikes against Al Qaeda, President Ali Abdullah Saleh was quoted as saying: "We'll continue saying the bombs are ours, not yours [...]";
- Iran: US Intelligence believes Iran has obtained missiles from North Korea;
- Singapore: Singapore's founding father, Lee Kuan Yew, called North Koreans "psychopathic types" and leader Kim Jong-il a "flabby old chap" who craved public worship.

These leaks of sensitive information, which are still going on, are telling us that the age of transparency has arrived and it is pervading every aspect of our lives. While people all over the world are in shock about the leaks, it is primarily the governments, corporations, and the super-rich that are most affected by this information leakage crisis. Many are violently reacting to these events: some by pleading with people not to release certain information, others by threatening to eradicate all those involved in WikiLeaks. It is not hard to imagine that such leaks could sharply escalate international tensions between countries with a history of bad blood, or even trigger wars. It is also very possible that the leaks could lead to violent stock market movements that could bring about the financial collapse of major corporations and set the world economy into a deep recession.

The message of this global-shift scenario is clear: information is power. With this world-shaking situation where vital information can no longer be held only by certain minority groups, and is made available to anyone who has Internet access, the power-holders of our world are changing. It is up to us to decide what we want to do with this vital information. For good or for bad, the choice is ours.

Climate and Geophysical Chain Reactions

As we have mentioned, the earlier estimates by scientists as to when earth's climate would reach a point of no return, or tipping point, have been drastically revised. For example, it was once predicted that the rising of sea levels by 1.5 metres would be reached by the end of 2100. That date has now been shifted to 2040. We are already seeing early refugees of climate change in the Republic of Kiribati, where two of the islands have gone under the sea. Tuvalu, another island in the South Pacific, has had waves washing the island's main roads, leaving coconut trees partly submerged, and patches of cropland unusable because of salt water flooding. Today, climate-related catastrophes are already a fact.

On 10 December 2009, an iceberg almost twice the size of Hong Kong—87 square miles—and weighing 200 billion tonnes began drifting toward southwest Australia. According to glaciologist Neal Young, an iceberg in those waters is uncommon, and as global warming continues, more can be expected. As it travels northwards, this iceberg is melting and adding to the rising sea levels.

A team of scientists who wrote the article "Casualties of Climate Change," featured in *Scientific American*'s January 2011 issue, reported that the frequency of natural disasters has increased by 42% since the 1980s, and the percentage of those that are climate-related has risen from 50% to 82%.

The work of geophysicists Mioara Mandea and Nils Olsen has confirmed unusual changes in Earth's molten core. Scientists speculate that these changes affect ocean tides, climate-water exchanges, and monsoon seasons. It is also possible that the imbalance between the rotational inertia and gravity is causing the movement of the continental crustal plates, resulting in earthquakes and tsunamis.

The words *faster than expected* are printed in bold in the November 2012 issue of the *Scientific American* magazine. Scientists thought that if planetary warming could be kept below two degrees Celsius, catastrophic rises of sea-levels and searing droughts could be avoided. The latest data from around the world show that in the Arctic Ocean more sea ice is disappearing, large areas of permafrost across Alaska and Siberia are spewing out more methane—a more powerful greenhouse gas than carbon dioxide—, ice shelves in West Antarctica are breaking up and sliding into the sea more quickly than once estimated, and extreme floods and droughts are occurring around the globe. All these data are clearly showing us that the Earth *is* changing faster than expected.

According to James E. Hansen, director of NASA's Goddard Institute, sea levels might ascend by as much as five meters within this century, which would inundate coastal cities from Miami to Bangkok. With the increase in heat, drought, and flooding, food production around the world would be severely affected, making massive famines inevitable. The three most important feedback systems—polar ice, atmospheric carbon (including permafrost methane), and oceanic temperatures and currents—affect almost all major systems on the planet. Winters get colder and longer (deep freezes), certain parts of the world more humid and damp, while others are confronted with searing heat waves. Insect growth rates are affected and hundreds of thousands of acres of trees are killed. A tipping point in any of the three systems would rapidly and dramatically change the habitability of our world.

Is the 2012 Prophecy a Non-Event?

The deterioration across multiple systems that the world has experienced in the last few years—socially, environmentally, and economically—have pushed us to look even more at our current situation and the likelihood of what the future holds. It has not been surprising to find that for every proposition made about what is happening to our world, there is an opposing view and counter argument. Many people spend their time searching for information and knowledge to debunk the theories that have been proposed by scientists and spiritualists. One school of thought argues that climate changes have very little to do with human activities. Another group of scholars have ridiculed leading authors like John M. Jenkins, Jose Arguelles, and Carl J. Calleman on their interpretation of the Mayan end-date. Other researchers attempt to disprove the notion of astrophysical changes and cosmological influences.

The web-links provided in the References and Resources section at the end of this book will dispense more detail on these conflicting points of view. In general, debunkers are trying to calm people or warn them of ill-conceived ideas that some authors or researchers have put forth. Many of these debunkers are saying that the upcoming years will be a *non-event*, which means that nothing out of the ordinary will happen. Alternatively, since it's impossible to predict the future intelligently, they ask, "What's the point in worrying about it?" Many of these caustic critics have based their conclusions strictly on what they can measure (science), while others are calling the spiritual advocates "false prophets."

Upon careful investigation of both sides, we have concluded that science alone cannot provide a coherent understanding of what is happening to our world, because the data are too complex and contradictory. On the other hand, spiritual knowledge without a verifiable framework is blind. It was Einstein who said, paraphrasing Immanuel Kant, "Science without religion is lame. Religion without science is blind."

Given the precarious nature of our world, we must radically overhaul our existing methods and assumptions in order to bring forth a new world where life can be ecologically sustained. We have entered an era in which a new balance is born between matter and spirit. After centuries of being separated, these integral elements of our Universe are reunited. Physics, biology, and biochemistry show us that our world is deeply interconnected. Science is proving what the sages from the East have said for thousands of years: there is no "us," there is no "them."

In the field of quantum physics, matter is highly interconnected, and space is not a vacuum. In fact, space is full of virtual particles, fleeting in and out of existence. This is similar to the Buddhist concept of *shunyata*, which states that space is empty, yet full of potential. These scientific and spiritual reciprocal developments are leading to a new paradigm, in the same way in which our paradigm was shifted when Newton discovered and calculated gravity.

It is perhaps surprising to many that Newton was first and foremost an alchemist, according to J. W. Sullivan, author of Newton's definitive biography. Newton wrote more than one million words about alchemy, which is an ancient practice and philosophy of transmuting lead into gold, which dates back to 3500 BCE. After rummaging through old alchemical texts, Carl Jung concluded that alchemy is really about the psychological processes of finding one's true self. This theory led to his famous postulation on the existence of a collective unconscious, a psychic system of a group consciousness which is identical in all individuals. This hypothesis is similar to the concepts of Ervin László's Akashic Field, David Hawkins's Collective Database, and Rupert Sheldrake's Morphic Field, all of which include spiritual knowledge.

The answers to the major questions of our world, therefore, can only be found within each individual who makes time for his or her own discovery. After all, this matter concerns the future of everyone and everything on this planet.

In recent years, many have been arguing over the possibility of a non-event instead of the end of the world; we, the authors, propose three ways to view this scenario:

a) nothing happens; the world proceeds as usual. But the point is: do we continue living our lives in a business-as-usual way, even though we have witnessed the shocking deterioration of our world into one that is inhumane, inhabitable, and unsafe? If nothing happens in the next few years, does that mean nothing life-threatening will happen in the years to come? Our decision and ensuing actions on the questions stated above are going to impact the future of our children and the fate of our planet;
b) the "shift of the ages" is already happening; it is just not what most of us were expecting: devastating climate changes, plant and animal extinction, depletion of food and water supply, a rise in human conflict, and a decline in human health are all signs of the shift of the ages. In other words, instead of one catastrophic change, critical incremental changes are taking place, and these changes are evoking a shift in planetary consciousness. Millions of people around the world are waking-up to a planetary and collective calling. Little by little, corporations are starting to be socially and environmentally responsible, and eco and ethical practices are being meaningfully implemented. Environmental and humanitarian groups, consisting of individuals with no previous academic and work experience, are making positive rippling changes in poor villages and endangered habitats. Technological advances are providing promising healing to our lands, oceans, and atmosphere. The questions that remain are: will these positive changes proliferate? Are these life supportive changes sustainable? How many people are aware of our challenges and are willing to support them actively?
c) an epochal energetic shift will sweep undetected across Earth leading up to 2017; this all-important energetic shift is so subtle that none of the scientific machines we have can detect it. Therefore, the majority, perhaps 90% of the population on Earth, will not sense anything. This intelligent energy that surrounds Earth has been called many names:
    - The Field, by Lynne McTaggart, author of the international bestseller *The Intention Experiment*;
    - Nature's Mind, by Dr. Edgar Mitchell, astronaut and scientist, founder of the Institute of Noetic Sciences;

- Divine Matrix, by Gregg Braden, author of the *New York Times* bestselling *The God Code and Fractal Time*;
- Noosphere, by Teilhard de Chardin, a well-known French Jesuit, palaeontologist, and geologist, author of *The Phenomenon of Man*; and
- Global Consciousness, a project hosted by Princeton University. This is an international scientific effort to measure the interaction of human consciousness with random event generators to see if it produces non-random patterns.

The details of this intelligent energy—or universal consciousness—will be discussed in Part 3. For now, it is sufficient to say that this complex energy has been evolving since time began, and we have come to a point where it is offering us the opportunity to make a quantum leap in human collective consciousness. That is, a wiser and more balanced way of thinking, feeling, and co-existing—so that ecological solutions for planetary healing, peace, and sustainability can be conceived and implemented.

The Question of All Questions

The question that scientists of all disciplines and spiritual people of all traditions are asking now is: "What is actually going to happen to us?" For the first time in human history this question is asked not only by a particular group of people, by religious believers, or by one particular nation; it is asked by the citizens of our planet. The concern about "us," has expanded beyond our family, society, nation, race, and religious faith; it addresses our future as a species and cohabitants on earth.

Questions are the answers. Questions channel our attention in specific, important points of focus, and they mobilise our abilities and resources to find the answers to act upon. The kinds of questions we ask determine what we look for from moment to moment. This helps us recognise the solution we need; therefore, without the proper questions, we might not recognise the solution, even if it is right in front of us! Obviously, some questions yield better answers than others. So the way we structure our questions is crucial.

Equally (if not more) important is to examine the intention behind our questions. They are connected to our assumptions and mindsets, which control what we accept or reject, attract or expel, take responsibility or blame others for, give life to or destroy, preserve or change. We will talk about the power of intention in later chapters, but for now, it is important that we hold these intentions in mind as we explore some vital questions regarding our future:

- to be open to new ideas, even if they challenge our beliefs and values;
- to invite and expand positive possibilities, while centring ourselves to our inner-wisdom;
- to choose to take responsibility for the affairs of our world; we are intrinsically connected and therefore, how we live makes a difference;
- to consciously choose to be life-supportive rather than life-depletive;
- to think, feel, and act from the perspective of *we*, which includes *me*; our collective life-supportive action is our best hope for the future;
- to be prepared for the necessary discomfort and inconvenience which a positive change might bring along.

Age of New Consciousness & New World

As has been foretold by ancient spiritual traditions (Greek, Hindu, Mayan, Hopi, etc.), the scenario in which the Golden Age begins will bring about peace, harmony, and cooperation and the new universal consciousness will be realised. Our era of separation (between matter and spirit) will be followed by an age of oneness, in which the awareness of the divine presence will once again return to everyday life. This unifying consciousness is a deep awareness and appreciation that we are all one, part of an organically interrelated living system. This transformation begins in our inner-world and it will help us unite our consciousness and collectively change our outer-world.

The idea of a change in global consciousness is far-fetched in most people's map of reality and, therefore, warrants an explanation. If one were to plot a curve representing this rate of change, one would find the curve soon approaches the vertical; mathematicians call such a point a "singularity" (mentioned earlier; see the section Coming to Zero Time, in Chapter 1, above). This rapid change that has been accelerating exponentially for billions of years is going to come to an end soon. Most experts in this area agree that this singularity in time will occur by 2050. While it is logical to assume that it will be brought on by superfast and intelligent computers, Peter Russell, a world-renowned scientific futurist, believes technological progress will only be part of the development of the singularity. As self-organising systems become more and more complex, they jump class in consciousness. When we look closely, all systems are currently converging—politics, climate, environment, genetics, economics, and society. Nature tends to form whole systems out of different micro-systems; the whole is greater than the sum of its parts.

From the agricultural to the industrial and to the current informational age, we see an evolutionary trend toward less matter and more light/energy, as in fibre optics, nanotechnology, and quantum energy generation. It is therefore very possible that the next stage of evolution on earth is a quantum leap in human consciousness.

As our external world worsens, this will fuel our inner worlds to resonate to global consciousness. With the combination of exponential advances in technology and quantum leaps in collective consciousness, it is possible that we clean up our mess in an unprecedented short time. A new world of peace, harmony, and unity will emerge with global shifts in priorities, a redefinition of trade and compensation, redistribution of wealth and resources, acceleration in development of renewable energies, rapid reduction of poverty, global cooperation in rejuvenating the planet, and other positive changes.

A danger that comes with embracing this scenario is that people think that the world transformation is on its way. People might hold the attitude, "Nothing needs to be done; it's all happening anyway." Such an attitude is naive, a form of escapism, deficient in thinking, and downright disempowering. This Golden Age is going to unfold according to our collective intentions and actions. This Golden Age is asking us to become co-creators of a new way of life. But, how many of us are awakening to this invitation? More importantly, how many people in the world are living their lives in ways that support the manifestation of this Golden Age?

Pause: Metamorphosis

The Collins Dictionary defines *serendipity* as "the gift of making fortunate discoveries while searching for other things." It fittingly describes my extraordinary journey of writing this book. Just about four days ago, as I was writing this chapter, I received an email from an associate of mine, who recently agreed to assist me in doing research for this book. In his email, David An Wei, a native of China, eagerly shared with me a teaching he had received from his spiritual teacher on the meaning of 2012:

> The shift may not be in 2012 and it may not even happen,
> based on the principle of uncertainty in quantum physics.
> It really depends on the state of our consciousness, as it would
> determine what we unfold. But if the shift does happen,

it would be very good for us, as we have been
in the three-dimensional (3D) world for so long.
Our transformation from a 3D to a 4D world is like a
caterpillar (2D) turning into a butterfly (3D). When a butterfly
tells a caterpillar (which is himself/herself) that he/she can be
a butterfly in the near future, a 2D caterpillar can only
perceive fleeting, unrecognisable flashes of the butterfly
(2D creatures can't perceive higher dimensional entities),
and therefore may not believe what the butterfly is saying...

Out of the blue—and very suitably—David gave me a spiritual perspective that was very appropriate for me. You see, for three years now, I've been coming across butterflies and moths everywhere. There was a period of eight months when I had frequent encounters with palm-size butterflies and moths. At first, they were just, well... butterflies. What's the big deal, right? Then, I started seeing them in public toilets, right on top of the urinal; then in front of the lift-doors of my office building, flying past my client's office window, on the 20th floor, landing on my cabby's windscreen on a jammed highway, and in many other such unusual places. But the weirdest of them all was when a butterfly was lying on top of my kitchen cooker-hood, staring right at me. There were two other family members in the kitchen, and they had not seen it! It was at that moment that I realised that there was a message for me.

The Symbolism of Butterflies

Almost universally, butterflies are honoured as a symbol of transformation. Through the process of metamorphosis, the caterpillar naturally transforms itself into a completely new entity. This transformation parallels the alchemical process of turning lead into gold and, in terms of ancient cosmology, the transformation of the Iron Age into the Golden Era. Here are some of the essential qualities and themes of the butterfly, which are great metaphors for what is happening to us, humans:

I. the constancy of change: from an egg to a caterpillar, a chrysalis, and a butterfly; we need courage to leave our old ways of life behind and accept the change cycles within ourselves and the environment;
II. transitions represent redefinitions; life has its stages and cycles; the butterfly embraces the changes by redefining itself. Inside the chrysalis it faces stagnation, darkness, pain, and uncertainty. These inner struggles serve to provide the strength and confidence we need to actualise the next life;

III. breaking free; each phase offers the butterfly an opportunity to break away from the old existence in order to wake up to a new, richer, and more meaningful one; and the same goes for us. This requires us to give up the comfort and security of the life that we know so well;

IV. spirit; in some cultures, a butterfly is a messenger from the spirit/dream world. It also symbolises the spiritual dimension of our physical being. It reminds us to stay constantly connected to our spirit;

V. a second birth; after a death-like process in the chrysalis stage, a beautiful butterfly emerges. In metaphysics, this symbolises the end of matter and the beginning of spirit. It is about living again in a new form. For us today, it means ending individualistic materialism and living again with a collective spiritual ideal;

VI. serving a bigger and higher purpose; by transforming itself and by accessing the dimension of flight, the butterfly fulfils the vital function of pollination and reproduction. Thus, it serves its purpose in the circle of life. In order to actualise our higher potentials we must face the unknown, and it is through faith that we transcend fear.

Butterfly Relationship to 2012

Many of us can relate to the symbolism of the butterfly in our lives and the world we live in today. According to David An Wei's spiritual teacher, not everyone will make it to the higher dimensions (butterfly). The idea that ascendance is available to everyone, but is not given or automatic, is supported by many esoteric and metaphysical teachings. In other words, we have to work on raising our consciousness, and our collective consciousness will determine which scenario we will unfold.

Metaphorically, some people may remain in the lower dimension (caterpillar), and be like dinosaurs, that served the purpose of evolution and were finally replaced by new species. A significantly large portion of the population might be in the cocoon stage: they might feel isolated and experience a high level of inner chaos, resulting in doubt and uncertainties. Their surrendering to the divine change coupled with the strength of their faith would determine their eventual reality. There's a smaller percentage of population who have been preparing themselves, breaking away from the heavier and denser lower dimensions of materiality and living the higher-dimensions of lightness and illumination. They are currently living in accordance with the world they are envisioning: they are being the change they want to see. This group, along with the larger group of "cocooners," would tip the scale and bring forth the desired world. In Part III we will explore what it takes for one to be transformed.

PART II
THE FIRST PATH—STRIFE AND DESTRUCTION

The Hopi pictograph in Chapter 1 shows that we are coming to a point where we must choose between the first path—a jagged line indicating chaos and devastation—and the second path—a smooth line leading to peaceful abundance. If the people in the pictograph continue walking at the same level on which they are when they reach this point of no return, they go onto the first path. They must choose to change levels in order to step on the smooth line.

This section of the book takes a deep look at the route we are advancing on at the moment, and where it might be leading us. This includes the state of our world today: environmentally, financially, socially, spiritually, and in terms of health and well-being. We know from our own experience of doing this research that looking so deeply into the current state of things, and into all the problems we face, can be a painful and sometimes discouraging process.

It is tempting to want to believe that things aren't as bad as they appear to be or that there is nothing we can do about them anyway, so we might as well not think about it. It is also easy to go into deep fear when we see how serious the situation really is.

In spite of this, we have reached the conclusion that we need to confront these facts, however unpleasant, so that we can see the dire urgency of making the necessary changes to alter the path of destruction that we are already on. In this section, we intend to present to you our distillations of what is happening in the world, with key scientific references, and the choices we have in order to shape the future of our world. There is no lack of knowledge today: we are just a few mouse-clicks away from all that we want to know. We have provided many references for books, journals, newspapers, as well as web-links in the "References and Resources" section.

Our research into ancient and modern wisdom has convinced us that there *are* important actions we can take—this is the good news. By making choices that support life, we can ensure that the changes in the near future will be positive instead of disastrous. We can also ensure that we don't reach a point where we would be forced to find the best ways to make it through difficult times. We believe we can only do this if we squarely face problems and make conscious choices for change based on reliable information.

The information in the following chapters has convinced me that we are nearing a tipping-point of some kind; this could be within two-to-three decades, at most, and very possibly within the next few years (2013-2017). As a species, I think we have to make choices about the dangers we are facing, and we have to make them now, so that catastrophes can be avoided. Time is of the essence, because choices start to slip away as we come closer to a tipping point.

CHAPTER 3
WHAT'S HAPPENING TO OUR WEATHER?

WEATHER LANGUAGE

Have you noticed the presence of weather words in our daily lives? We say things like, "When it rains, it pours," "The boss thundered into the room," "You are the sunshine of my life," "The winds of change are shifting…" and so many more. Clearly, in some intimate and deep ways, the climate and the environment shape the human psyche and the way it creates reality. It is not inappropriate, then, to approach someone and ask them, "How's your weather today?" because collectively, as the dominant thriving species on the planet, we might be the ones creating the weather that we are now experiencing.

CRAZY WEATHER

In the cold light of day, we can no longer ignore the clear and loud messages from our weather. On 8 February 2009, *The Sunday Times* (Singaporean newspaper) questioned, "What's happening to the weather?" Down under, just five hours from Singapore, Australians were experiencing 46.4 degrees Celsius in Melbourne, with intense wildfires and rising death tolls. On this same land mass, in the northern state of Queensland, flood waters continued to rise, destroying 3,000 homes, a fifth of the region's sugarcane fields, and stranding tens of thousands of cattle. The same article noted, "Parts of Europe went into deep freeze, America's groundhog declared six more weeks of winter, while Argentina and China were in the throes of a drought." Britain was experiencing the coldest winter in 20 years, China and Argentina were suffering from the worst drought in 5 decades, and Moroccans in North Africa were fleeing from floods. What is the cause of all this extreme weather that we are experiencing all around the world?

WEATHER SCIENCE

The Intergovernmental Panel on Climate Change (IPCC) stated clearly that, "Warming of the climate system is unequivocal, as is now evident from observations of increases in global average air and ocean temperatures, widespread melting of snow and ice and rising global average sea level." The IPCC was jointly set up by the World Meteorological Organization and the United Nations Environment Programme to provide an authoritative international scientific assessment of climate change. They were awarded a Nobel Peace Prize in 2007 for their critical contribution to the study of climate change.

The IPCC's Climate Change 2007 Synthesis Report states that the primary cause of global warming is greenhouse gas emissions created by human activities. The studies on this subject were very complex, and it is not easy to identify a single cause and explain it in a simple manner. Earth's atmospheric concentrations of carbon dioxide, methane, and nitrous oxide have increased markedly, and it is predominantly due to man's agricultural activities and burning of fossil fuels. More importantly, there is extensive agreement among IPCC scientists that with the current climate change mitigation policies and related sustainability development practices, global greenhouse gas emissions will continue to rise over the next few decades, despite the devastation to our climate they have already caused.

This is largely due to the economic expansion in China and India. The International Energy Agency, based in Paris, boldly stated that around 60% of the global increase in carbon dioxide emissions in 2005–2030 will come from China and India.

Earth's Air Blanket

Earth's atmosphere is a layer of gases surrounding the planet that is retained by Earth's gravity. This mixture of gases becomes thinner until it gradually reaches space. The atmosphere protects life on Earth by absorbing ultraviolet solar radiation, warming the surface through the greenhouse effect, and reducing temperature extremes between day and night. Through this natural greenhouse effect, some of the heat from the Sun is retained in Earth's atmosphere. Without the greenhouse effect, Earth's temperature would be below freezing. However, Earth's greenhouse effect is getting stronger as we add more greenhouse gases to the atmosphere.

This increase in greenhouse gases is largely caused, as we have mentioned above, by the burning of fossil fuels and agricultural practices that release these gases and other air pollutants into the atmosphere.
The amount of carbon dioxide in the atmosphere has increased by 25% since the middle of the 19th century, and it is expected to double by 2050! The levels of other greenhouse gases, like methane and nitrous oxide, are increasing as well. With more greenhouse gases in the air, heat gets trapped on its way out of Earth's atmosphere, and this is increasing the temperature of our planet. Thousands of scientists around the world are scrambling for solutions to reverse global warming.

Toxic Expansion

From now on, China and India are taking the centre stage when it comes to worsening the planet's ecology, along with the United States (currently ranked third). Energy experts and international climate officials know that within 25 years the Chinese will have melted the Himalayan ice-cap with incalculable consequences for themselves and the world. Without taking into consideration precipitation factors, more than two billion people (one-third of the world's population) rely on the Himalayas for potable water.

Even if the rest of the world were to limit carbon dioxide emissions to current levels for the rest of the century, China's growth alone will ensure that world temperatures would rise by up to three to four degrees by 2100. China is burning so much coal, so inefficiently, that by 2030, its carbon dioxide emissions will equal Europe's, and it will have emitted more into the atmosphere in 25 years than Europe and America have done in the past one hundred years.

In its 2008 report, the World Energy Outlook stated that the world's energy system is at a crossroads: the current global trends in energy supply and consumption are obviously unsustainable. To prevent catastrophic and irreversible damage to the global climate ultimately requires a major de-carbonisation of the world energy sources.

Vinegar Rain

When fossil fuels are burnt, sulphur dioxide and nitrogen oxide are released into the atmosphere. These chemical gases react with water, oxygen, and other substances to form mild solutions of sulphuric and nitric acid. The effects of acid rain, combined with other environmental stressors, leave trees and plants less able to withstand cold temperatures, insects, and disease. The pollutants may also inhibit trees' ability to reproduce. Acid rain has many effects on ecology, but none is greater than its impact on lakes, streams, wetlands, and other aquatic environments. It makes water acidic and causes it to absorb the aluminium that goes from soil into lakes and streams. This process makes waters toxic to crayfish, clams, fish, and other aquatic animals. Some species can tolerate acidic waters better than others; however, in an interconnected ecosystem, an impact on some species eventually impacts many more through the food chain, including non-aquatic species, such as birds.

Global Impact of Vinegar Rains

Acid rains are already having a negative impact on the entire world. In Brazil, fish are disappearing in reddish water, trees are turning to skeletons, and human health is immensely affected. Poland is one of the heaviest burners of coal in the world, and there are now complex illnesses in cities close to coal-burning factories; in Krakow, the golden roof of a chapel is quickly dissolving. In Czechoslovakia, children are suffering from breathing problems at a high rate due to acid rain and much of the fresh water is too acidic to drink. Scientific research carried out by Dr. Leon Rotstayn, from Australia's Commonwealth Scientific and Industrial Research Organisation (CSIRO), and his colleague, Professor Ulrike Lohmann, from Canada's Dalhousie University, said that pollution from Western countries may have caused the droughts that ravaged Africa's Sahel region in the 1970s and 1980s. Millions died in those droughts, which hit Ethiopia hardest in 1984.

Vinegar Rain Thickens in Asia

Acid rain is now emerging as a major problem in the developing world, especially in some parts of Asia and the Pacific region, where energy use has surged and the utilisation of sulphur-containing coal and oil is very high. An estimated 34 million metric tonnes of sulphuric and nitric acid were emitted in the Asia region in 1990, over 40% more than in North America.

Acid deposition levels were particularly high in areas such as southeast China, northeast India, Thailand, and the Republic of Korea, which are near major urban and industrial centres. The China Meteorological Administration (CMA) statistics show that of the 155 acid rain monitoring stations located across the country, in August 2006, 19 recorded acid rains on each rainy day and 51 recorded acid rains on half of the rainy days. In Beijing, 80% of the rainy days were acidic. The regions most affected by acid rain spread from the northeast to the southwest of China. Soils and water that soak up acid rain transfer the pollution to crops, and eventually into people's corneas and respiratory tracts. China is the world's biggest sulphur dioxide polluter, with 25.5 million tonnes of sulphur dioxide discharged in 2005.

Singapore's *The Straits Times* newspaper reported on 14 September 2009 that "20 species of animals, plentiful in the Bukit Timah Nature Reserve in the 1980s, including frogs, crabs, and fish, are slowly being wiped out," according to a four-year study led by Associate Professor David Higgitt of the National University of Singapore. The team led by Professor Higgitt noted that the stream, which covers five hectares of land, is more acidic after rainstorms. The acidity of the water comes from industrial pollutants and lightning that dissolve in rain water, which then falls into streams and other bodies of water.

DARKER OCEANS

Changes in the chemical composition of the atmosphere trigger alterations in climate. Climate change has already reached a critical point. The western Arctic temperatures are at a 400-year high. Satellite pictures taken in September 2005 proved that the extent of the Arctic ice cap was 20% below the long-term average for that month. If this condition persists, the Arctic will be ice-free before 2030. The warming process feeds on itself: as ice disappears, the surface of the sea becomes darker, absorbing more heat; less ice is formed, which means that the sea becomes still darker, absorbing still more heat. This rapid reduction of the Arctic ice cap is changing the world's weather. While Europe experiences Siberian cold, most of the planet is subject to rising temperatures. If no definitive action is carried out to reverse the warming trend within the next 25 years, the damage to the planet will become *irreversible*. There will be an almost total destruction of coral reefs, most fynbos shrublands in South Africa will be ruined, the alpine flora of Europe, the Amazon and Borneo rain forests, many parts of Australia will dry out, and most of China's broad-leafed forests will disappear.

DISEASES CAUSED BY CLIMATE CHANGE

In 2005, the World Health Organization (WHO) estimated that climate change led to more than 150,000 heat-related deaths and 5,000,000 illnesses annually, and this toll could easily double by 2030. Tony McMichael, professor of epidemiology at the London School of Hygiene and Tropical Medicine, said that diseases that breed in warm and wet conditions will likely become more prevalent in the next 20 years. This would mean widespread incidences of diarrhoea, and of cardio-respiratory and infectious diseases. There's an inseparable connection between the quality of our air, environmental temperature, and our health.

According to the World Animal Health Organization, a Paris-based agency with 174 member countries, from a survey done in 2009 on 126 of its member states, 58% had already identified one climate-change disease that was new to their territory. The world's experts in animal health said that climate change is proliferating viral disease among farm animals and some microbes are potentially lethal to humans. Three of the most mentioned diseases were Bluetongue—spread among sheep by midges, Rift Valley fever—a livestock disease that can also infect humans through handling infected meat, and West Nile virus—which is transmitted by mosquitoes from infected birds to both animals and humans. These diseases increasingly take their toll on human and animal lives.

Breathing Problems

According to the *WHO Air Quality Guidelines* (2005), air pollution, both indoors and outdoors, is a major environmental health problem affecting everyone in developed and developing countries. In a conservative estimate, urban air pollution has contributed to over 500,000 deaths in Asia. People living in Guangzhou (China) have a greater chance of getting lung cancer from polluted air than from smoking cigarettes. Pollution—particulate matter, ground-level ozone, nitrogen dioxide, and sulphur dioxide—has caused many severe lung-related problems. Chronic exposure to particulate matter contributes to the risk of developing cardiovascular and respiratory diseases, as well as lung cancer. Epidemiological studies have shown that symptoms of bronchitis in asthmatic children increase in association with long term exposure to nitrogen dioxide. Produced primarily by burning fossil fuels, sulphur dioxide causes coughing, mucous secretion, aggravation of asthma, irritation of the eyes, and it makes people more prone to infections of the respiratory tract.

Scientific data gathered from prehistoric times indicate that the oxygen content of the atmosphere was well above today's level of 21% of the total volume. The percentage of oxygen in the air has decreased in current times, largely due to the burning of coal, which began in the 1850s. The oxygen content of the atmosphere today wavers at 19% over impacted areas, and is down to 12–17% over major cities. This level is insufficient to keep body cells, organs, and the immune system operating optimally; cancer and other degenerative diseases are very likely to develop. If the oxygen level drops to 6–7% of the volume of air, life can no longer be sustained.

Interconnectedness

As our planet's ecosphere and ecosystems are intimately connected, it is impossible to isolate the precipitating patterns, such as the melting of mountain and polar ice-caps, the rising sea-levels, hurricane and cyclone occurrences, plant and wildlife extinction, agricultural disasters, human health dangers and many other negative effects. *The Millennium Ecosystem Assessment Synthesis Report* (2005) put together by 1,300 experts from 95 countries, emphasised that approximately 60% of the ecosystem services that support life on earth were at *serious risk* of being damaged by pollution, overuse, and mismanagement. This includes ecosystems essential to human living, such as water, agriculture, and fisheries. The lack of a sustainable climate management system is at the core of what is causing the weird weather we have been experiencing, and probably a main factor that could lead to ecosystem collapse in the future. And this was in 2005!

It is scientifically clear now that a climate change in one area of the planet does affect other areas—often quite rapidly. Forest fires in Borneo contribute significantly to air pollution in Singapore. Extreme droughts affecting the agriculture production of the Midwestern US shape food prices in Hong Kong. Flooding of cornfields in China starves children in Africa. A small whirlpool in the South Pacific Ocean can cause torrential rain and flooding in Indonesia. The interconnection of living systems is becoming clearer every day: we have *one environment*, we inhabit *one planet*, and we are *one people.*

PAUSE: WISDOM FROM "THE ELEMENTS"

I inhaled a breath of "fresh air," slowly and deliberately; my mind cleared and I asked myself, "What does all this mean?" As I pondered the question, my attention drifted to my slightly blocked nostril, and indeed, the clue was right under my nose—the air that I breathe. Breathing isn't something we consciously do, it happens by itself. The intelligence within our bodies does it on its own, and because we are not aware of our breathing, we take it for granted. We can live without food for a month, without water for a week, but without air—no more than three minutes.

We are caught up with what we need to do almost every moment of our lives, mostly pushed by the judgements and fears of our past and pulled by our desires for our future. In our quest for a better life, we have unintentionally destroyed, contaminated, and destabilised vast landscapes and oceans. We have depleted, misused, polluted, and destroyed the essential elements of life—water, wood, earth, fire, and metal—which support our existence. These elements are the five essential elements in the ancient Chinese healing philosophy. They are understood as different types of energy in a state of constant interaction, movement, and exchange. According to this healing philosophy, changes in the climate, time of the day, and geographical locations affect humans biologically. Nature is, therefore, regarded as a fractal of the human body, and vice-versa.

BODY AND THE UNIVERSE

In the ancient Chinese healing system, known today as Traditional Chinese Medicine (TCM), *metal* corresponds to lungs, *fire* to heart, *earth* to spleen, *wood* to liver, and *water* to kidneys. These elements and parts of the body work intelligently together, having their own inherent wisdom. Each element is not only associated to an organ system, but it also corresponds to a sensory organ, to emotional states, to physical and emotional symptoms, and to a general orientation toward life.

It is interesting to notice the correlation between metal, which corresponds to our lungs in TCM, and the issue of global greenhouse gas emissions that are contaminating the air. From a microcosm-macrocosm perspective, which is the recognition of the same traits appearing in entities of different sizes, one can say that our planet is wheezing, coughing, and having a severe runny nose. This *dis-ease* is accompanied by intermittent bouts of cold and fever. Analogously speaking, one can also say that the planet is like a 21-year-old walking into a TCM clinic with these symptoms and asking what the TCM doctor can do for him. The patient admits that he is a heavy smoker—up to 90 unfiltered cigarettes a day. Upon closer examination, the TCM doctor discovers that half of his lung has collapsed. What would the TCM doctor do?

Building on this analogy, one can say that the planet is addicted to smoking (burning of fossil fuel/minerals). Nicotine helps to stimulate unnaturally faster movement of the energy (*chi* or *qi*) in the body. One might also say that "smoking" makes a country more vibrant and productive, therefore the number of states that want to emulate this vibrancy and productivity is increasing. The smoker (industrialised countries) pollutes the atmosphere, making the air impure and warmer, sometimes dryer, and sometimes wetter. With half of the earth's forests gone, the carbon and oxygen exchange is adversely affected. In TCM, the key approach to restoring health is through balance.

YING AND YANG: COMPLEMENTARY OPPOSITES

In Chinese history, the theories of the Five Elements and the two forces—yin and yang—co-existed since as early as 3000 BCE. *Yin* represents everything in the world that is dark, hidden, passive, receptive, yielding, cool, soft, and feminine. *Yang* represents everything in the world that is illuminated, evident, active, aggressive, controlling, hot, hard, and masculine. Everything in the world can be identified with either yin or yang. Together, they provide the intellectual framework for much of the Chinese studies of cosmology, psychology, philosophy, biology, and medicine. Illness is seen as a disturbance in the balance of yin and yang or the Five Elements. Thus, treatment or interventions depend on the accurate diagnosis of the source of the imbalance. Smoking weakens the lungs' function of commanding the overall movement of qi and blood along the body's meridians, by creating heat in the lungs and stomach, which in turn causes fire in the heart and likely stagnation of the liver's qi. *The Yellow Emperor's Classic of Medicine* (2497 BCE), an ancient text that forms the basis of Taoism and serves as the highest authority in Traditional Chinese Medicine, emphasises the importance of maintaining balance between the yin and the yang to attain good health:

> If Yang is overly powerful, then Yin may be too weak.
> If Yin is particularly strong, then Yang is apt to be defective.
> If the male force is overwhelming, then there will be
> excessive heat. If the female force is overwhelming,
> then there will be excessive cold.

The Wisdom of the Tao

To be in alignment with this all-important theme of balance, we must strive to exist in harmony with the entire Universe, cooperating with the principles of the whole. While the sky, ocean, and trees are in a state of wholeness, our ego insists that we are separate, distinct, and superior. The moment we see ourselves as being superior to others or to our world, we are interfering with the wholeness. To further appreciate this wisdom, let us tap into the *Tao Te Ching (The Book of the Way)*, one of the most profound books in philosophy:

Verse 39

In harmony with the Tao,
the sky is clear and spacious.
Earth is solid and full.
All creatures flourish together,
content with the way they are,
endlessly repeating themselves,
endlessly renewed.
When man interferes with the Tao,
the sky becomes filthy.
Earth becomes depleted.
The equilibrium crumbles.
Creatures become extinct.
The Master views the parts with compassion,
because he understands the whole.
His constant practice is humility.
He doesn't glitter like a jewel
but lets himself be shaped by the Tao,
as rugged and common as stone.

—Lao Tzu, the father of The Way (Tao)

Lao Tzu firmly believed that wholeness can only be sustained with humility and reverence. With these qualities, we live a reality in which we are aspects of the whole. We see and feel the interconnectedness of everything. Our body is a convenient analogy of the Universe. Although it is one entity, it consists of trillions of individual, intimately connected cells. Just one cell with a self-centred and disconnected relationship with the whole makes all the other cells suffer. This is much like the individual who disrupts the Tao by polluting the sky, exploiting the soil, oceans, forests, and the animals.

The first step to any change is awareness. Begin to feel a connection to all life, rather than the separateness that our egos prefer. Our lives need to have a relationship with nature, and that relationship is kept alive with our humility and reverence for life. We must begin to see ourselves in everyone we encounter, in every creature on our planet, in forests, oceans, and the sky. The more we connect and relate to these elements, the more we want to stay in a state of cooperation, rather than separation and competition. Whenever we pay attention to our breathing, we are absolutely present, and the positive change we seek starts from where we are in each moment.

CHAPTER 4
WHAT'S HAPPENING TO OUR WATER?

ABUNDANTLY AVAILABLE

"Have you brushed your teeth?" "Wash your feet before you pray." "Please remember to flush." "Go take a shower!" Which parent has not encouraged in his or her child the frequent use of water? What's there to consider when it is so readily available via the tap, shower, and toilet? But will it always be so?

Four-fifths of our earth's surface is water. However, salt water represents 97.5% of all water on earth, the remaining 2.5% being fresh water. Two-thirds of the fresh water is frozen in the polar ice-caps and the rest is mainly underground or in soil moisture. That leaves less than 1% of the fresh water on earth accessible for direct human use and this is the water in rivers, lakes, and reservoirs. Do we value our fresh water? Is it accessible to all humans on this planet?

WATER UTILISATION AND DISTRIBUTION

Water usage differs significantly between developing countries and developed ones. Developing countries use 90% of their water for agriculture, 5% for industry, and 5% is consumed in urban areas. Developed countries use 45% for agriculture, 45% for industry, and 10% is left for urban-area consumption. An average human needs 5 litres of water a day for drinking and cooking and 25 litres for personal hygiene. An average American uses 350 litres a day, of which 80 litres only for flushing the toilet! An average European or Japanese uses 130–150 litres a day. Contrast this with the fact that 48% of Africans lack access to water that is safe for drinking and cooking; many have to walk for two miles to get safe water!

At the beginning of the 21st century, the earth, with its diverse and abundant life forms, including humans, is facing a serious water crisis. The seven billion people on this planet are using nearly 30% of the world's total accessible renewable supply of water. This usage is expected to rise to 70% by 2025. Yet one-third of people lack basic water and sanitation and are now facing near catastrophic water shortages. That number may increase to two-thirds by 2025. The worst-hit areas will be Africa, the Mediterranean Basin, the Middle East, and south and central Asia. The World Health Organization estimates that over five million people die annually from diseases caused by unsafe drinking water, lack of sanitation, and insufficient water for hygiene.

The latest report by the Leadership Group on Water Security in Asia (2009) headed by Mr Tommy Koh, Singapore's Ambassador at Large, states that today, one out of six people (more than a billion worldwide) does not have adequate access to safe water. The UN projects that by 2025, half of the countries in the world will face water stress or outright shortages. By 2050, as many as three out of four people on planet Earth could be affected by water scarcity.

Acute Shortages in Asia

Asia is particularly susceptible to water-related problems. With more than half of the world's population, this continent has less fresh water than any continent other than Antarctica—only 3,290 cubic metres per person per year. With a growth of almost two-thirds of the global population expected to occur in Asia (500 million within the next 10 years), water requirements would rise sharply. Presently, 48% of all people live in towns and cities; by 2025 this number will rise to 60%. The logic of urbanisation is clear: those countries that urbanised most in the past four decades are generally those that have the largest economic growth. Urban areas logically provide the economic resources to install water supplies and sanitation, but they also escalate and concentrate waste. Without a good waste management system, urban areas are among the world's most health-endangering environments.

Scientists agree that reduced access to fresh water will lead to a chain of consequences, including impaired food production, the loss of livelihood security, large-scale migration within and across borders, and increased economic and geopolitical tension and instabilities. Over time, these effects will have a dramatic impact on security throughout the world.

Global Efforts to Increase Water Access

The Water for Life International Decade was established by the United Nations General Assembly and launched on World Water Day in 2005 (on 22 March). It calls for efforts to increase access to water and sanitation for everyone on earth. It aims to fulfil international commitments made on water and related issues by 2015. It coincides with the Education for Sustainable Development International Decade, which is also set for 2005-2015. This is the second time that water issues have been highlighted, as the UN declared the period 1981-1990 to be the Drinking Water Supply and Sanitation International Decade, with the aim of providing safe drinking water and adequate sanitation systems for all people by 1991. An excerpt from the UN World Water Development

Report (2003) clearly states that the outlook of the global fresh-water shortage is gloomy:

> Reform and liberalization of the water sector,
> better valuation of water, and more private sector involvement,
> could bring forward new technology and [...] could enable us to
> muddle through. This however, is to take an optimistic view.
> The realist would have to say that, on the basis of the evidence
> put forth by this first WWDR, the [water-access] prospects for
> many hundreds of millions of people in the lower income countries,
> as well as for the natural environment, do not look good.

IMPACT OF ECONOMIC DEVELOPMENT AND POVERTY

Water use is dependent on population size and economic development. As the levels of income growth increase, domestic water use rises. Poverty, on the other hand, results in overworking the land, contaminating rivers and lakes, and lowering the water table. Agriculture is responsible for 87% of the total water used globally. In Asia it accounts for 86% of the total annual water withdrawal, compared with 49% in North and Central America, and 38% in Europe. Rice growing, in particular, is a heavy consumer of water. Another problem is that rapid erosion of many agricultural lands and chemical use in agriculture increase the expansion of algae that choke lakes and waterways. Refuse and 100,000 chemical compounds are dumped into rivers and seas, poisoning marine life, especially in coastal regions. Every day, 250,000 tonnes of sulphuric acid fall as acid rain in the northern hemisphere (which includes Asia). In short, most of the water we are using today is contaminated, whether it is used for agriculture, industry, or domestic consumption.

BP Oil Spill Damages Ecosystems

On 20 April 2010, the largest accidental marine oil spill in the history of the petroleum industry occurred in the Gulf of Mexico: an explosion on the Deepwater Horizon oil drilling rig (owned by Transocean's Triton Asset leasing GmbH and leased to BP) led to a series of human and mechanical errors that allowed natural gas under tremendous pressure to shoot onto the drilling platform, causing a blast and fire that killed 11 crewmen and injured 17 others, sending 4.9 million barrels of oil into the water in a flow that was unabated for 87 days. Two million gallons of toxic chemicals known as *dispersants* were used for the clean-up soon after, which, when combined with crude oil, were found to be 52 times more toxic than oil alone.

The consequences were devastating to what used to be one of the richest seafood grounds in the United States, affecting many ecosystems; the spill has greatly disturbed marine life and hundreds of miles of beaches, has eroded marshlands, and caused billions of dollars in damage. Although scientists believe that it may take a decade before the full effects of the disaster are apparent, fishermen found shrimp with tumours, eyeless fish, and crabs with holes in their shells; organisms on the seafloor and plankton—vital to the aquatic food chain—were destroyed, creating a big imbalance in the food web, which ultimately affected and still affects the fishing industry as well as consumers; 6,100 birds, 600 sea turtles, and 153 dolphins were found dead, and many more were injured, as the accident hit at the peak breeding season for many species of fish. The Gulf Coast states that rely heavily on commercial fishing and outdoor recreation to sustain their local economies suffered a great decline.

In a bigger picture, the usage of fossil fuels is causing temperatures to rise, which, in turn, is triggering many of the natural calamities we are confronted with today. According to *The Guardian*, UK, "[...] these extreme extraction processes [deep-water drilling] are the fossil fuel industry's last-ditch efforts to stretch their financial dominance as far into the 21st century as they possibly can."

Climate Change and Water

IPCC's observational records and climate projections released in 2008 provided evidence that freshwater resources are vulnerable and can be strongly impacted by climate change. This has wide-ranging consequences for human societies and planetary ecosystems. Global warming, observed over several decades, has been linked to changes in the large-scale hydrological cycle, such as the increase in atmospheric water vapour content, the change in precipitation patterns (intensity and extremes), dwindling snow cover and widespread melting of ice, changes in soil moisture, and runoff.

Extra Dry and Extra Wet

Warm air holds more water vapour, so a hotter planet is one where the atmosphere contains more moisture. For every degree Celsius increase in the atmosphere, there is approximately 7% more water vapour in the air closer to Earth's surface. This will not necessarily turn into more rain, because of the dynamics of precipitation, which occurs when the atmosphere, a large gaseous solution, becomes saturated with water vapour and the water condenses, i.e., it precipitates. Two processes, possibly acting together, can lead to air becoming saturated: cooling the air (dry) or adding water vapour to the air (wet). "The basic argument would be that the transfers of water are going to get bigger," explains Isaac Held, a scientist at the National Oceanic and Atmospheric Administration's Geophysical Fluid Dynamics Laboratory at Princeton University (United States). A good general rule of thumb is that the wet areas are going to get wetter, and the dry areas—drier. Between 1991 and 2000, more than 665,000 people died in 2,557 natural disasters, of which 90% were water-related. Of these water-related disasters, floods represented about 50%, water-borne and vector-borne diseases—about 28%, and droughts—11%.

Australia Is Drying out

In 2009, the people of the Murray-Darling Basin were facing one of Australia's most devastating droughts of seven continuous years. Once a dry and semi-fertile land, this region was turned into Australia's granary through a massive water-management programme, which dammed rivers, filled reservoirs, and tapped water for irrigation and other human needs.

The early European settlers followed the habits of their homelands without studying the natural ecosystem. They chopped down some 15 billion trees, unaware of what it would mean to disturb an established hydrological cycle by uprooting vegetation well adapted to arid conditions. The New Australians introduced sheep, cows, and water-hungry crops (cotton and rice) which are foreign to a desert ecosystem. The manipulation produced unintended and serious consequences. Irrigation caused soil-salinity levels to skyrocket, which in turn poisoned wetlands and rendered large stretches of acreage unfit for cultivation. The endless ploughing to encourage Australia's new bounty further accelerated the degradation of its soil.

The long seven-year drought (2002–2009) brought farmers to their knees. They sold most of their cattle because they couldn't stand lying in bed every night hearing their animals screaming in hunger. Hundreds of farmers committed suicide, drowning in hundreds of thousands of dollars of debts, seeing their crops turn into dust and their meadows receding into desert scrubland. It is hoped that the world may learn from Australia's lesson on manipulating the limits of natural resources and weathering the destructive climate changes.

Other countries and regions with severe drought include Spain, Cyprus, Africa, Russia, northern and central China, the Middle East, Cambodia, Afghanistan, Pakistan, the United States, and Central and South America. These places jointly account for two-thirds of the world's food production. Food prices will be climbing upwards sharply in the years to come.

Flooding around the World

The Global Disaster Alert and Coordination System reported that a massive and severe flood occurred in eastern Nepal and northeast India, covering 163,700 square kilometres, between 18 August and 24 September 2008.

Annual flooding during the monsoon season, which began in June, was compounded when the Koshi dam in southern Nepal broke on 18 August, causing extreme flooding downstream, in the Bihar region of northern India, where hundreds of rural villages were destroyed.

Many people were killed by drowning, collapsed houses, and electrocution. There was a huge loss of crops. In the districts that were worst hit, nearly two million persons were trapped and ten million affected. The Indian government reported this as the worst flood in the area in 50 years. In Nepal, an estimated 85,000 people were displaced and had to live in schools and colleges.

According to Dartmouth Flood Observatory, located in New Hampshire, United States, during the period April 2008-April 2009, there were 429 major incidences of floods around the world. Some countries' flooding was catastrophic, as it impacted millions of people in China, India, Nepal, Myanmar, and Columbia. The African continent had the highest incidence of flooding (26), which displaced hundreds of thousands of people, and claimed 390 lives. The country with the second highest incidence of flooding was the United States (22).These floods displaced hundreds or thousands of people, and claimed 59 lives. In the same period, the Philippines and Indonesia were hit by 9 to 10 major floods, with 1,180 and 80 deaths and 440,000 and 91,500 people displaced, respectively.

Undoubtedly, the two major floods of 2010 in Pakistan and Australia are deeply imprinted in our minds. In Pakistan, the flood began in July of 2010, killing 2,000 people, destroying 700,000 homes, and affecting 20 million people. Ten million people were forced to drink unsafe water. The economic impact was estimated at $43 billion. From drought to flood, Australia's Queensland was three-quarter flooded in December 2010. More than 3 million were affected, with 35 deaths and 9 missing persons. "This flood devastated crops, tourism, retail, and manufacturing [...] and coal exports," said Australian Treasurer Wayne Swan. The cost of repairs was estimated to be over $5.58 billion.

In most flooding disasters, collateral damages are incalculable. The massive cost of rebuilding affected places and relocating people, and the time needed to re-establish people's income often negatively impacts the country's economy.

Dehydration and Water-Borne Diseases

Flooding comes with other distressing problems: dehydration, diarrhoea, and vector-borne diseases. The irony is that even though flood victims are surrounded by water, they cannot drink it because it is contaminated. Floods make potable fresh water even scarcer. With global warming, water-borne diseases flourish in hot weather. Diseases such as cholera, leptospirosis, malaria, and typhoid and dengue fever have spread increasingly around the world, especially in flooded areas.

Health Status of Rivers

The World Commission on Water (WCW) reported that more than 50% of the world's major rivers in both rich and poor countries are seriously depleted and polluted, poisoning surrounding ecosystems and threatening the health of millions of people. The most seriously affected rivers are in certain regions of India, China, western Asia, the Middle East, the former Soviet Union, and the western part of the United States. River systems that suffer the greatest damage are those whose water has been used to support intensive irrigation and industrialisation. The primary cause for the rivers' bad condition is overuse and misuse of land and water resources in river basins. In addition, many rivers are also being depleted because demands for water are increasing sharply.

Although the technology to make water cleaner and its use more efficient is widely available, the main problem is the lack of coordinated management of watersheds, both within and across national boundaries. According to the World Wildlife Fund, the ten most endangered rivers in the world are the Salween, the La Plata, the Danube, the Rio Grande, the Ganges, the Murray-Darling, the Indus, the Nile, the Yangtze, and the Mekong.

Global Warming and Rising Sea-Levels

Since 1900, the sea level has risen on average between 1 and 2 millimetres per year (10 to 20 centimetres per century). Due to global warming, the sea level rises as heat from the land and lower atmosphere is transferred into the oceans, which causes sea water to expand (thermal expansion). Also, the glaciers and ice sheets are melting, contributing to increasing sea levels, which are expected to be one metre higher by 2050.

In many places, a one metre rise in sea level would mean that entire beaches and significant chunks of coastlines would disappear. Tens of millions of people living in the low-level coastal areas of southern Asia, including the coastlines of Pakistan, India, Sri Lanka, Bangladesh, and Myanmar, would suffer from it.

On 24 November 2009, *The Straits Times* reported that Antarctica is losing ice faster than expected, according to a study done by scientists at the University of Texas. Since 2007, the East Antarctic ice sheet has been losing 57,000,000,000 tonnes a year. If melted completely, it contains enough water to push up global watermarks by six-to-seven meters.

As we have seen, the first refugees of global rising waters have already appeared—in the island nation of Kiribati, made up of 33 atolls, with a population of about 115,000. "Already we have whole villages being washed out, there is no running away from the reality that the seas are rising," says Anote Tong, President of Kiribati. "We have to face the reality that we have to relocate," he continues. Australia and New Zealand are helping 75 Kiribatians to migrate each year. This is a snippet of the things to come: more and more climate change refugees will need to be evacuated. The question is: which country's responsibility is it to take them in?

We have already seen more and more water-related disasters in recent years: hurricanes, monsoons, typhoons, flooding, and drought. On the one hand, we have a severe shortage of fresh water, and on the other, we are drowning in rising sea water.

Water Inflation and Virtual Water

People are now required to pay high prices for spring water, which has natural filtration, is tasty, and is good for health. We are faced with steeply increasing demands for water as the population grows rapidly and freshwater sources are diminishing and are further reduced by pollution and contamination. Global investors are starting to see fresh water as a commodity—just like gold or oil—a commodity that is under-appreciated, under-valued, and that is growing scarce.

Britain is the world's sixth largest importer of water, and its "water footprint" is threatening global water reserves. The average Briton uses a massive 4,645 litres of water every day, of which only 150 litres per person are used for drinking and washing; the rest represents the hidden or virtual gallons of water used in the production of their food and clothing.

The virtual water content of a product or service refers to the total water used in the various stages of the production chain. The water is said to be "virtual" because, using a simple example, once the wheat is grown, the real water used to grow it is no longer actually contained in the wheat. Once a person adds up all the virtual water he consumes in the products that he buys along with the daily use of tap water, he will have a better idea of what his water footprint is. Water footprints are used to give nations a better consumption-based indicator of water use.

Pause: Watery Dream

Coming to the end of this chapter, I cannot help but recall a vivid dream I recently had (as of this writing). As it was so unusually realistic, when I woke up, I could remember the details and the sequence of the events in it, which is extremely rare for me. Briefly, in this dream I saw in a distance of about one kilometre, a water avalanche of 30 stories high crushing office buildings. I ran with my son Dylan (who is autistic) into the attic of a three-storey shophouse, with water fiercely gushing the stairs behind us. It was so real and traumatic for me that when I woke up, I had to rush to the window to see if there were any anomalies around my neighbourhood! I quickly jotted down the dream and got some help deciphering it. Among many important messages, the dream signalled to me that I had to reprioritise my life in order to write this book.

Perception of Our Relationship with Water

With water rapidly deteriorating in quality and quantity all over the globe, a range of transnational entities are scuttling around to assume ownership of and distribution rights for the world's freshwater resources. There are a number of compelling economic and political viewpoints for merging the control of water under such transnational entities, and it is very likely that global water schemes consistent with these viewpoints will be implemented. Perhaps the real issue is not to back up or disapprove of the political and economic arguments supporting such water schemes, but to examine the association and connection between humans and water, which is ultimately not related to government regulations, corporate investments, water management plans, or even technological breakthroughs. Our self-imposed difficulties and limitations with water arise primarily out of our own misguided and limited perception of water, and only secondarily out of the institutions that seek to exploit it. A significant shift in our perception may be the most beneficial step that we need to include in our overall strategy for mitigating our global challenges with water.

By contrast, ancient and indigenous cultures considered water to be a sacred gift, making the concept of owning water absurd. Why should one species claim ownership and control over a gift from nature that has always been shared among all life forms?

Water Symbolism

Since the beginning of recorded history water has held a special place in the spiritual world and physical lives of people. Myths from around the globe recognise that creation was preceded by an original state of chaos, which was defined as a state of formlessness, often identified as a watery abyss or primordial sea. Such a sea apparently reflected the fact that ancient people recognised creation as the emergence of form from formlessness. Cultures and civilisations such as the Sumerians from Mesopotamia, Egyptians, Greeks, and Maoris from New Zealand share the same myth.

Water was revered not only because it was required by physical bodies but also because it was connected to the divine. The beliefs of modern people often contradict the wisdom of the ancients. Historian and mythologist Joseph Campbell suggests that ancient spirituality focused on putting people in accord with their own human nature and the natural world, while people today advocate subduing one's human nature and controlling the natural world.

> Men gradually lost the knowledge and experience of the spiritual nature of water, until at last they came to treat it merely as a substance.
>
> —*Theodor Schwenk, German flow scientist*

Shamanic Wisdom of Water

Animism is the world's most ancient set of spiritual beliefs regarding the nature of reality, and is now seeing a global revival. Its presupposition is that everything in the Universe is alive, highly conscious, and intelligent. Animistic cultures believe in giving utmost respect to the spirits in everything—clouds, soil, water, and animals, because all of these participate in the greater sphere of life.

While animism refers to the beliefs of the members of a community, *shamanism* is an area of expertise developed within that society by certain individuals. It is not a religion because it has no defined hierarchy or set of dogmas; rather, it's a set of special techniques for harnessing and directing energy. Without the ancient practice of shamanism, it is unlikely that humans would have survived prehistoric times, as it includes many practical techniques designed to solve everyday problems and to assist survival in the world's most cruel and difficult environments.

The practice of shamanism is astonishingly similar from continent to continent, even though in many instances there has been little or no contact between the shamans of the Amazon, the steppes of Asia, Africa, Australia, Europe, Polynesia, and all of North America. Shamans from all over the planet say that this similarity is a consequence of their cross-cultural use of deep trance to gain access to the ever-present symbol (*axis mundi*) that expresses "a point of connection between the sky and the earth where the four compass directions meet." Metaphorically, it is the universal tree of life, an invisible highway leading to all locations in the world.

Dr. Jose L. Stevens conducted a study in which he mapped out the shamanic understandings of the nature of water from three different cultures and corresponding locations in the world. One culture was from the Peruvian Amazon—the Shipibo tribe, another one was from central Mexico—the Huichol tribe, and lastly, there was the Tuvan tribe of Mongolia. Although these cultures were geographically remote from one another, all of them have amazingly similar beliefs about the nature and power of water:

- regard water as the giver of life and hold it with utmost reverence;
- view water as alive and guided by intelligence in the unseen realm;
- understand water to be a medium for other energies;
- communicate with water to resonate with "her;"
- use water to heal, purify, and cleanse on important occasions;
- form an alliance with water for assistance and power;
- acknowledge that water has a deep structure that reflects the patterns in nature.

Bridging Spiritual Water to Science

Whether represented by a deity, element, geometry, or life-sustaining force, water has been identified in diverse cultures as a key participator in creation. In some yet-to-be-fully-understood way, water is a kind of bridge for different energies between the seen and the unseen realms. The mechanism by which water performs this feat is through the process of sympathetic vibration and resonance. In other words, water's vibrational repertoire is complex enough to act as an energy and information transducer. Many scientists today believe that information, rather than matter or force, is fundamental to creation in the Universe, i.e., physical properties are a result of information transfer.

Rudolf Steiner, an early 20th century Austrian philosopher, wrote extensively about an etheric realm: it is the vital but unobservable, life-giving, and form-shaping force of the physical world. Building on Steiner's work, flow scientist Theodor Schwenk postulated that all organic information is based upon etheric forces. These forces use the medium of water, which vibrates in resonance with them and permits the passage of formative impulses to the material world. Note that water acts as the conduit between the ether and the material world via the process of vibration. In other words, water is the observable counterpart of the unobservable ether. Water serves as the mediator of the life force in biological structures. Biochemist and Nobel Laureate Albert Szent-Gyorgyi maintained that water plays a vital role in the assembly, activation, functioning, maintenance, and recycling of biological structures. Without water, few biomolecules can exist in a recognisable or life-sustaining form.

Life force exists within organisms (along pathways known as *meridians*), surrounding organisms, and in the space between matter. The life force can be sensed, but not measured, and it profoundly affects the health and vitality of organisms. This life force interacts with water. In addition to the water-ether relationship already discussed, various researchers at the HeartMath Institute of California (United States) have postulated that water's physical properties and molecular network can be altered by people's psycho-physiological states or subtle energy fields. In short, *water responds to human intentions*.

The Wisdom and Power of Water Revealed

In the 1990s, Japanese scientist Masaru Emoto used samples of water from all over the world, froze the water and photographed the unique crystals as the ice started to melt. When water freezes, its molecules systematically connect and form the nucleons of a crystal; it becomes stable when it has the structure of a hexagon. Then it starts to grow and a visible crystal appears. Different water samples from difference sources formed different crystals. Some were beautiful, some were deformed, and sometimes no crystals were formed.

In tap water, with chlorine used to sanitise it, the structure found in natural water was destroyed. Within natural water, springs, underground rivers, glaciers, and upper reaches of rivers—complete crystals formed.

Dr. Emoto collected visible proof that water reacted to its experiences and stored that information; this demonstrated that water has memory. Each molecule of water carries information, and that data becomes a part of our body when we drink it!

Water exposed to classical music revealed well-formed crystals, while water exposed to heavy-metal music resulted in fragmented and malformed shapes. Water with positive expressions written on the paper wrapped around the bottles, like "Thank You," "Love," and "Gratitude," formed beautiful crystals. Water exposed to negative expressions did not form crystals. Dr. Emoto says, "The vibration of good words has a positive effect on our world, whereas the vibration from negative words has the power to destroy." He discovered that molecules of water are also affected by our prayers, thoughts, words, and feelings, which means that *our consciousness affects water!*

Quantum physics shows that the entire Universe is in a state of vibration: everything generates its own frequency, including humans. Vibrations attract and interact with each other. Resonance occurs when something emits a frequency and another thing responds with the same frequency or a multiple of it. The vibrations we emit create the kind of world (morphic field) we live in. If we emit hate, frustration, or sadness, we'll find ourselves in situations that make us hateful, frustrated, or sad. Simply put, the things that we choose to focus on or feel in our inner world, are absorbed and mirrored back to us by the water and energies that surround us. When we emit peace, love, and gratitude we find ourselves surrounded by them. The choice is ours.

Chapter 5
What's Happening to Our Land?

Concretely Speaking

"We need to build our foundation on solid ground." "We are going to take root here." "This is a landmark of success for our community." Stop for a moment and think about these expressions that we use every day; notice how they reflect a sense of being supported, of going back to our roots, of needing to be sure-footed and connected to our ground or land. It is on the ground that we build our homes, grow our food, tap into minerals and resources, and get connected to nature and all living things. In fact, we call our planet *Earth*. On this vast planet with resources we had believed unlimited, the notion of our earth being used-up, sick, or dying has been unimaginable for most of us... until now.

Arable Land

The total land surface area on earth is approximately 149.8 million square kilometres, out of which 20% is covered by snow, 20% mountains, 20% dry land, 30% good land that can be farmed, and 10% land that doesn't have topsoil. The total arable land is about 30.3 million square kilometres, of which 71% is in developing countries. This scarce farmable land is currently supporting more than 7,000,000,000 people on this planet. The United Nations projects that there will be 9 billion people on our planet by 2045. The most alarming news is that our land is degrading at an accelerating rate.

Severity of Soil Degradation

The Food and Agriculture Organization (FAO) of the United Nations has published on its website what it calls "National Soil Degradation Maps," showing the severity of human-induced soil degradation. Almost every country in the world has a certain degree of some form of soil degradation, varying from "light to moderate" to "severe to very severe." Land degradation refers to the reduction in the productive capacity of land due to human actions. Areas on earth with severe and very severe soil degradation are Vietnam, India, China, Indonesia, the Philippines, the United States, Brazil, Russia, and many more. These countries also happen to be the world's top producers of rice, wheat, soybeans, maize, potatoes, vegetables, and fruits—categories of food which are essential to sustain human life in a minimal way.

Soil degradation is primarily caused by:

a) agricultural demands: rapid depletion of soil nutrients due to excessive farming and poor crop rotation; arable lands are constantly pushed beyond their ability to supply;
b) deforestation and overgrazing: the rapid clearing of forests through burning, logging, and livestock grazing, has rendered a massive amount of land unproductive and vulnerable to wind and water erosion;
c) agricultural mismanagement: salinisation, acidification, and poor soil and water management are making land weak and un-arable;
d) industrialisation and urbanisation: urban growth, road construction, mining, and industry are major factors contributing to land degradation in many regions; valuable agricultural land is lost;
e) natural disasters like droughts, floods, and landslides have also greatly contributed to soil degradation.

In 1992, Oldeman, Hakkeling, and Sombroek conducted the Global Assessment of Soil Degradation (GLASOD) and classified about three million square kilometres (roughly the size of India) of the earth's arable soil as "strongly degraded" and "extremely degraded." Extremely degraded soils cannot be restored. The issue of severe soil deterioration has obviously increased over the past 20 years. A more recent study conducted in 2008 by the FAO confirms this: land degradation is increasing in severity and extent in many parts of the world, spreading rapidly to more than 20% of all cultivated areas, 30% of forests, and 10% of grasslands.

The consequences of this major problem include reduced productivity, migration of animals and humans, food insecurity, damage to basic resources and ecosystems, and loss of biodiversity through changes of habitats at both species and genetic levels. The situation is going to worsen rapidly unless governments and related scientific bodies around the world start taking effective action to arrest and reverse it.

What this means is that on the one hand, we have a steady increase in consumption by the human population, and on the other hand, we have a sharp decrease in the quality of global agricultural soil. As Jason Clay irrefutably noted in his book *World Agriculture and Environment*, "The two trends are on a collision course."

Biocapacity and Renewable Resources

The Living Planet Report 2008, funded by the World Wildlife Fund, states that more than three quarters of the world's people live in nations that are ecological debtors—their national consumption has outstripped their country's biocapacity. This new conservation term, *biocapacity*, refers to the capacity of a given biologically productive area to generate an on-going supply of renewable resources and to absorb its spill-over wastes. According to the report, most of us are living our current lifestyles by drawing upon the ecological capital of other parts of the world.

The progressive destruction of ecological balances was not clearly recognised until the ground-breaking book *The Silent Spring* was published in 1962. Chemically-strengthened and mechanised agriculture increases the yield per metre and makes more metres available for farming, but it also increases the growth of algae that suffocate lakes and obstruct waterways. Chemicals such as DDT are effective pesticides, but they poison entire populations of animals, birds, and insects.

Over the past 35 years alone we have decimated one-third of the planet's wildlife population. Today, globally speaking, about 1,130 species of mammals and 1,194 species of birds are considered endangered. This has an incalculable impact on the biodiversity of the planet, on which we (humans) are dependent.

BIODIVERSITY: INTER-SPECIES DEPENDENCY

Biodiversity reflects the number, variety, and variability of living organisms, as well as how these shift from one location to another over time. It includes diversity within species, between species, and among ecosystems; in sum, it is the diversity of all life on earth.

Ecosystems provide resources for the basic necessities of life, such as food, clean air, and water. The loss of biodiversity affects ecosystems, making them more vulnerable to environmental and climate changes and less able to supply humans with valuable "services." Whether we live near a forest, a sea, or in the hub of the city, our livelihood depends on the services provided by the earth's natural ecosystems. The changes in biodiversity at the ecosystem level further affect global warming. The equation is simple: less plant life, less photosynthesis, and therefore less oxygen.

Below is a summary of the global biodiversity status, gathered from the United Nations Environment Programme, the World Conservation Union, the US Academy of Sciences, and the World Resources Institute:

- agriculture, population growth, pollution, and global warming are killing species at an alarming rate;
- global security is based on the sustainability of the ecological and social systems; the five major ecosystems that are rapidly

disintegrating are: agriculture, coasts, forests, fresh water, and grasslands;

- more than 60% of coral reefs—essential to marine ecosystems—are in critical condition;
- more than 75% of major marine fish species are dwindling due to overfishing and pollution;
- the world's forests—nature's carbon absorbers—are rapidly disappearing; this also means the loss of habitat for hundreds of animal species;
- biodiversity disintegration has severe impact on the rural poor, amounting to 1.5 billion people, who depend on the natural systems—forests, grasslands, oceans, rivers, fresh water and so on—for their livelihood;
- by 2025, with the growth of more than 20 megacities and 60% of the world's population living in cities, the rate of biodiversity extinction will be further accelerated;
- biodiversity supports human health; 10 of the world's top-selling drugs come from natural sources; future drugs for diseases like cancer, diabetes, hypertension, and others may be lost if we do not preserve biodiversity.

DISAPPEARING FORESTS

National Geographic's November 2008 issue featured an article entitled "Borneo's Moment of Truth." Throughout human history, Borneo has had its natural resources plundered by businessmen and governments around the world.

Since the mid-19th century, scientists have continued to discover new species in Borneo, proving that its rain forest is one of the most biologically diverse places on earth. However, in a short period of just 20 years (1985–2005), nearly a third of the rainforest that stood in Borneo disappeared. This works out to approximately 162 thousand square kilometres being cleared primarily for oil palm plantations, coal and timber mining, and industrial and urban development. According to wildlife conservationists, converting an entire wildlife area to a monoculture plantation (e.g. oil palm) is biological suicide.

Since the Second World War, half of the world's forests have been cleared. Many types of plants and animal species have disappeared and will continue to disappear. And Borneo is not an isolated case: we can now see parallels to Borneo's ecological disaster in many other parts of the world as well.

Carbon Released from Disturbed Lands

Borneo has also given the world a new concern: a specialised ecosystem called a "peat swamp forest," which covers 81.4 thousand square kilometres, an area larger than the entire Czech Republic. There, trees grow on highly organic soil built from centuries of waterlogged plant materials, which represent a massive store of the world's carbon. Stripped of its trees and drained, tropical peat in these forests decays and releases its carbon into the atmosphere, and as it dries, it becomes extremely susceptible to burning, thus releasing even more carbon and other pollutants. Massive annual fires set deliberately to clear previously forested land for new oil palm plantations have burnt out of control and filled Borneo's sky with smoke, closed airports, and caused severe respiratory problems for millions of people as far away as mainland Asia. Singaporeans and Malaysians have first-hand experience of this issue, almost on an annual basis.

A similar problem of carbon release from soil is found in an area of permafrost spanning one million square kilometres in Siberia, which has started to melt. Russian researchers found that what was—until recently—a barren expanse of frozen peat is turning into a broken landscape of mud and lakes. The area, the world's largest frozen peat bog, has been producing methane since it formed at the end of the last ice age, but most of the gas has been trapped under the permafrost. Calculations show that the melting peat bog could release around 700 million tonnes of carbon into the atmosphere every year, about the same amount that is released annually from all of the world's wetlands and agriculture put together. This would double atmospheric gas levels, leading to a 10–25% increase in global warming.

Waste Disposal

Midden is a word often used by archaeologists to describe any kind of ancient site containing waste products relating to daily human life. In the modern day equivalent, 90% of our urban waste includes food scraps and paper, plastic covers, bags and bottles, aluminium and tin cans, textiles, wood/timber, glass, and scrap metals. Many of these items are recyclable and reusable, but they are wastefully transported to landfills in most countries.

The Fresh Kills Landfill in New York was formerly the largest landfill in the world at 9.7 square kilometres. The site's volume eventually exceeded the Great Wall of China, with 650 tonnes of garbage added daily. Jessica Williams's book *50 Facts that Should Change the World* informs us that Americans produce enough plastic wrap to cling-film the state of

Texas, offices use enough paper to build a four metre high wall between Los Angeles and New York, and they throw away enough aluminium cans to rebuild the country's entire commercial air fleet. Every three weeks, they discard 1.26 billion plastic bottles—enough to reach the moon—and here we are only talking about one country!

Developing nations are catching up. China produces and discards more than 45 billion pairs of disposable chopsticks every year and cuts down 25 million trees to do it. In the Bangladeshi capital Dhaka, more than 10 million plastic bags are dumped every day, clogging the city's drains. Among the ASEAN countries, Malaysia, with a population of 22 million, generated an estimated 5,475,000 tonnes of solid waste in 2001. Singapore, with its population then of only 4.4 million, generated a comparable 5,035,415 tonnes of waste in the same year. Developed nations over-consume and overload the planet with harmful waste, while millions in the 41 heavily indebted poor countries (567 million people) starve to death.

Currently, the largest landfill on earth is, ironically, in the ocean, in the North Pacific Subtropical Gyre, a slowly moving, clockwise spiral of currents. The area is an oceanic desert, filled with tiny phytoplankton, but few big fish or mammals. The area is not just filled with plankton, though: it is loaded with modern waste—millions of pounds of it—, mostly plastic. In our battle to preserve goods against natural deterioration, we have created a class of products that defeat even the most creative and insidious bacteria. Plastics are now virtually everywhere: we drink out of them, eat off them, sit on them, and even drive in them. Plastics are like diamonds: they last forever.

Plastic doesn't biodegrade; in the ocean, it *photo-degrades*, a process in which it is broken down by sunlight into smaller and smaller pieces, all of which are still plastic polymers, eventually becoming individual molecules of plastic, still too tough for anything to digest. The United Nations Environment Programme estimated in 2006 that every square mile of ocean hosts 46,000 pieces of floating plastic. In some areas, the amount of plastic outweighs the amount of plankton by a ratio of 6:1, and the fish that humans consume feed on plankton. The damaging impact of plastic on fish, ocean ecosystems, and humans is incalculable.

WHAT, NO MORE TOP-UPS?

Based on the *BP Statistical Review of World Energy 2008*, at the rate of 80 million barrels per day, the natural oil reserves will last 42 years. That is, fossil fuel will disappear before 2050. We need petrol to fuel cars, motorbikes, airplanes, and ships; we need crude oil to provide energy for electricity generation and as a raw material in plastics, solvents, fabrics, and detergents. It's not an overstatement to say that without it, society, industry, and the world would come to a halt.

According to geologist Dr. King Hubbert, who has worked for Shell, a theory of decline was first suggested in the 1950s. The crux of the matter is not to ask "When will oil run out?" but "When will production start to become too expensive?" The first oil fields discovered are the big ones that can be exploited cheaply. When they are exhausted, the industry is forced to turn to smaller fields, where extraction costs are much higher.

This is already happening: currently in Alberta (Canada), oil companies are squeezing sand for oil. With electric shovels of five stories high, they are digging bitumen-laced sand from the ground. After washing off the sand, the tarry bitumen is cracked and converted into synthetic crude oil in upgrading facilities. This is then sent to oil refineries. To extract each barrel of oil from a surface mine, the industry must first cut down the forest, then remove an average of two tonnes of peat and dirt above the oil-sands layer, and then two tonnes of the sand itself.

Processing a barrel of crude oil from the oil sands emits as much as three times more carbon dioxide than letting one gush from the ground in Saudi Arabia. According to Simon Dyer, an ecologist of the Canadian Pembina Institute, "The fact that we are willing to move four tonnes of earth for a single barrel really shows that the world is running out of easy oil."

The impact on the environment and biodiversity is inevitable. Earlier, in April 2008, five hundred migrating ducks mistook a polluted pond for a hospitable stopover, landed on the oily surface, and died. Whitefish from the Athabasca Lake (in Saskatchewan and Alberta provinces of Canada) are often covered in unusual red spots; natives no longer eat them. Many have suspected the phenomenon had to do with the toxic chemicals released during bitumen production that seeped into rivers and lakes.

Nature's Energies

The race is on, now, for alternative renewable energies. Renewable energy generated from natural resources—sunlight, wind, geothermal sources, tides, and waves that are naturally replenished. According to the *World Energy Outlook 2008*, modern renewable technologies are growing rapidly, overtaking gas to become the second largest source of electricity, behind coal. Altogether, renewable energies are growing faster than any other source worldwide, at an average rate of 7.2% per year. Currently, most of the increase occurs in the power sector.

Governments around the world are putting considerable emphasis on the development of hydrogen as a primary fuel for vehicles. In a fuel cell, hydrogen is consumed by a pollution-free chemical reaction. The fuel cell simply combines hydrogen and oxygen chemically to produce electricity, water, and waste heat, nothing else. Hydrogen-fuel-cell-powered cars are one of the best alternatives to polluting, gasoline-powered cars for several reasons: (i) the cars are completely emissions-free, (ii) hydrogen is renewable and abundant, (iii) the cars are compatible with cold weather, (iv) the fuel cells are compact and lightweight, (v) the cars are about three times more efficient than gasoline-powered cars, (vi) the cars will have incredible mile ranges, and (vii) the tanks can be refuelled quickly.

As with any change, there are inconveniences and concerns. There are some challenges that come with using hydrogen fuel: it is a difficult gas to contain, it is still expensive to produce, and there are transport and storage issues. Despite these concerns, the move toward a hydrogen-based economy has started. Samsung and Sanyo are working on producing fuel cells for phones and laptops. BMW, Honda, and Mercedes have produced small numbers of concept cars, while nine European cities have conducted testing of hydrogen-powered buses. Perhaps it is too early to tell whether hydrogen will be the solution to the world's long-term energy requirements, but the clock is ticking. We may have only two decades to find an alternative source of ecological energy.

HUMAN CONSUMPTION: ECO-FOOTPRINT

Humanity needs what nature provides, but how do we know how much we are using? The Ecological Footprint, conceived in 1990 by William Rees and Mathis Wackernagel at the University of British Columbia, measures how fast we consume resources and generate waste, and compares it to how quickly nature can absorb our waste and generate new resources.

The Global Footprint Network has calculated the footprints of 150 nations from 1961 to the present, based upon 5,400 data points per country per year. As of July 2009, humanity uses the equivalent of 1.3 planets to obtain the resources it uses and to absorb its waste. This means it now takes Earth one year and four months to regenerate what humanity uses in one year. If the current population and consumption trends continue, by the mid-2030s, we will need the equivalent of two Earths to support us.

The ecological footprint of a person is calculated by taking into account all of the biological materials consumed and all of the biological wastes produced by that person in a given year. All these materials and wastes are then individually translated into an equivalent number of global hectares. The ecological footprint of a group of people, such as a city or a nation, is simply the sum of the ecological footprint of all the residents of that city or nation.

An ecological deficit/reserve refers to the difference between the biocapacity and the ecological footprint of a region or country. An ecological deficit occurs when the footprint of a population exceeds the biocapacity of the area available to that population. The chart below shows the top 12 countries with an ecological deficit:

Ranking	Country	Population Millions	Ecological Deficit
1	Kuwait	2.7	8.4
2	United Arab Emirates	4.5	8.4
3	Israel	6.7	4.4
4	Spain	43.1	4.4
5	United States	298.2	4.4
6	Japan	128.1	4.3
7	Greece	11.1	4.2
8	Singapore	4.3	4.1
9	Belgium	10.4	4.0
10	Switzerland	7.3	3.7
11	United Kingdom	59.9	3.7
12	Italy	58.1	3.5

Source: Global Footprint Network, 2005

Note: If everyone lived the lifestyle of an average Singaporean, we would need 4 planets.

Turning resources into waste faster than the waste can be turned back into resources puts humanity in global ecological overshoot. The result is collapsing fisheries, diminishing forests, depletion of fresh water, and the rapid build-up of pollution and waste; these create problems like global climate change. The overshoot also contributes to resource conflicts and wars, mass migrations, famine, disease, and other human tragedies. This situation has a greater impact on the poor, as they cannot buy their way out of the problem by obtaining resources from somewhere else.

How can we all live well and within the means of one planet? This is the most pressing question of the 21st century. If we are serious about sustainable development, there is no way around this question. If we do not design ways to live within the means of one planet, then living in a world with starvation, sickness, war, and a rapidly declining habitat is a future certainty.

Pause: Mother Earth, Father Sky

While I'm writing this chapter, I can clearly see outside my window a thin layer of "skimmed milk" five blocks and beyond in front of me. Moments of synchronicity are rampant in the writing of this book: the haze is back. According to Lieutenant-Colonel N. Subhas of the Singapore Civil Defence Force, there were 265 bush fires in 44 days in 2009, the highest number in the past decade. This was due to an unprecedented two-month-long dry spell and the northeast monsoon dry winds. These are just scientific reasons, but what are the bush fires trying to tell us? The strangest thing about this is, as I ask around, most people in Singapore are not aware of the rampant number of bush fires within our tiny 693 square kilometre island, ranked as the second most densely populated country in the world. We have lost touch with our environment. We walk on roads and pavements, take lifts, drive cars, stroll in shopping malls, and rest in cement homes.

While deeply contemplating this in February 2009, I received the *Discovery Channel Magazine* and I was drawn to a story entitled "Bee-Gone." As I read on, I was saddened by the article, yet at the same time curious about the possible message it brought. In search of the meaning of bees in our lives, I stumbled onto the knowledge of animal totems, which led me to the wisdom of Native Americans.

The Native Americans arrived in America about 20,000 to 30,000 years ago from north-eastern Siberia, through Alaska. Though they evolved into different tribal groups, they have common beliefs that connect them. They believe in wholeness and the interconnectedness of every living and non-living thing in their lives.

They believe in change and that it flows in seasons. If they cannot see how a particular change is connected to their lives, it usually means that their own standpoint is affecting their perception. They believe that the physical world and the spiritual world are two aspects of one reality. Breaking a spiritual principle will affect the physical world, and vice versa. A balanced life is one that honours both.

Native American Proverbs

To further appreciate Native American wisdom, let us delve into their proverbs. Succinctly described, a proverb is an old common saying which briefly expresses some practical truth, based on long experience and observation. Often, a proverb is the "swift horse" that brings us to the discovery of new insights.

Regard heaven as your father, earth as your mother,
and all things as your brothers and sisters.

~~~

When we show our respect for other living things,
they respond with respect for us.

~~~

Every animal knows more than you do.

~~~

Take only what you need and leave the land as you found it.

~~~

When a man moves away from nature,
his heart becomes hard.

~~~

All things are connected. Whatever befalls earth befalls the sons of earth.
Man did not weave the web of life. He is merely a strand in it.
Whatever he does to the web, he does to himself.

~~~

Bees Are Disappearing

"Whatever he does to the web, he does to himself" is a description of our current stage of evolution. The story of "Bee-Gone" is a good example. Bees around the world are vanishing. US beekeeper David Hackenberg announced the fact that 75% of his bees disappeared between 2006 and 2007, when he moved his base from Pennsylvania to Florida. Beekeepers from New Zealand to Slovenia also said the same thing. A similar phenomenon related to bees has appeared in Sulawesi, Indonesia, and Sichuan, China; their fruit harvest has been reduced significantly as a result.

With an increased demand in recent years for certain crops that require bee pollination, beekeeping entered a new era: bees are now pollinators for hire. Entire colonies pollinate one crop and are then carted thousands of kilometres on trucks to do their next job.

Professor Maria Spivak, a bee expert from the University of Minnesota (United States), said that bees are dying in droves because of the way climate and plant lives are changing, and the way viruses and bacteria are mutating. In 2011, Joseph DeRisi at the University of California, San Francisco, discovered four new previously unknown viruses that exist in bee hives. The viruses remain a serious threat, with about a third of all bee colonies affected, and no cure in sight. If insecticides are used, they don't discriminate between good bugs and bad ones. Scientists fear that this problem could result in global fruit shortages.

Frogs Are Becoming Extinct

We are also witnessing a mass extinction of other species: frogs, toads, newts, and salamanders. Though they were hopping around and croaking for millions of years before dinosaurs, a third of the world's amphibian species have most likely disappeared and almost half are rapidly declining in population.

According to Joseph Mendelson, Curator of Herpetology at the Atlanta Zoo (United States), it is the largest scale extinction event that has happened in human history. Habitat destruction, the introduction of exotic species, commercial exploitation, and water pollution are working in concert to decimate the world's amphibians.

A major cause for this possible annihilation is a fungal infection, chytridiomycosis (chytrid for short). Global warming is increasing cloud cover in tropical mountains, creating the cool, moist conditions that the fungus loves. Disease is the bullet killing the frogs, but climate change is pulling the trigger. Chytrid is now reported on all continents where frogs live, in 43 countries.

When amphibians disappear, there will be consequences up and down the food chain. With no tadpoles to feed on them, algae will clog clear streams. Birds and snakes that once fed on frogs will starve or be forced to seek new habitats. Furthermore, a promising pharmaceutical resource will be lost. The skin of amphibians is filled with a rich brew of anti-microbial chemicals called *peptides*, which have already been used to block transmission of HIV in lab tests and to produce a painkiller far more powerful than morphine, yet with none of its side effects.

Which Species Will Live?

In 2008, the Wildlife Conservation Society gathered to decide on a near-to-impossible objective: to save a small number of animals. The researchers had spent months studying thousands of declining bird and mammal species around the world and had chosen several hundred species for the organisation to focus on. Very often, at different points in their discussion, the enormity and complexity of the process would hit the scientists. As entire groups of species were determined valuable, but not valuable enough to save, scientists would quietly shut down, shoulders slumped and eyes glazed: "I'm just overwhelmed," they would say. They had to remind themselves constantly of the severity and importance of what they were doing, and that they were confronted with a loss of animal species at an unprecedented scale.

Due to the drastic changes in climate and environment and their precipitating factors, which include the mutation of bacteria and viruses, huge numbers of animals on land, air, and sea are dying. Many thousands of species are already extinct. With environmental pollution, rapid disappearance of natural habitats like forests and corals, and animal poaching and exotic meat consumption, many animal species are not able to survive and reproduce.

Not only is the task of deciding which animal species to save overwhelmingly hard, but scientists are finding great difficulty in getting sufficient support and funding from governments and corporations to save them. All animals are deemed important because scientists have yet to fully understand the significance of the role of each animal in their respective ecosystem. And many of them hold potential cures and solutions to human diseases and other life problems.

In March 2011, the well-respected science journal *Nature* published the online article "Has the Earth's Sixth Mass Extinction Already Arrived?" from which we find out that the five major wipe-outs of species that have occurred over the past 540 million years were naturally-induced events. By contrast, our current threat is man-made, inflicted by habitation loss, over-hunting, over-fishing, the spread of germs, and by climate change caused by greenhouse gases. According to paleobiologists, until the big expansion of mankind some 500 years ago, mammal extinctions were very rare—approximately 2 species died out every million years. But in the last 5 centuries, around 80 out of 5,570 mammal species have become extinct, and many other mammals, marine and other types of animals are on the critically-endangered-species list.

If we look at our current situation through Native American wisdom, Mother Nature is no longer talking to us, she is shouting. We must listen and change our ways.

Wisdom from Animal Totems

According to American Indians, there was a time when humanity recognised itself as part of nature. People used images of nature to express this unity and to instil a transpersonal experience. In the past, shamans and priests were the keepers of the sacred knowledge of life. These individuals were tied to the rhythms and forces of nature. They were capable of walking the threads that link the invisible and visible worlds. They helped people remember that all trees are divine and that all animals speak to those who listen.

The study of nature totems helps us to understand the divine manifestation within our natural lives. A totem is any natural object, being, or animal with whose phenomena and energy we feel closely associated during our lives. The understanding of animal totems helps us to learn the language of nature as it speaks to us every day, read, and apply what it says, and in doing so help ourselves to develop greater reverence for all life and higher wisdom in our own.

Let us learn from the animal totem of bees. They are highly productive, remain focused in their activities, and do not get distracted from their goals. If a bee shows up in our lives, it is asking us to examine our productivity. Are we doing all that we can to make our lives more fertile, or are we attempting to do too much? Are we keeping our desires in check so that they can be healthily productive? Bees' legs are one of their most sensitive organs. A bee tastes through its legs: this reminds us to slow down, smell the roses, and taste the sweet nectar of life.

Bees carry the power of service. They are important pollinators of many plants. As a bee lands upon a flower, collecting its nectar, pollen also attaches itself to the legs of the bee. The pollen is then transferred to other flowers, creating a fertilisation process. The movement of bees from one plant to another symbolises the interconnectedness of all living things. The bee is a messenger that holds the secrets of life and service.

Frogs have important messages for us, as well. Many shamanic cultures, especially North and South American, link the frog with rain and control of the weather. Their voice is said to call forth the rains. Frogs are associated with abundance and fertility. This is because after the rain, frogs come up to dry land and feed on insects and worms that have come out of the rain-soaked land. This is connected to fertility, for rain makes things grow.

If frogs hop into our lives, this may be a signal for us to call forth new rain in our lives or perhaps new rains are coming, as maybe the old waters are becoming dirty and stagnant; so frogs can teach us how to clean up. Frog energies can bring rain for many purposes: to cleanse, to heal, to help things grow, to flood, to stir. The frog is also a totem of transformation: from eggs to tadpoles, then to adult frogs; this signifies the awakening of one's creativity. To awaken creativity, it is helpful to know the life stage we are in at any particular moment. By closely studying the characteristics of the frog, we can discover the different stages of our lives. The frog invites us to jump into our own creative power, to reinvent our lives.

Eerily, the messages from bees and frogs are relevant to our current challenges. In our quest for more in life, we have degraded our lands, decimated forests, and destroyed animal habitats. We have also lost touch with our natural inclination to be in harmony with nature. In the pursuit of things that we want (not need) we have forgotten the sheer joy of doing our work in service to others. We must call forth a new tide of change and forsake our old unhealthy ways of living.

CHAPTER 6
WHAT'S HAPPENING TO OUR FOOD?

We are what we eat not just physically, but emotionally, mentally, and energetically as well. The food we eat contains natural chemicals that affect the brain's biochemistry, which gives rise to our moods and behaviours. We have also formed a relationship with food based on our past experiences, which may work for or against our health and well-being. How many of us long for the food we enjoyed in our childhood? The movie *Ratatouille* highlights this when the food critic is transported back to his childhood kitchen and the memory of the taste of his favourite dish brings a turning point in his joyless adult life. But do we eat to live, or live to eat?

FOOD SECURITY: DO WE HAVE ENOUGH TO EAT?

Food is one of the basic physiological needs for survival. Having enough to eat is surely a basic human right. Article 25(1) of the Universal Declaration of Human Rights (1948) states that, "everyone has the right to a standard of living adequate for the health and well-being of himself and his family, including food." The United Nations' Food and Agriculture Organization (FAO) defines the existence of food security as when all people, at all times, have physical and economic access to sufficient, safe, and nutritious food to meet their dietary needs and food preferences for an active and healthy life.

World Hunger

However, according to the FAO, in the early 2000s, 850 million people went hungry every day, two billion suffered from chronic malnutrition, and 18 million died each year from hunger-related diseases. The World Health Organization warns that "hunger is the gravest single threat to the world's public health." Yet annual world food production is enough to feed everyone, if it were equally distributed. There are huge surpluses of food in the developed West, which are even sometimes destroyed to keep prices up! *The New York Times* reported on 9 April 2008 that the government of the United States paid farmers to idle their cropland; 8% of cropland was idled in 2007, representing a total area bigger than the state of New York. Developed nations also give $300 billion in subsidies annually to protect their farmers, making it impossible for developing nations to compete in the world market. For example, each European Union cow is subsidised by $2.50 per day, more than what three quarters of African people have to survive on. In other words, more money is given to cows than to starving humans.

Soil Erosion

Virtually all our food comes from cropland (99.7%), which is eroding by more than 10 million hectares (37,000 square miles) every year. "Soil erosion is second only to population growth as the biggest environmental problem the world faces," said David Pimentel, professor of ecology at Cornell University, in 2006. Due to erosion, over the past 40 years 30% of all arable land has become unproductive. The United States is losing soil 10 times faster, and China and India are losing soil 30 to 40 times faster than it is being replenished.

Food Prices Rise

Rising global population and increasing consumer demand have pushed prices up. From 2006 to 2008, the average world price for rice rose by 217%, wheat by 136%, maize by 125%, and soybeans by 107%. Asia and Africa were particularly affected, with many protests and food riots in late 2007 and early 2008 over the lack of basic food staples. Agricultural subsidies in developed nations also contribute to high global food prices. In 2007, developing nations had to pay 25% more for food imports.

The Bane of Biofuels

Another systemic cause of rising prices is that more farming is channelled to producing biofuels, with oil prices shooting over $100 per barrel. An estimated 100 million tonnes of grains per year are being redirected from food to fuel, and the total worldwide grain production was just over 2,000 million tonnes in 2007. This further dichotomises rich and poor nations, as filling a tank of an average car with biofuel uses as much maize as an African consumes in an entire year (maize being their principal food staple)! While we would think nothing of filling up our tanks x number of times a year, would we consider sponsoring x number of humans with one year's supply of maize each?

Industrialisation of Agriculture

Agriculture, compared to hunting and gathering, led most to the development of human civilisation with food surpluses. The Industrial Revolution in the late 18th and early 19th centuries gave rise to a much higher agricultural production, with new machines for ploughing, seeding, threshing, and digging drainage channels. In the 20th century, agricultural production increased further with the use of synthetic nitrogen fertiliser, rock phosphate, pesticides, and more mechanisation. Industrialised agriculture with high yield varieties is extremely water-intensive, with US agriculture consuming 85% of all its freshwater resources. Only 60% of the water for irrigation comes from surface water supplies, the other 40% coming from underground aquifers that are being used up.

From 1945, the Green Revolution spread technologies used in industrialised countries to developing countries, including irrigation projects, chemical pesticides, herbicides and fertilisers, improved crop varieties (such as Asian rice), and breeding methods. World grain production increased by 250% between 1950 and 1984. In India, annual wheat production at 10 million tonnes in the 1960s rose to 73 million tonnes in 2006.

The Green Revolution shifted subsistence-oriented polycultures to monocultures of cereal grains oriented toward export, animal feed, and biofuel. Extensive use of pesticides became necessary to limit high pest damage due to monocropping (the production of a single crop over a wide area). Pesticides in rice production poisoned the fish and weedy green vegetables that co-existed in the rice cropland and provided nutrition to the farmers. Biodiversity International stated that less people suffer from hunger, but many are affected by malnutrition due to changes in dietary habits. In children under five years of age, 60% of annual deaths are related to malnutrition.

Pesticides, Insecticides, and Herbicides

Pesticide use has jumped 50-fold since 1950, with 2.3 million tonnes now being used every year, of which 75% is used in developed countries. One study found that without pesticides, crop yields are reduced by 10%. Insecticides which have been used for many years include heavy metals such as lead, mercury, arsenic, and plant toxins such as nicotine. Many are toxic to humans, and they have the potential to alter ecosystems significantly. They can poison pollen and nectar and kill bees, and the loss of pollinators will result in a reduction of crop yields. Over 98% of sprayed insecticides and 95% of herbicides drift from their target species reaching non-target areas, contributing to water pollution and soil contamination. In Europe, recent EU legislation has banned the use of highly toxic pesticides.

The WHO and the UN Environment Programme estimate that around three million agricultural workers are severely poisoned by pesticides and about 18,000 die every year. In China, around 500,000 people are poisoned by pesticides, of whom 500 die every year. Many studies have pointed to pesticide exposure being associated with respiratory problems, memory disorders, dermatologic conditions, cancer, depression, neurological deficits, miscarriages, and birth defects. A study of the Harvard School of

Public Health has discovered a 70% increase in chances of developing Parkinson's disease if exposed to even low levels of pesticides. Many crops, including fruits and vegetables, contain pesticide residues even after washing and peeling. Such concerns have led to the organic food movement.

Antibiotics and Drugs

In the United States today, close to 70% of the total antibiotics and related drugs produced are fed to cattle, pigs, and poultry, according to the Union of Concerned Scientists. Antibiotics are given not just to treat sick animals, but to speed up their growth. The European Union has banned some of the animal antibiotics used for animal growth promotion. Animals in confined feeding operations are routinely given antibiotics to prevent stress-related illnesses, all contributing to increasing antibiotic-resistant diseases in humans. Bacteria from farm animals, such as salmonella and campylobacter, have been causing antibiotic-resistant cases of food poisoning in people. Humans have long been exposed to antibiotics in meat and milk. Research now shows that people may be ingesting them from vegetables as well.

Hormones for Animals

Hormones increase the profitability of the meat and dairy industries. They make young animals gain weight faster, reducing the waiting time and amount of feed eaten before they are slaughtered. Hormones increase milk production in dairy cows. Synthetic forms of oestrogen have been used since the early 1950s to increase the size of cattle and chickens. One of the first synthetic oestrogens made to fatten chickens, DES, was found to cause cancer and was phased out in the late 1970s. Early puberty in girls (aged eight or younger) and breast enlargement in young girls and boys have been traced to steroid hormone residues in beef and poultry.

Animal Diseases

Factory farming of animals has led to their having health problems, with epidemics of mad cow disease, bird flu (avian influenza), and foot-and-mouth disease. Antigenic shift resulting in a virus that crosses the species barrier between birds and humans is often traced to locations where humans, chickens, and pigs live in close proximity. Because the new virus is different from the strains that combined to produce it, there is no natural immunity to it. This enables the new virus to spread rapidly, causing extensive sickness and death.

Bovine Spongiform Encephalopathy (BSE)

Commonly called mad cow disease, BSE is a fatal neuro-degenerative disease in cattle. It is caused by herbivore cattle being fed the remains of other cattle in the form of meat and bone meal and feeding infected protein supplements to young calves. In the UK, 179,000 cattle were infected, and 4.4 million cattle were slaughtered in the resultant eradication programme. About 482,000 BSE-infected cattle had already entered the human food chain before controls were introduced in 1989. Humans who eat BSE-contaminated products (including hamburgers) contract a new variant, called Creutzfeldt-Jakob Disease, which killed 164 people in Britain and 42 people elsewhere by February 2009. The number is expected to rise over time due to the long incubation period of the disease.

Bovine Leukaemia Virus (BLV)

According to *Hoard's Dairyman* magazine (Wisconsin, USA), Volume 147, Number 4, 2002, 89% of the cows in the United States are infected with BLV, a virus closely related to HTLV-1 which is a human tumour virus. Robert Cohen, an expert in the dangers of milk, says that virtually all animals exposed to BLV develop leukaemia. Relatedly, in Russia and Sweden, areas with uncontrolled BLV have been linked with increased incidences of human leukaemia.

Bird Flu (Avian Influenza)

As China and other countries in eastern Asia grew wealthier, there was greater demand for meat, especially for chicken. Billions of chickens in crowded battery farms were prey to any flu virus. The highly pathogenic H5N1 virus struck in 1997 in Guangdong, China, and outbreaks have since been occurring across the world. Massive slaughtering campaigns have failed to eradicate H5N1 in poultry. While China has tried vaccinating intensively reared poultry, the virus's persistence in vaccinated birds drives it to evolve into new forms.

The virus has spread to wild bird species in Asia, and along with bird trade, it has spread to Europe and other places. It is now the biggest animal disease outbreak ever recorded. However, all influenza viruses have the ability to change, and they are mutating so fast that scientists around the world are terrified and defenceless.

According to Helen Branswell, a Nieman Fellow for Global Health Reporting at Harvard University, the 2009 pandemic of H1N1 flu in Mexico underscored that the greater threat may come from pigs, not birds, because it is typically easier for pig viruses to make the jump to people. What made the pandemic so alarming was that it contained genetic material from birds, pigs, and humans, something totally different from what the human immune system has experienced before. Although the 2009 outbreak was mild, we may not be so lucky the next time.

Genetically Modified Food

Genetically modified (GM) foods are produced from crops that have specified traits created by genetic engineering. First marketed in the early 1990s, they include soybean, corn, canola, and cotton seed oil. The main GM food crops, soybean and maize, have been genetically modified to produce bigger yields or the same yield for less input (less herbicide, insecticide, and fertiliser). More than 70% of the GM plants grown today are herbicide resistant. The largest share of the GM crops planted globally is owned by Monsanto: 246 million acres throughout the world in 2007, a growth of 13% from 2006. While Europe has largely rejected GM crops, the United States has embraced it so much that most meals consumed there will have some GM content.

The safety of GM food has been questioned, with concerns about new allergens and the spread of antibiotic-resistance and transgenic organisms. Open field trials have led to GM crops spreading to non-GM crop fields nearby. In 2001, hundreds of food products were recalled in the United States as they contained traces of GM corn (StarLink) suspected to cause allergies. While some groups advocate prohibition of GM foods, others call for mandatory labelling. Another controversy is the sale of GM seeds that produce plants which yield sterile seeds, making patent claims more financially significant. Farmers have to buy new seeds every year.

GM food has moved from plants to animals.

In September 2010, the FDA approved GM salmon as safe to eat. Many scientists have opposed this move, calling these salmons "Frankenfish." They cited unpredictable results (GE salmon had more physical deformities than non-GE salmon) and exacerbating allergies in humans. Fish allergies are one of the eight most common allergies in the United States.

In January 2011, GM pigs that could digest phosphorous were announced in Canada, and UK scientists created GM chickens that do not spread bird flu. Novel genes that do not exist in nature are introduced, and these scientists believe they are harmless to the animals and the people who eat them; however, opponents claim they have not been adequately tested for safety.

Food Processing and Distribution

We are living in a world today where lemonade is made from artificial flavours and furniture polish is made from real lemons.

—*Alfred E. Newman*

Food Additives

People in industrialised countries eat six to seven kilograms of food additives every year. In 2000, the food industry spent $20 billion on making food look better, taste nicer, and last longer. Less than 10% of food additives are used to preserve food; 90% are cosmetic additives: flavourings, colourings, emulsifiers to make food feel smoother, thicker, sweeter, and some of them are allergenic or carcinogenic. These additives have neurotoxic effects and contribute to hyperactivity, allergies, asthma, headaches, nausea, diarrhoea, learning problems, concentration difficulties, etc.

Sugar and high fructose corn syrup can lead to high blood sugar, hypoglycaemia, hyperactivity, yeast problems (candida), food cravings, increased blood fats, obesity, diabetes, and dental cavities. Artificial sweeteners like saccharin and aspartame were proved to have caused cancer in lab animals. Excessive salt can lead to fluid retention and increased blood pressure.

Monosodium Glutamate (MSG)

MSG, a man-made flavour enhancer, is an excitotoxin, a type of drug which damages brain cells. MSG was first created in 1908 in Japan and brought to the United States during the Second World War, where it was used to make wartime fare taste better. Usage has doubled every decade since and hundreds of thousands of tonnes of MSG are added to our food every year.

MSG is found in most processed foods, such as soups, snacks, junk food, and fast foods, and also in infant formulas and vaccines. Independent research has identified many adverse reactions to MSG, including headaches/migraines, nausea/vomiting, stomach cramps/irritable bowel syndrome, retinal degeneration, asthma attacks, chest pain/palpitations, anxiety/panic attacks, mental confusion/disorientation, learning disabilities, reproductive disorders, obesity, and lesions on the brain (especially in children).

Commonly known by trade names like Ajinomoto, Accent, or E621 in Europe, MSG is contained in many additives, including every hydrolysed protein product. Thus, MSG is largely hidden in food labels.

More information on people recovering from problems defying medical diagnosis after eliminating MSG from their diet can be found at http://www.truthinlabeling.org.

Food from Around the World

Supermarkets today stock an incredible variety of food from all around the world. International food trade is increasing faster than the world's population and food production. Food is flown in planes and freighted by land. Air-freighted food is costly and environmentally damaging, emitting greenhouse gas and consuming fuel. The Kyoto Protocol excludes emissions from international air and sea freight under its purview of limiting the global emission of greenhouse gases.

Research shows that the further food travels, the more its vitamin and mineral content deteriorates. Animals transported in tightly packed containers without adequate food or water often die, the meat may be bruised or dry, and transporting animals contributes to the spread of diseases, such as foot-and-mouth disease, BSE, or mad cow disease.

FAST FOOD

Fast food restaurants started in the United States in 1912, with franchising introduced in 1921. This food is prepared and served quickly, designed as finger-food to be eaten on the go. Generally, menu items are made from processed ingredients prepared at a central supply facility, then distributed to individual outlets, where they are reheated, cooked (often deep-fried or microwaved), or assembled quickly.

Fast food restaurants represent the largest segment of the food industry. In 2006, the global fast food market increased by 4.8%, with 80.3 billion transactions valued at over $100 billion. In India, the growth is of 40% per year. McDonald's has over 31,000 restaurants in 120 countries, Subway has over 29,000 in 86 countries, Burger King has over 11,100 restaurants in 65 countries, and so on. Starbucks, with 13,000 outlets, aims to have 40,000 stores worldwide, overtaking McDonald's.

Fast food chains have come under fire over high caloric content, trans fats, and portion sizes. There have even been lawsuits against them for making people obese. The first study on the impact of fast food on Asians (including around 50,000 Singaporeans), conducted by the National University of Singapore and the University of Minnesota, found that Singaporeans eating Western fast food more than four times a week had an increased risk of death from heart disease by 80%! Those who ate fast food more than twice a week had an increased risk of type 2 diabetes by 30% and an increased risk of death from heart disease by 56%. (Source: *The*

Straits Times, 14 July 2012, "Fast food can be life-threatening," by Melissa Pang.)

Trans Fats

Trans fats, or partially hydrogenated oils, are found in fast foods, particularly French fries and chicken nuggets, in baked goods, and processed snacks. They are used because they can be repeatedly fried at high temperatures without breaking down chemically. Trans fats raise the proportion of bad cholesterol in the blood, accumulating fat in the arteries. They increase the risk of arterial inflammation and irregular heartbeat.

Studies have shown that consuming artery-clogging trans fats leads to more abdominal fat, which dramatically increases the risk of heart disease and diabetes. Steen Stender, of the Gentofte University Hospital in Copenhagen, Denmark, and lead author of the analysis in *The New England Journal of Medicine*, calls it "the killer fat."

Researchers analysing fast food say that daily consumption of five grams or more of trans fats raises the risk of heart attack by 25%. They found that half of 43 large-sized fast-food meals, 24 from McDonald's and 19 from KFC, examined in the study—purchased in outlets around the world—exceeded the five-gram level.

Super Size Me

Super Size Me: A Film of Epic Portions (2004) is a documentary film written, produced, directed by, and starring Morgan Spurlock, an American independent filmmaker. For 30 days Spurlock ate only McDonald's food. The then-32-year-old gained 24.5 lbs. (1.75 stone, 11.1 kg), a 13% body mass increase, experienced mood swings, sexual dysfunction, and liver damage. Doctors begged him to stop before the 30 days, due to the effect the food was having on him. It took him 14 months to lose the weight he gained.

IMPACT OF LIVESTOCK AND MEAT CONSUMPTION

Livestock's Long Shadow, by The Livestock, Environment and Development (LEAD) Initiative (2006), supported by the Food and Agriculture Organization (FAO) of the United Nations of the United Nations, the World Bank, and others, names the livestock sector as the second or third most significant contributor to local and global environmental problems. This sector employs 1.3 billion people, accounting for 40% of agricultural gross domestic product (GDP). Global meat production is projected to double from 229 million tonnes in the year 2000 to 465 million tonnes by 2050, and milk—from 580 to 1,043 million tonnes.

LEAD's report states that:

- livestock production accounts for 70% of all agricultural land (including feedcrop production) and 30% of the planet's land surface;
- this sector has caused mass deforestations, especially in South America, where 70% of the previous Amazon forest land is now pastures and land for feedcrops; the forests are mainly cleared by burning, releasing CO_2, and contributing to global warming;
- land degradation has occurred in 20% of the planet's pastures and rangelands, with 73% of rangelands in dry areas affected, through overgrazing, compaction, and erosion caused by livestock action;
- the livestock sector is responsible for 18% of greenhouse gas emissions, a higher percentage than the transportation sector; it is also responsible for 37% of the emissions of anthropogenic methane, which has 23 times the global warming potential (GWP) of CO_2, and 65% of anthropogenic nitrous oxide, with 296 times the GWP of CO_2, mainly from manure;
- the same sector is the cause of 64% of anthropogenic ammonia emissions, contributing to acid rain and acidification of ecosystems;
- it accounts for over 8% of global human water use, mainly for irrigation of feedcrops; this contributes to water pollution, degradation of coastal areas and coral reefs, drying up floodplains, lowering water tables, and the replenishment of fresh water;
- this sector is a leading player in the reduction of biodiversity, with loss of species running 50 to 500 times background rates.

Fished Out

According to the FAO, more than 50% of the world's main stocks are fished out or depleted. This is caused by large fishing ships, called *bottom trawlers*, pulling up to a million pounds of fish in a single haul!

The World Wildlife Fund estimates that about 300,000 dolphins, whales, and porpoises are killed by fishing gear annually. Many species are in danger of becoming extinct. About one-third of the global fish catch is ground up into fishmeal and fed to livestock. This includes feeding farmed fish, where two to three kilograms of wild fish are required for every kilogram of farmed fish produced. The ocean pens where farmed fish are raised are also known for leaking large quantities of antibiotics and additives, damaging the nearby environment.

Fish (wild or farmed) also contains no fibre, and is high in saturated fat and cholesterol. Fish absorb a lot of the contamination humans deposit in their waterways such as mercury, PCBs, DDT, dioxins, lead, and other chemicals.

Climate Change

The Food Climate Research Network, UK, studies how the food system contributes to greenhouse gas emissions (around one-third globally), and promotes ways to reduce it. Their four-year study *Cooking Up a Storm*, from September 2008, showed that to reduce climate change people should eat less meat and consume less milk—from the current weekly 1.6 kilograms to 500 grams of meat, and 4.2 litres to 1 litre of milk. The report said that total food consumption should be reduced, especially items with low nutritional value, such as sweets, chocolates, and alcohol. It encouraged buying local, in-season products, cooking in bulk, avoiding waste, and walking to the shops.

Waste Management

Livestock in the United States produce 130 times as much manure as the human population. However, there is almost no waste management infrastructure for farmed animals, no sewage pipes, treatment, or federal guidelines regulating it. This waste seeps into rivers, lakes, and oceans killing wildlife and polluting air, water, and land. It contains ammonia, methane, hydrogen sulphide, carbon monoxide, cyanide, phosphorus, nitrates, and heavy metals. It also consists of more than 100 microbial pathogens, including salmonella, cryptosporidium, streptococci, and giardia.

FEEDING THE WORLD ECOLOGICALLY

Jonathan A. Foley, Director of the Institute of the Environment at the University of Minnesota, states that even with the most efficient meat and dairy systems, feeding crops to animals reduces the world's potential food supply. Together with a team of international experts, they came out with a five-step global plan to double food production while greatly reducing environmental damage. In brief the steps are:

a) halting expansion of agriculture's footprint (implementing incentives and policies to protect tropical forests and savannas);
b) closing the world's yield gaps (increasing the yield of farms with genetics and smarter farming technologies);
c) using resources much more efficiently (increasing crop output per unit of water, fertilizer, and energy);
d) reducing food waste (on both consumer and producer ends);

e) shifting diets away from meat.

On the last point, Foley clearly asserted that global food availability and environmental sustainability can be dramatically increased by using more foodcrops to feed people directly and less to fatten livestock. Globally, humans could get up to three quadrillion additional calories every year—a 50% increase from current supply—by switching to all-plant diets. With fair trade and efficient distribution networks we can end world hunger today.

HEALTH PROBLEMS

When diet is wrong, medicine is of no use.
When diet is correct, medicine is of no need.
—*ancient Ayurvedic proverb*

Many studies have shown that our health improves with more plant-based foods compared with animal-based foods. Eating meat has led to higher risks of heart disease (due to saturated fat and cholesterol), cancer (vegetarians have 25–50% lower cancer rates), obesity, food poisoning (faecal matter in meat can result in deadly E. coli bacteria), antibiotic resistance, toxicity (from chemicals and pathogens, pesticides, and herbicides), learning disabilities, impaired immune systems (dioxin exposure), and mercury and other types of poisoning (from fish).

Twenty-Year Diet and Disease Study

The China Study, by T. Colin Campbell, Ph.D., and Thomas M. Campbell II, studied the relationship between diet and disease, giving critical life-saving information on nutrition. The father and son team based their study on a 20-year research conducted by the Chinese Academy of Preventive Medicine, Cornell University, and the University of Oxford. It covered 367 variables from 6,500 adults in over 2,500 counties in China and Taiwan. It is the most comprehensive study of its kind ever undertaken. Some key findings included:

- people who ate the most animal-based foods got the most chronic disease, whereas people who ate the most plant-based foods were the healthiest and tended to avoid chronic disease;
- breast cancer was connected with higher dietary fat—animal-based foods, especially milk;
- higher levels of bad LDL cholesterol were associated with Western diseases;

- the more animal protein you eat, the more prone you are to heart disease;
- a dietary shift to eating more animal protein and animal fat has shown marked increases in the rate of type 2 diabetes;
- high fibre intake, exclusively found in plant-based foods, was associated with lower rates of cancers of the rectum and colon, and lower levels of blood cholesterol;
- the consumption of animal-based foods, especially cow's milk (high in calcium and acid-producing), is associated with a higher risk of autoimmune diseases, such as multiple sclerosis (MS), rheumatoid arthritis, lupus, type 1 diabetes, and rheumatic heart disease;
- dairy products are high in calcium but increase metabolic acid, which draws calcium from the bones, and the amount of calcium in the urine is increased; osteoporosis and high fracture rates are attributable to consuming animal protein;
- cognitive dysfunction and dementia (vascular dementia and Alzheimer's disease) are associated with high animal protein and fat consumption;
- our habit of eating hot dogs, hamburgers, and French fries is killing us.

Dr. Campbell's study concluded that there are virtually no nutrients in animal-based foods that are not better provided by plants. Dr. Campbell, the leading global expert in plant-based diets, firmly believes that genes do not determine disease on their own. Genes function only by being activated or expressed, and nutrition plays a critical role in determining which genes, good and bad, are expressed. His study firmly establishes that heart diseases, diabetes, obesity, and many other diseases can be reversed by a healthy diet.

Vegetarian Diet

Nothing will benefit human health and increase the
chances for survival of life on earth
as much as the evolution to a vegetarian diet.

—*Albert Einstein*

Issues such as global warming, hunger, animal cruelty, GM foods, food production technology (factory farming, hormone and antibiotic use, food irradiation, prevalence of salmonella and E. coli) are succinctly covered by John Robbins in his book, *The Food Revolution: How Your*

Diet Can Help Save Your Life and Our World. Though he was the heir to the Baskin-Robbins ice-cream empire, he rejected it to crusade for people to adopt a vegan diet, free of meat, milk, and eggs.

Jonathan Safran Foer's book, *Eating Animals*, was based on one and a half years of going to farms all over the country to experience first-hand animal husbandry and meat production. Here is some of the animal cruelty he highlighted:

- for the past half century, there have been two kinds of chickens—layers and broilers, engineered with distinct genetics; egg-laying hens (layers) are in cages of around 67 square inches stacked 3 to 9 tiers high, in windowless sheds; male layers are destroyed as chicks, by being sucked through pipes onto an electrified plate, tossed in large plastic containers to suffocate, or sent fully conscious through macerators!
- chickens (broilers), which can live 15–20 years, are grown twice as large in half the time, and typically killed at six weeks for meat;
- modern trawling leads to massive catches with massive bycatch—unwanted sea creatures caught by accident and thrown overboard dead or dying; the average shrimp trawling operation throws 80–90% of the catch overboard as bycatch;
- a factory sow will be kept pregnant most of the time; for the 16 weeks of pregnancy she is confined in a small "gestation crate" where she cannot even turn around and must lie in excrement; chafing causes pus-filled sores, and her bone density decreases due to lack of movement; a sow nurses nine piglets on average; when they are weaned, a hormone injection gets her ready to be artificially inseminated in three weeks;
- the treatment of baby animals, and later their slaughter is another horrifyingly painful story.

Safran Foer describes cruelty as "not only the wilful causing of unnecessary suffering, but the indifference to it." He suggests vegetarianism as a lot of small (easy) choices and advises to "eat consciously as few animals as possible, ideally none. More than 99% of animal products are produced under factory farm conditions." Natalie Portman, the Academy Award winner actress, adds, "*Eating Animals* changed the way I choose what to eat. The book reminded me that what we choose to eat defines not only our physicality, but also our humanity."

More Than 59 Billion Animals Slaughtered

In the May 2011 issue of *National Geographic* magazine, a one-page article provided shocking data. In 2009, we ate about 293 million cows, 398 million goats, 518 million sheep, 693 million turkeys, 1.1 billion rabbits, 1.3 billion pigs, 2.8 billion ducks, and 52 billion chickens! A total of approximately 59 billion sentient beings are tortured and slaughtered every year to sustain a mere population of 7 billion humans. And these staggering numbers exclude animals from the sea, reptiles, and amphibians. Not to mention that more than 30% of global food crops—that we clear forests to cultivate—are fed to these animals. Large amounts of the faeces generated are dumped into our lands and rivers untreated. The damage to the environment is incalculable. We have to come to the full realisation that the fate of our world is literally in our mouths when we sit down to eat every meal.

Making a big difference in our world can come in small, consistent steps. The citizens of Ghent, Belgium, took that step on 14 May 2009. They support their city's Veggie Day: every Thursday, everyone has meatless meals. Ghent officials instituted this programme to support sustainability; being vegetarian one day a week for one year is equivalent to taking half a million cars off the road (in CO_2 savings).

FOOD SCARES AND FOOD SCAMS

More and more food contamination, food poisoning, and outright food scams have been hitting the headlines. "In the UK, 6 million people or 10% of the population have a case of food poisoning every year," reported the BBC on 26 March 2007. *The New York Times*, on 14 May 2009, stated that "increasingly, corporations supplying processed food are unable to guarantee the safety of their ingredients because they don't even know who their actual ingredient suppliers are."

Most of the estimated 76 million cases of food-borne illnesses every year go unreported or are not traced to the source. Annually, these result in 325,000 hospitalisations and 5,000 deaths, meaning that 13 people die every day because of food-borne illnesses. Some examples include:

- salmonella scares—from eggs in the UK (20 January 2000), chicken in UK (25 October 2005), chocolate in the UK (2006/2007), tomatoes in the United States (10 January 2008), pistachios in the United States (30 March 2009);
- dioxin (carcinogenic chemical) scares—from Belgian meat (1999), eggs in the UK (5 August 2003), pork in Ireland (December 2008), mozzarella cheese in Italy (28 March 2008);

- according to the Chinese Ministry of Agriculture, in 2008 the Chinese melamine milk scandal affected more than 54,000 children with kidney stones and agonising complications, with at least 6 infant deaths; this led to farmers dumping milk and killing cattle after companies stopped buying from them; more than 170 tonnes of tainted milk powder were found in February 2010 having been repackaged for sale instead of destroyed!

The Truth about Milk

Cow's milk contains 59 active hormones (injected to accelerate growth and milk production), scores of allergens, fat, cholesterol, herbicides and pesticides (from their feed), dioxins, up to 52 powerful antibiotics (to prevent cows from falling sick), blood, pus, and faeces (due to the way cows are being treated). One of the hormones is insulin-like Growth Factor One (IGF-1), which is also a fuel for ALL cancer cells in the human body, especially breast, prostate, and colon cancers.

Professor Jane Plant, a then 42-year-old scientist, was struck down with breast cancer. In her research for a cure she came across an interesting discovery and quit milk and dairy food completely. After seven weeks, she was declared free of cancer by every doctor. Her book, *Your Life in Your Hands* (2000), emphasises: "to cure breast cancer, avoid all milk products."

Health problems most prominently associated with milk are:

- cancer: breast, colon, and prostate;
- upper gastrointestinal disorders: gastroesophageal reflux, peptic ulcer, colic;
- lower gastrointestinal disorders: colitis, chronic constipation, irritable bowel syndrome;
- respiratory conditions: nasal stuffiness, runny nose, sinusitis, asthma;
- bone and joint problems: rheumatoid arthritis, lupus, psoriatic arthritis, osteoporosis;
- skin rashes: atopic dermatitis, eczema, seborrhoea, hives;
- nervous system: multiple sclerosis, Parkinson's disease, autism, allergic-tension fatigue syndrome;
- blood disorders: abnormal blood clotting, iron deficiency, anaemia, thrombocytopenia, eosinophilia;
- lifestyle diseases: obesity, heart diseases, strokes, diabetes.

Cows get their calcium from plants that contain magnesium, which helps the body absorb and use calcium. Cow's milk has insufficient magnesium for us to absorb the calcium in it. In fact, the high protein in milk strips the human body of calcium through urinary excretion. The irony is that countries with high milk, meat, and dairy consumption have the highest rates of osteoporosis. These high acid foods must be neutralised when consumed, and bones dissolve themselves to release alkaline materials to neutralise the high acids!

In *The China Study*, Dr. Colin Campbell disclosed that casein, which comprises 85% of the protein in cow's milk, promoted cancer in ALL stages of its development. Casein also creates problems such as mucus in the nose, throat, lungs, and intestinal tract, and immune system responses causing hives, itching, stomach bloating, vomiting, and rashes. The cow protein lactalbumin plays a significant role in the development of diabetes mellitus. The world famous Dr. Benjamin Spock shocked the medical world years ago by stating that no child under the age of one should consume cow's milk.

Three glasses of fresh whole milk contain the same amount of cholesterol as fifty-three slices of bacon. Condensed milk contains five times more cholesterol than fresh whole milk. Homogenising milk makes it last longer, but it breaks up the large fat molecules (which could not get through the intestinal wall) into smaller molecules, which get into the bloodstream, bringing fat-borne toxins (lead, dioxins, pesticides and herbicides, etc.) to organs.

The above information is largely unknown to the general population as the milk industry is one of the oldest and largest industries in the world. In the United States alone, the milk industry generated $25.9 billion in 2008. They use very aggressive marketing tactics to sell dairy products, with "research" showing the benefits of milk. The marketing budget runs into hundreds of millions, with a significant portion targeting children, parents, teachers, and school food service professionals.

SOFT DRINKS

A study on 60,524 people in the Singapore Chinese Health Study, extended over 14 years found that people who drink 2 or more sweetened soft drinks a week have an 87% higher risk of pancreatic cancer. People who drink mostly fruit juice, instead of soft drinks, do not have the same risk. Dr. Mark Pereira of the University of Minnesota, who led the study, wrote in the journal *Cancer Epidemiology, Biomarkers and Prevention*, that the findings would apply elsewhere. He said that, "the high levels of sugar in soft drinks may be increasing the level of insulin in the body, which we think contributes to pancreatic cancer cell growth." Insulin is made in the pancreas, and helps to metabolise sugar. Pancreatic cancer, which kills 230,000 people globally every year, is one of the deadliest forms of cancer.

Pause: We Are What We Eat

The phrase "You are what you eat" came from French lawyer and politician Jean Anthelme Brillat-Savarin in 1826. He gained fame as an epicure and gastronome. In one of his famous gastronomic writings, *The Physiology of Taste*, he said, "Tell me what you eat and I will tell you what you are." Brillat-Savarin was trying to emphasise that the food we eat has a strong bearing on our state of mind and health; it affects what our body assimilates and benefits from, which has direct effect on the cellular chemistry of our body, the bloodstream that nurtures the cells, and in turn, our overall physical, mental, and emotional well-being.

Penny Kelly (US naturopath), in her book *From the Soil to the Stomach* (2005), stated that most medical doctors do not know much about nutrition, and the truth is, the majority of chronic illnesses and degenerative diseases people are struggling with are directly linked to poor and missing nutrition.

In our world today, more and more people are getting sick, dying early, and holding themselves together with a regiment of drugs that often leads to further dependence on heavier drugs later on. It is the realignment of our lives to the ancient adage that "the mind and body are one," that we can turn our health around.

We must fully realise that the mind *does* affect the body for good or ill. And in turn, the body (what we eat, drink, breathe, and smoke) affects the mind in the same ways, for good or ill.

The Impact of Food Production and Consumption Choices

In the last 20 years, books have been frequently and intensively written about the impact of our food choices; Professor Michael Pollan of Berkeley University has probably been the one leading the way.

In *The Omnivore's Dilemma* (2006), Dr. Pollan investigates our "eating disorders" by tracking food from its origin to ultimately into our dinner plate. He suggests that there is an essential tension between the logic of nature and the logic of human industry. He further adds that the way we eat represents our most profound engagement with our natural world, and our production and consumption obscures vital ecological relationships and connections.

In the book *In Defense of Food* (2008), Dr. Pollan offers a seven-word mantra with three rules: "Eat food. Not too much. Mainly plants."

In 2009, his *Food Rules* gave us 64 rules to healthier and happier eating. Pollan's books are not only about nutrition. In fact, he is pointing at a bigger picture for us, which includes not only our health, but also our conscience and our responsibilities toward humanity and the environment.

There is a systemic relationship between the food we eat and our health, economy, environment, humanity, and spirituality. At this juncture, the book has clearly established the devastating health consequences and the environmental impact of our food production and consumption choices.

Below, on the next page, is an illustration of the systemic relationship between these overlapping and interpenetrating contexts.

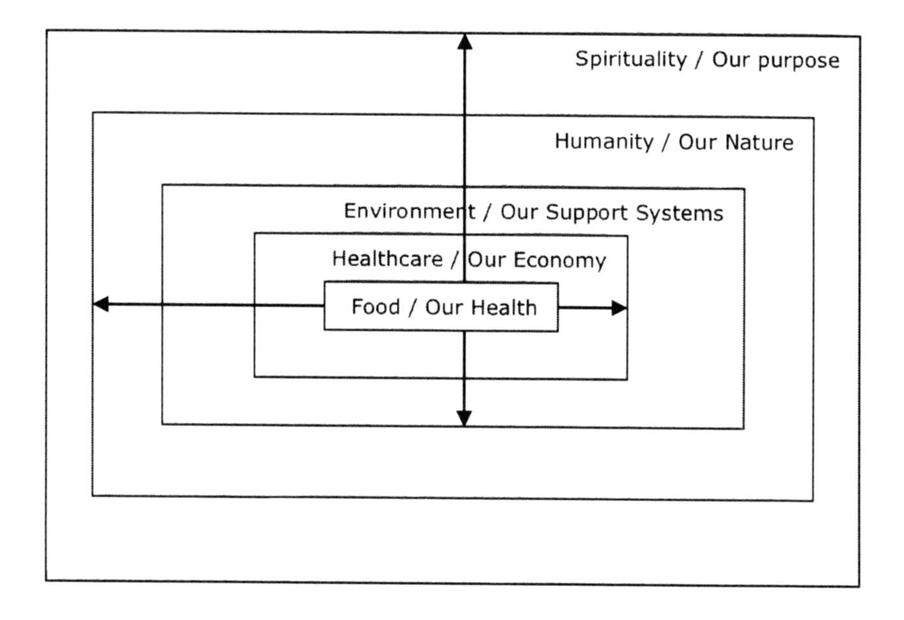

Impact on Economy

The Centers for Medicare and Medicaid Services (CMS) reported in 2007 that the United States spent $2.24 trillion in healthcare, the most in the world. According to the Milken Institute, an independent economic think-tank, the rising rate of chronic diseases is a very large contributor to the rising healthcare cost in the United States. The chronic diseases focused on in the Milken study in 2007 were cancer, diabetes, hypertension, stroke, heart disease, pulmonary conditions, and mental disorders. The combined cost of treatment expenditures and loss of economic output for the United States was $1.3 trillion for the seven diseases. They clearly affirmed that most of the medical costs are avoidable by lowering weight, eating better, and getting exercise.

Similar studies around the world also show that healthcare costs are rising every year, especially in developed, wealthy, and fast developing countries. According to the findings of Standard and Poor in January 2012, healthcare spending in a number of advanced economies will almost double as a proportion of GDP over the next 40 years. For example, Germany, the United States, UK, and France would typically rise from 6.3% GDP in 2010 to 11.1% in 2050. Paul Ryan, a US Congressman, stated that Medicare is supposed to hold the promise of healthcare security for millions of citizens. However, the United States faces an impending $38 trillion unfunded healthcare liability, and each year they put off dealing with this looming massive problem, Medicare falls deeper into debt and closer to collapse.

Left unchecked, unmitigated, and unreformed, the rising costs of healthcare in many wealthy countries would cripple major economies and trigger deep recessions around the world.

Impact on Environment

Following the rapid growth of the fast food industry, consumption of beef and chicken has risen tremendously. Fast food chains have been under attack from environmental groups because of the environmental impact of meat consumption. Intensive breeding of livestock and poultry leads to deforestation, land degradation, and the contamination of water sources and other natural resources. The same goes for clearing land for cash crops, such as coffee and tea. All these production choices have severely impacted biodiversity on the planet.

From the research of Paleobiologist Anthony Barnosky of the University of California (2011), earth creatures are on the brink of the sixth mass extinction, comparable to the one that wiped out dinosaurs 65 million years ago. He calculated that three-quarters of today's animal species could vanish in just 300 years. This disappearing of plant and animal life has direct effect on the environmental life-support systems that living things—including us—need.

More than one third of the world's total grain production is used to feed livestock (in the United States, 70%). For every pound of red meat, poultry, eggs, and milk produced, five pounds of irreplaceable top soil is lost. About 719 litres of water are required for meat breeding per animal per day, ten times what a normal Indian family uses in a day. It takes 25,000 litres of water to produce one kilogram of beef, 200 times more water than to produce a kilogram of potatoes. Overall, meat requires much more of our natural resources to be produced than comparative amounts of plants as food. Compounding on this problem of food shortage, large fields of crops have been converted to biofuel instead of being fed to the millions of starving humans.

Impact on Humanity

The Merriam-Webster dictionary defines *humanity* as the quality of being humane, which encompasses the qualities of compassion and benevolence. However, to understand the meaning of our humanity we must first answer the question, "What does it mean to be human?"

Perhaps a quote from Dr. Carl Sagan (1934–1996), the renowned American astrophysicist and cosmologist, could help us understand and appreciate the importance of asking ourselves this perplexing, yet vital question:

> Who are we? Where do we come from? Why are we this way and not some other? [...] Are we capable, if need be, of fundamental change, or do the dead hands of forgotten ancestors impel us in some direction, indiscriminately for good or ill, and beyond our control? Can we alter our character? Can we improve our societies? Can we leave our children a world better than the one that was left to us? Can we free them from demons that torment us and haunt our civilization? In the long run, are we wise enough to know what changes to make? Can we be trusted with our own future?

To expand on the meaning of humanity, one needs to look into the contributions of Dr. Albert Schweitzer (1875–1965), the famed German philosopher, theologian, medical missionary, and Nobel Peace Prize Laureate. He founded and built a hospital in Gabon, Africa, which still exists today. He introduced and justified a universally valid ethics of responsibility for all forms of life and is called by many of his contemporaries "the conscience of our age."

The keynote of Dr. Schweitzer's contribution to mankind was the idea of "reverence for life." He conceived this idea in 1925 serendipitously, realising that this term implies an ethics that leaves behind everything that has been formulated before. This ethics is not limited to the relationship with fellow human beings, but includes all forms of life. He firmly stated that until we extend the circle of our compassion to all living things, we, ourselves, will not find peace. And throughout history there have been many luminaries who supported Dr. Schweitzer's philosophy:

The greatness of a nation and its moral progress can be
judged by the way its animals are treated.

—Mohandas Gandhi

I am in favour of animal rights as well as human rights.
That is the way of a whole human being.

—Abraham Lincoln

I have no doubt that it is a part of the destiny of the human race,
in its gradual improvement, to leave off eating animals, as surely
as the savage tribes have left off eating each other.

—Henry David Thoreau

Non-violence leads to the highest ethics,
which is the goal of all evolution.
Until we stop harming all other living beings,
we are still savages.

—Thomas A. Edison

If you have men who will exclude any of God's creatures from
the shelter of compassion and pity, you will have men who
will deal likewise with their fellow men.

—St. Francis of Assisi

Life is life, whether in a cat, or dog, or man. There is no
difference there between a cat or a man. The idea of difference
is a human conception for man's own advantage.

—Sri Aurobindo

It is clear from the accounts of key historical figures that humans can relate to animal suffering. The vast majority of meat, eggs, and dairy products sold in modern cities around the world come from animals raised in intensive-confinement systems. Animals on these factory farms suffer immense pain and stress as a result of extreme confinement, bodily mutilations without pain relief, and completely denied opportunities to behave normally. They are dehorned, debeaked, and castrated without anaesthesia. Finally, after living a life of total suffering, they are trucked over great distances without food or water, through all weather extremes, to a horrifying death. This is the reality for the suffering tens of billions of animals every year, but such practices are kept out of our daily living. If they were carried out in front of us—say we order a chicken and it is slaughtered before our eyes—I think our humanity would kick in. We are not cruel or blood-thirsty by nature.

Recent studies on compassion and benevolence strongly suggest that we are wired to be good, kind, and compassionate. Such traits are part of our human nature, rooted in our brain and biology, and therefore ready to be cultivated for the greater good.

These studies include the works of Dr. Dacher Keltner (*Born to Be Good*), Jeremy Rifkin (*The Empathic Civilization*), Frans de Waal (*The Age of Empathy*), and Dr. Marc Bekoff (*The Animal Manifesto: Six Reasons for Expanding Our Compassion Footprint*).

Dr. Marc Bekoff is Professor Emeritus of Ecology and Evolutionary Biology at the University of Colorado and a scholar-in-residence at the University of Denver's Institute for Human–Animal Connection. According to Bekoff's research, many animals are far more empathic and fair than many people realise. While we recognise the rules of right and wrong behaviour in our own human societies, we are not accustomed to looking for them among animals. Bekoff's long-term studies on animal activities show that animals demonstrate behaviours such as fairness, empathy, forgiveness, trust, altruism, social tolerance, and reciprocity.

In his book *The Animal Manifesto*, he lists six reasons for expanding our compassion footprint, and with them he provides extensive evidence for his assertions.

Reason #1: Animals have the right to live with dignity and respect on this earth just as much as humans do. People and animals alike have intrinsic value by virtue of existing on this planet. We all share the same earth and this should be reason enough for us to stop dominating animals, but instead to coexist peacefully with them.

Reason #2: Animals are capable of thinking and feeling. These are not qualities that we attribute to them in a desire to believe they have these features, but they exhibit these traits in order to be able to survive. Numerous scientific studies have shown that animals think, feel, remember, strategize, and display emotions, just like humans.

Reason #3: Animals are compassionate and deserve to be treated with compassion. Bekoff observed animals in their natural environment, they have a natural tendency to work in cooperation with other animals and to exhibit compassion and empathy toward them. By being compassionate, we are expressing our true nature.

Reason #4: By feeling connected to animals, we experience feelings of care toward them. When we alienate them, disrespect arises. Our disconnection from animals has made us heartless (inhumane); we have turned them into "things and stuff" and subjected them to torture in battery farms. When we reconnect with animals as nature intended it, we will allow them to bring joy, sustainability and abundance back into our world.

Reason #5: We are not living in a world that treats animals with compassion. If we take a close look at how billions of animals are actually being treated by humans, we will see why we must behave more humanely with animals in order to fulfil our dream of creating a peaceful and life-supportive planet.

Reason #6: All living creatures and our world benefit when humans are compassionate. Bekoff wrote that our actions should always reflect kindness, empathy, and compassion toward all beings. When we live compassionately, life reciprocates.

This segment on humanity beseeches us to reconnect to all life around us, because when we do, we find our true humanity. When we align ourselves to our highest ideals we know that we are not a pain-inflicting and blood-thirsty species. From the information that we have uncovered thus far, we now know that what we choose to eat affects our health, economy, environment, and our humanity. We must make a choice to eat mainly plant-based food, practice ecological farming systems, support locally grown produce, eradicate preservatives and highly processed food, reduce our consumption of meat and dairy as much as possible, and be resolved in treating animals humanely.

Not only are we looking at our own present needs, we are also looking at the sustainability of our planet in the way we eat. A popular definition of sustainability is from the Brundtland Commission of the United Nations, on 20 March 1987: "Sustainable developments meet present needs without compromising the ability of future generations to meet their needs." So, to think about sustainable living is to think about the well-being of our children's children's children.

The ramification of our food choices is far reaching. It is clear now that the way we eat determines the kind of world we live in today and the kind of world we will leave to our future generations.

> We cannot abdicate our conscience to an organization, nor to a government. "Am I my brother's keeper?" Most certainly I am! I cannot escape my responsibility by saying the State will do all that is necessary. It is a tragedy that nowadays so many think and feel otherwise.
>
> —*Dr. Albert Schweitzer*

Impact on Spirituality

Ultimately, everyone on their journey of personal development strives to nurture the qualities of being more compassionate, loving, open, benevolent, enduring, and to rise above the fear of death. Such qualities aptly describe Saints and Avatars like Mother Theresa, Gandhi, Nelson Mandela, the Buddha, Jesus, etc. With the decline of membership in organised religions and the rise of secularism around the world, more and more people are exploring their own experience and definition of spirituality.

Secular spirituality emphasises humanistic qualities (such as the ones described above) and aspects of life and human experience which go beyond a purely materialistic view of the world. In more recent definitions, spirituality is often described as a way of living that nurtures thoughts, emotions, words, and actions that are in harmony with a belief that everything in the Universe is mutually connected and dependent. In aligning oneself to the belief that everything is connected, one includes all living things and the planet as a whole. In a broader sense, it means there is a reciprocal relationship between all things.

Modern spiritualists, usually not bound to any religious tradition, focus on enriching their inner life and on discovering their deepest values, many of them through practices like meditation, prayer, contemplation, and adhering to a particular form of diet. Such practices help to achieve and maintain simplicity by cutting down on living consumption and expanding the time to connect to the larger community, nature, the cosmos, and the divine realm.

It is now common knowledge that what we put or don't put into our bodies not only impacts our physical, but also our spiritual well-being. Many Eastern and Western religions have ancient laws forbidding certain foods or combination of foods to prevent contamination of the body, which is regarded as a temple of the spirit.

Many spiritual leaders and philosophers from different traditions and beliefs chose a vegetarian diet; Zoroaster, Confucius, Pythagoras, St. Francis of Assisi, Leo Tolstoy, William Booth, Rudolf Steiner, Mahatma Gandhi, Sri Aurobindo, Maharishi Mahesh Yogi, Sri Sri Ravi Shankar, Thich Nhat Hanh were or are all vegetarians.

Science has conclusively determined that a vegetarian diet promotes a much better health than any other diet; research studies have been conducted on long-lived people around the world and have shown that the Hunza people, living in the Himalayas, had an average life span between 90 and 120 years! The investigation concluded that the Hunza's longevity was based on clean water, exercise, and adherence to a diet of fruits, grains, and nuts. Interestingly, their diet matches the one described in the Bible (Genesis 1:29). With the birth of the Christian Vegetarian Association and the increasing number of publications by Christian researchers (e.g., Dr. Rex Russell, author of *What the Bible Says About Healthy Living*, Keith Akers, author of *The Lost Religion of Jesus: Simple Living and Nonviolence in Early Christianity*, etc.), more Christians are choosing to adopt a plant-based diet. They believe vegetarianism is the appropriate way to serve God, honour their bodies as temples of the spirit, and preserve God's original creation.

In the ancient Indian religion of Hinduism, Hindus believe that their diet affects the body, mind, and emotions, which can alter their states of consciousness. In Hindu scriptures, the Universe contains three primal forces which govern all creation and evolution processes. They are known as Sattvas (preservation forces), Rajas (creation forces), and Tamas (destruction forces). These primal forces are imbued in every aspect of the Universe, including animals, plants, humans, and all forms of subtle energies and vibrations.

In Hindu scriptures, Lord Krishna speaks of three different types of food:

- Sattvic foods are juicy, wholesome, and pleasing to the heart. They promote longevity, purity, strength, health, and contentment. Examples of such foods are fresh fruits, almonds, dates, sprouts, lettuce, barley, lentils, etc.
- Rajasic foods are bitter, sour, salty, hot, pungent, dry, and over cooked. They cause greed, jealousy, anger, and delusions. Examples of such foods are sour apple, guava, corn, buckwheat, potato, broccoli, fish, shrimp, chicken, etc.
- Tamasic foods are not fresh, but tepid, putrid, and impure. This kind of food leads to stagnation and degeneration. It promotes self-centredness, mood swings, and chronic diseases. Examples of such foods are avocado, wheat, garlic, onion, pumpkin, beef, lamb, pork, etc.

By eating a non-vegetarian diet, there is therefore an increase in Tamasic properties which are more likely to promote physical, emotional, and mental disturbances and imbalances. In a vegetarian diet there would naturally be a higher proportion of Sattvic properties, which are conducive for spiritual practice. Notice how a different religious context directs our focus again to vegetarian food as a better diet to support our physical, mental, emotional, and spiritual development.

In Eastern spiritual traditions bondage and liberation are states of the mind. The mind is a mass of vibrating energies, controlled by the constitution or condition of those energies. If the energies are heavy or stagnant, they impede silence and clarity, which are needed for spiritual development. Depending on certain kinds of thoughts or actions, one can darken the mind, making it thick and heavy, or one can brighten the mind, making it light and fluid. To attain such liberation, the mind must be purified and refined, and dieting and meditation are necessary and powerful ways to "brighten" the mind.

In ancient Yogic wisdom, it is believed that when we eat food, we absorb some of its consciousness. All the animals suffer immense pain when they are slaughtered, many of them still being conscious when their bodies are split open or thrown into boiling water.

From a scientific perspective, according to the *Journal of Animal Science*, researchers at the University of Milan confirmed that fear experienced by animals during slaughter significantly raises the stress hormones of adrenaline, cortisol, and other steroids in meat.

These hormones remain in the meat long after the animals are slaughtered, so when we eat their meat, we ingest these hormones ourselves.

The concentrated vibrations of terror, grief, and desperation transmitted through these hormones permeating terrorized meat are invisible and completely unrecognised by conventional science, yet they may be even more detrimental to us than physical toxins because they work of the level of our consciousness.

When our bodies are graveyards of animals, how can we achieve inner peace? And consequently, how can the world around us be peaceful?

Many vegans and vegetarians around the world testify to the fact that when they stopped eating meat, they noticed a distinct reduction in negative emotions like mood swings, anger, anxiety, and sadness. Thich Nhat Hanh, the renowned Vietnamese Buddhist monk, author and peace activist, sums it up simply and directly:

> When we eat an egg or chicken, we know that [...it] can also contain a lot of anger. We are eating anger, and therefore we express anger. [...] Be careful what you eat. [...] If you eat despair, you will express despair. If you eat frustration, you will express frustration.

Other research studies, by Dr. George A. Bray and Dr. Irwin H. Putzkoff, have shown that such pain hormones contribute to obesity, sexual impotence, and increased levels of bad cholesterol. The impact of stress on meat is well known to the Food and Agriculture Organization

(FAO) of the United Nations. Pale Soft Exudative meat—or PSE meat—occurs when animals are subjected to manhandling, fighting in pens, and bad stunning techniques, as the fright and stress cause a rapid breakdown of muscle glycogen. This lightens the colour of the meat and turns it acidic and tasteless, making it difficult to sell, so it is usually discarded or added to other meat products and made into sausages.

When it comes to talking about food choices, many people insist that we all make individual choices about the foods we eat, resisting the fact that they have been indoctrinated by governments and corporations. Will Tuttle, Ph.D., author of *The World Peace Diet* (one of the bestsellers in 2010 on Amazon) clearly pointed out that we never actually chose an omnivorous diet; we were fed this way since our herding culture (domestication of animals). For thousands of years, humans are told by parents, doctors, teachers, governments, and more recently by advertisements and the media, that animals are for us to eat and become strong.

Today, these forces continue to distort the truth about the negative impact of meat-eating on health and continue to manipulate consumers who still believe they are making free choices, while the terror and agony of billions of animals remain hidden from view. We have trained to dissociate and disconnect ourselves from the way food has been grotesquely manufactured before being cosmetically dressed-up on our plate. We have become unconscious accomplices to mass murder.

Dr. Tuttle also pointed out how universal spiritual principles (e.g. Hinduism, Zoroastrianism, Christianity) are routinely ignored and violated, including the five precepts of Buddhism: (1) against killing—animals are killed by the billions every year; (2) against stealing—animal babies, eggs, milk, and freedom are stolen from them; (3) against committing sexual violence or misconduct—animals are sexually abused through artificial insemination and painfully castrated without anaesthetics; (4) against deceiving—animals are deceived by barbed hooks, and by dark tunnels ending with electroshocks, stun guns, and sharpened blades; (5) against alcohol and drugs—animals are forced to take antibiotics, drugs, hormones, and psychotropic pharmaceuticals.

If we continue to inflict pain, suffering, and agony on other living beings, ignoring the fact that they have feelings, mother-child emotions, and a divine purpose, just like us, it is unlikely that there will ever be peace on earth. What we do to others, we do to ourselves, because the web of life connects us all.

Almost every day we hear people talk about how horrible the world is, how everything is getting worse, and how they feel powerless and sad in trying to make a positive difference in our rapidly deteriorating world.

Now, we can all do something about it daily when we sit down to eat. Many of our problems are directly and indirectly connected to our ignorance and fixation on meat consumption.

A groundbreaking study and book written by John Robbins, *Healthy at 100*, shows a vital commonality in the study of centenarians—the world's healthiest and longest-lived humans. He studied people who are between 90 and 130 years old and still healthy in Abkhazia (in the south of Russia), Vilcabamba (in Ecuador's Andes mountains), Hunza (in the northernmost tip of Pakistan), and Okinawa (the southernmost state of Japan). There were many different factors that contributed to these people's longevity, such as living at high or low altitudes, the amount of physical exercise they do every day, the quality of the water they drink, or the amount of food they consume daily. But one thing is common to all of them: they have a very low animal protein intake, 90–99% of their diet being based on plants.

Their diets are low in calories and high in good carbohydrates from whole grains, vegetables, and fruits. They mainly eat whole foods, without preservatives and very little processed or refined. They all depend on fresh foods, eating what is in season and locally grown, which means their food is not canned or shipped from far away. Their fat intakes is low, mostly comes from seeds and nuts, and occasionally fish. They take their proteins from plants—beans, peas, whole grains, seeds, and nuts.

The Buddhist, Taoist, Yogi, Hindu, and many other spiritual traditions share a common ideal: in whatever they do, they strive to increase compassion and alleviate human and animal suffering. Among other life-supportive practices, embracing a vegetarian diet—an uncomplicated solution—would help us to change our world positively through practising healthy and ecological consumption. Most of all, vegetarianism and veganism help us to propagate a consciousness of compassion and nonviolence in our world. When we treat life with compassion, life treats us back with compassion.

CHAPTER 7
WHAT'S HAPPENING TO OUR BODIES?

Deeply engrained into our psyche of reality, we refer to our bodies when expressing our thoughts and feelings through expressions like "Play by ear," "See eye to eye," "Stick one's neck out," "Pull one's leg," "Get something off one's chest," etc. There are many sayings regarding our health, some depicting what we should or should not do: "Early to bed and early to rise makes a man healthy, wealthy, and wise," "Laughter is the best medicine," "Health is more important than wealth."

> I see rejection in my skin, worry in my cancers, bitterness and
> hate in my aching joints. I failed to take care of my mind,
> and so my body now goes to hospital.
>
> —*Astrid Alauda, author*

> He who has a straight body is not worried
> about his crooked shadow.
>
> —*Chinese proverb*

> The body is your temple. Keep it pure and clean
> for the soul to reside in.
>
> —*B.K.S. Iyengar, founder of Iyengar Yoga*

In spite of this, we have not been treating our own bodies and those of others with much care; it is not at all surprising, thus, that we've also been mistreating other species and our environment, which we are dependent on for our lives. Let us take a look at recent scientific studies to discover the changes in our bodies.

RECENT CHANGES IN OUR BODIES

According to a study conducted in 2009 by the UK National Sizing Survey, the physiques of women have become bigger compared to the 1950s. Women used to have an hour-glass figure of 36-24-35 inches in the 1950s; now it has rolled in to a barrel-like 38-34-40. Other vital statistics were included: most women are 2 inches taller, 7 pounds heavier, and their feet have grown 3 sizes. These important findings do not just carry implications for how we look; they provide crucial information about our health. Overall improvements in hygiene, medicine, and nutrition in many parts of the world have increased our life-span from 70.9 years to 81.5 years. And this is not necessarily a good thing if we examine the lives of our elderly: they do live longer, but in poor health, heavily relying on medication.

Implications of Increases in Height

The improvements in nutrition and availability of more ample diets have largely contributed to the increase in height, explained Dr. Bernard Harris, professor of history of social policy at Southampton University. "Broadly speaking, people who are shorter are more likely to die early," according to Dr. Harris, who was involved in a large scale international study on the evolution of human physique.

Implications of Increases in Weight and Waistlines

In the 1940s, we were consuming 32% saturated fats, and currently they form 40% of our diet. Our intake of sugar has more than doubled, and we consume more processed and junk food. Our increased consumption of foods that contain corn syrup and preservatives has impeded the production of healthy hormones that help to regulate our appetite and fat storage. The change in diet and lack of exercise explain our expanding waistlines.

In a study published by *The New England Journal of Medicine*, a big waistline is deemed more hazardous for health than just being overweight because the fat cells carried around the stomach pump out chemicals that can damage the insulin system, raise blood pressure, and increase cholesterol levels.

Implications of Increases in Breast Size

Professor Michael Baum, a specialist in breast cancer and Professor of Surgery at University College Hospital, London, states that, "Fat is laid down on breasts as much as thighs or bottoms, and we are experiencing an obesity epidemic, so the increase in women's measurements isn't that surprising." The increase in certain hormones in women could contribute to the changes in breast size, as women on hormone replacement therapy for menopause have evidently increased one cup size. Speculation has been made on the increased usage of hormones in dairy cows that found their way into fresh milk and beef.

According to a health feature on CNN in August 2010, more girls are reaching puberty earlier, even at ages seven or eight. It is a public health concern as studies have shown that girls who start puberty earlier are more likely to develop breast and uterine cancer later in life.

Implications of Increases in Hip Dimensions

Based on the same UK survey, our hips have not increased in size as much as our waists have. This is partly due to our oestrogen levels which determine where fat is stored. When hormone levels are balanced, most women store fat around the hips, as opposed to the current trend whereby most is stored around the waist.

In a study by the WHO on waist-hip ratio (WHR), men and women with "apple-shaped" bodies (more weight around the waist) face more health risks than those with "pear-shaped" bodies (more weight around the hips). Oestrogen levels in women affect where fat is stored and if their hormone levels are unbalanced, they end up storing more fat at the waist rather than the hips, where they should. It is unclear now why women's oestrogen levels are unbalanced. It is speculated that it may be due to the widespread use of contraceptive pills, environmental pollution, as well as unhealthy diets and an increasing number of lifestyle stressors.

TECHNOPHYSIO EVOLUTION

Coined by Nobel Prize-winning economist Dr. Robert Fogel, *technophysio evolution* is a uniquely modern form of rapid physiological development derived from humanity's ability to control its environment and create technological innovations to adapt to it. For about three decades Dr. Fogel and his small team of esteemed colleagues have tirelessly researched what the size and shape of the human body say about economic and social change throughout history and vice versa. His book, entitled *The Changing Body: Health, Nutrition, and Human Development in the Western World Since 1700*, has clearly shown that in most parts of the world, the height, weight, and longevity of the human body have changed more substantially and rapidly in the past 300 years than over many previous millennia.

With scientific rigour, Dr. Fogel and his team synthesised mountains of data from demography, economics, medicine, biology, and sociology. Some of the major conclusions of their research include:

- increases in average height in men and women can be taken as a reliable measure of the success of different societies, specifically the productivity of their economies, the fair distribution of their resources, and wider access to scientific advances in healthcare and the workplace;
- poor nutrition affects labour productivity because it diminishes cognitive ability and the capacity to undertake and gain from education. Populations with high malnutrition rates suffered from low productivity not simply because of diminished physical strength, but also diminished cognitive ability;
- the health and nutrition of pregnant mothers and their children contribute to the strength and longevity of the next generation.

If babies are deprived of proper nutrition in the womb and early in life, they will be feebler and more prone to diseases later on. These weakened adults will in turn produce weaker offspring in a self-intensifying manner;

- over-nutrition has become the primary health problem in the United States and other westernised nations. The availability of calories combined with modern sedentary lifestyles has increased the number of obese people. They are susceptible of having chronic diseases such as heart problems, stroke, hypertension, and some types of cancer.

Dr. Fogel's research has given us a clearer picture of the intricate cause-and-effect relationships between economic growth, nutrition, and healthcare. To prevent decline in height, weight, and longevity, emphasis must be put on food production, which is tied to economic growth. It is also important to focus on the control of infectious diseases through better public healthcare and education. His research also tells us that nutrition has a direct impact on the intelligence of a country's workforce and the health of their future generations.

OBESITY

The WHO states that obesity rates have risen three or more times in the past twenty years, often faster in developing countries. There are about 1.6 billion overweight and obese people in the world, and at least 2.5 million deaths associated with these conditions annually. Over 300 million people are obese (BMI≥30), but 115 million live in developing countries. According to a report titled *Obesity: Halting the Epidemic by Making Health Easier—at a Glance, 2009*, by the Centers for Disease Control and Prevention, over one-third of the adults in the United States (more than 72 million people) are obese, as are 16% of children. Obesity rates have doubled for adults and tripled for children since 1980. The US Surgeon General report states that obesity is responsible for 300,000 deaths every year in the United States.

Physical, Psychological, and Social Consequences

The American and many other societies have become obesogenic, with over-consumption, eating unhealthy foods, and physical inactivity. The availability of food, incessant advertising, yo-yo dieting, and sedentary lifestyles have added up over the years. By watching television for two hours per day, a person will see over 20,000 food advertisements for sweets, chocolate, snacks, and fast food every year, many of them during TV programmes for children. Obesity-related diseases include type

II diabetes, heart disease, stroke, high cholesterol, high blood pressure, hypertension, sleep apnea, osteoarthritis, liver and gallbladder disease, infertility, and some types of cancer. Obesity is expensive. The United States spends over $100 billion on obesity every year, which account for 27% of the rise in medical costs between 1987 and 2001.

A report titled *Psychological Impact of Obesity on School-Aged Children*, by the Texas Department of Health in June 2004, found that "Obese children are often stigmatised by their peers, suffer social isolation [...], have lower self-esteem, emotional distress and anxiety, [some have] depression, suicidal thoughts, [or] bulimia. Long-term impact [includes] lower educational attainment, disparate treatment in the workplace, and lower rates of marriage."

Beautifying Surgeries

Cosmetic or aesthetic surgery has flourished tremendously over the years. According to the American Society of Aesthetic Plastic Surgery, nearly 12 million cosmetic surgeries were performed in the United States alone in 2007, up 7% from 2006, 59% from 2000, and 457% from 1997. Nine out of ten cosmetic procedures were performed on women. Over $13 billion were spent by Americans in 2007, $8.3 billion for surgical, and $4.7 billion for nonsurgical procedures. In Europe, the second largest market for cosmetic procedures, this is a $2.2 billion business.

The top five surgical cosmetic procedures in 2007 were:

- liposuction to remove fat;
- breast augmentation ("breast implants");
- blepharoplasty (the "eyelid surgery");
- abdominoplasty (the "tummy tuck");
- breast reduction.

The top five nonsurgical cosmetic procedures in 2007 were:

- botox injections;
- hyaluronic acid (including hylaform, juvederm, perlane/ restylane) procedures;
- laser hair removal;
- microdermabrasion;
- IPL laser treatment.

Other popular procedures include:

- buttock augmentation or lift;
- facial changes: reshaping the nose, ears, chin, cheeks, lips, and "face lifts," and
- hair transplants.

TV shows such as *Extreme Makeover*, *Nip/Tuck*, and *Dr. 90210* have sensationalised the whole plastic surgery industry. American TV star and singer Heidi Montag had plastic surgery a second time on 20 November 2009 (when she was 23 years old), undergoing ten procedures in one day. Addicted to plastic surgery, she said, "I'm beyond obsessed."

The risks of plastic surgery have been more seriously looked into following the death of model Solange Magnano, 38, former Miss Argentina, on 29 November 2009, arising from complications from buttock-enhancement surgery. As reported on CNN on 9 December 2009, Nigel Mercer, president of the British Association of Aesthetic Plastic Surgeons, described the cosmetic surgery industry an "unregulated mess" and warned that "no cosmetic surgery is totally risk free [...] even botox and fillers [...] there's a chance of infection, bruising or bleeding with any procedure."

In recent years, plastic surgery has become more prevalent with younger people in pursuit of a more ideal body. According to *time.com* (7 May 2008) younger Internet users flock to websites that dwell on personal appearance, such as ones that are dedicated to plastic surgery, weight loss, skincare, and bodybuilding. The death of an 18-year-old high school cheerleader, Stephanie Kuleba, on 22 March 2008, as a result of complications during an elective breast surgery, highlights the unforeseen dangers.

Psychosomatics journal reported in August 2004 that at least four studies in different countries have found that women who get breast implants commit suicide two to three times more often than women of comparable ages in the general population. Many who seek multiple cosmetic procedures aren't pleased with the results, and may suffer from body dysmorphic disorder (BDD), a preoccupation with an imagined deficit in appearance that causes distress in life. Men are also affected. In Singapore, a 44-year-old CEO, Franklin Heng, died from punctures to his intestines during a liposuction surgery in December 2009.

Body dysmorphic disorder is a mental illness where the person obsesses over a (perceived) flaw in his or her appearance. The person sees major defects despite other people's reassurances. Sufferers may withdraw from the world, not wanting to be seen by anyone, many have eating disorders, and others seek (numerous) plastic surgeries.

How do we see ourselves? What is our level of self-acceptance with our physical appearance? Do we deem that being normal is ugly? Would Michael Jackson have been as successful without cosmetic surgeries, or did his changed appearance strengthen the public's opinion of his creative genius?

JUNK SEX

According to Dr. Lisa Love, a relationship coach and author, more people are having a higher frequency of sex and with more partners, just as more people are eating sugar and junk food. Though junk food tastes good, it provides a brief superficial high that lacks the capacity to nourish the body. Similarly, the sexual experiences most people are addicted to today only give short-lived superficial highs that feel good at the time. But, without embodying life-affirming values, sexual experiences quickly dissipate from the initial high—felt during orgasm—into an empty, heart-breaking, often self-disgusted, and even guilt-laden experience.

What people don't understand is that more sex doesn't produce more satisfying sex, even if they spice it up with different sexual positions, sexy outfits, sex toys, etc. Frequent sex doesn't lead to a true sexual fulfilment if life-affirming values such as love, deep appreciation, thoughtfulness, and commitment are cut out.

HIV/AIDS and HPV

More sex with different people has its risks: sexually transmitted diseases (STD). Both the Human Immunodeficiency Virus (HIV) and the Human Papillomavirus (HPV) are at the top of the WHO list of epidemic STDs. The highest increase in STDs is among young people of 15 to 24 years of age. According to the *United Nations Programme on AIDS* (UNAIDS) 2008 report, there are 30.8 million people in the world living with AIDS. Globally, around 11% of HIV infections are among babies who acquired it from their mothers, 10% result from using drugs by injection, 5-10% are due to sex between men, and 5-10% occur in healthcare settings. AIDS acquired from sex between men and women accounts for the largest portion, between 59–69%. Though there are several main routes of HIV transmission, the most prevailing one is unprotected penetrative sex with someone who is infected. Most of the time, the infected person is not aware as he/she has yet to develop the symptoms.

About a third of the strains of HPV are sexually transmitted and can cause warts in the genital area. The greatest danger of the HPV infection is that about 10 of the HPV viruses are linked to cervical cancer, which affects 470,000 women every year. It is the second largest cancer killer in women. According to the latest report of the WHO, HPV types 16 and 18, which are linked to cervical cancer, have spread all over the world.

Asia has the highest incidences of cervical cancer, followed by Eastern Russia, Europe, Africa, South, Central, and North America. In the United States alone, it is estimated that 20 million Americans were infected with the broad spectrum of sexually transmitted HPV in 2009, and

5.5 million more are infected every year. It is estimated that 50% to 70% of the sexually active people will get an HPV infection during their lives.

According to Medimix, a global healthcare marketing research company, over 90% of the 1,022 physicians surveyed suggest that boys and girls should be vaccinated as early as ages 9–13. More than a thousand general practitioners in France, Germany, Italy, Spain, the UK, Australia, and Canada participated in the survey and most agreed that the government of each country should mandate the new HPV vaccine. Nonetheless, controversies about the safety and morality of the vaccine continue.

SUBSTANCE ABUSE

Smoking

There are 1.1 billion smokers in the world today. Around 10 million cigarettes are purchased every minute, 15 billion sold every day, and over 5 trillion are produced annually. One cigarette contains eight to nine milligrams of nicotine, and most smokers take in one to two milligrams of nicotine per cigarette. The rest is burned off. The nicotine in four to five cigarettes is enough to kill an average adult if ingested whole. Every eight seconds one human life is lost to tobacco use; there are approximately five million deaths annually. Second-hand smoke contains more than 50 cancer-causing chemical compounds. Trillions of filters filled with toxic chemicals from tobacco smoke make their way into our environment as discarded waste annually.

Drugs

Illegal drug abuse is a serious public health problem resulting in about 40 million serious illnesses or injuries in the United States every year. The US 2007 National Survey on Drug Use and Health found that around 8% of persons aged 12 or older were illicit drug users at the time, consuming marijuana/hashish, cocaine/crack, heroin, LSD, hallucinogens, inhalants, or prescription-type psychotherapeutics used non-medically. Drugs distort the way a user perceives the world, disrupting the ability to think, communicate, and recognise reality, causing emotional swings and feelings of disconnection and being out of control.

Alcohol

According to the Institute of Alcohol Studies, alcohol is the third leading cause of disease and injury in developed nations, causing nearly 10% of all ill health and premature deaths in Europe. On a worldwide level, alcohol causes 20–30% of cancer of the oesophagus and liver, cirrhosis of the liver, epilepsy, homicide/murder, and motor vehicle accidents. The American corporation WebMD states that 30% of US adults have experienced alcohol abuse or alcoholism.

Life-Coping Addictions

People who are addicted to over-the-counter medication and prescription drugs start out by wanting to alleviate some unwanted conditions such as pain, sleeplessness, anxiety, allergies and so on, or for fun, to lose weight, or to study better. Self-medication can easily lead to increasing dosages and maybe even illegal substances. The US National Institute on Drug Abuse states that the abuse of prescription drugs to get high has become more and more prevalent among teens and young adults; and the abuse of prescription pain killers ranks second only to marijuana as the most rampant illegal drug problem.

The death of Michael Jackson has spotlighted this problem. In 2008, the US Food and Drug Administration put out an alert that prescription drugs were the fourth-leading cause of death in the United States. As people get unwittingly addicted to the drugs they rely on to get through their day, the addiction insidiously robs them of their true personal power.

Distrust of pharmaceutical groups is very likely to rise as they have been caught with large health-care frauds. GlaxoSmithKline paid a $3 billion fine in July 2012 for marketing anti-depressant Paxil—which was never approved—as it raised suicide rates in children, for marketing Wellbutrin for sexual dysfunction and weight loss when it was only approved for depression, and for hiding clinical trials data that anti-diabetic Avandia could lead to heart attacks. Other drug-makers fined in recent years include Abbott, Merck, AstraZeneca, Eli Lilly, and Pfizer, for marketing infractions and drugs which caused serious health problems. (Source: *The Straits Times*, 14 July 2012, "Big Pharma fraud: Jail time could be the cure," by Andy Ho.)

Cardiovascular Diseases & Diabetes Epidemics

Don't dig your grave with a fork and knife.

—old English proverb

Cardiovascular diseases (CVDs) are the number one cause of death worldwide. They are disorders of the heart and blood vessels and include heart attacks and strokes. In 2004, 17.1 million people died from CVDs—29% of all deaths. Over 80% occur in low- and middle-income countries. The WHO estimates that an additional 23.6 million people will die from CVDs by 2030, with large increases in the eastern Mediterranean and south-east Asian regions. High blood pressure or hypertension increases the risk of CVDs. Heart diseases and strokes can be prevented through a healthy diet, regular physical activity, maintaining a healthy body weight, lowering cholesterol, and avoiding smoking tobacco.

Diabetes affects over 230 million people worldwide, nearly 6% of the adult population. India has the world's largest diabetes population of about 35 million, followed by China and the United States. Around 3.5 million people died from diabetes in 2007, and it is one of the major causes of premature death worldwide. The WHO estimates that the number of people with diabetes will grow to 370 million in less than 20 years if action is not taken. The incidence of this disease is increasing faster in developing economies, especially in the Middle East, sub-Saharan Africa, and India. Type 2 diabetes forms 90–95% of diabetes cases, 80% of which are preventable by changing one's diet, increasing physical activity, maintaining a normal body weight, avoiding tobacco use, and improving the living environment.

Heart diseases and diabetes are estimated to reduce GDP by 1–5% in low and middle-income countries experiencing rapid economic growth, as many people die prematurely.

Environmental Hazards

People, especially children, now face a more toxic, polluted and hostile environment compared with 20 years ago. The food we eat, the water we drink, the air we breathe, the chemicals used in our homes, the sounds (noise) we hear, the soil children play in, the vaccinations received, even toys and playground equipment are polluted with heavy metals, chemicals, toxins, and harmful gases. These can affect the immune, reproductive, endocrine, and respiratory systems, and cognitive development.

Watch Annie Leonard's "The Story of Cosmetics" and her call to get toxic products off the shelves, at http://storyofstuff.org/cosmetics.

Sodium Lauryl Sulphate and Sodium Laureth Sulphate

SLS and SLES are chemicals commonly used in shampoos, hair conditioners, toothpastes, body washes, bubble baths, etc. These chemicals started out as industrial degreasers and garage floor cleaners. They are very harsh detergents for humans, and have been shown to affect eye development, cause cataracts, affect the skin, corrode hair follicles, build up in the heart, liver, and brain and cause problems in these areas.

Toxic Chemicals in Household Cleaning Products

According to the television series of documentary programmes *The Nature of Things* (CBC-TV 2002), the levels of chemicals in the indoor air on a cleaning day can be hundreds, even thousands of times higher than in the outdoor air in the most polluted cities. Household cleaning products contain many toxic chemicals, to which even low-level exposure can be dangerous in the long-term. Many people are now affected by chemical sensitivities. Common cleaning products include:

- air fresheners and deodorisers: nerve-deadening chemicals that coat the nasal passage to interfere with our sense of smell; neurotoxins and carcinogens;
- all-purpose cleaners: ammonia (damages kidney and liver), neurotoxins, eye and skin irritants, carcinogens, and hormone-disrupting parabens;
- carpet cleaners: carcinogens affecting the central nervous system, causing dizziness, sleepiness, nausea, tremors, and disorientation; neurotoxin, eye, skin, and respiratory tract irritants;
- dishwashing detergents: carcinogens, eye and skin irritants, cause difficulty in breathing;
- disinfectants: phenols that damage DNA, liver, kidney, and nervous systems; carcinogens, respiratory toxins which contribute to forming antibiotic-resistant bacteria;
- laundry detergents: petrochemical ingredients causing skin irritation and respiratory reactions.

Another concern is that common disinfectants promote the growth of antibiotic-resistant superbugs, as reported by the National University of Ireland in the January 2010 issue of the journal *Microbiology*. Overuse of antibiotics is known to cause antibiotic resistance, but disinfectants could also produce the same effect. Lead researcher Gerard Fleming said that bacteria have adapted by pumping out both the disinfectant and the antibiotic from their cells as well as by the ability of DNA to mutate, making them resistant to antibiotics. He advised leaving disinfectants on a surface for long enough to kill bacteria, and to vary the disinfectants to prevent organisms from becoming less susceptible to the disinfectant.

Dioxins, PCBs, and PAHs

Industrial pollution (environmental chemicals) has been shown to disrupt hormones and cause reproductive abnormalities, birth defects, and type 2 diabetes. Cancer, especially breast cancer, has been linked to dioxins and PCBs (polychlorinated biphenyls), which are not metabolised when consumed and accumulate in body fat and breast milk. 90–95% of the exposure to these chemicals comes from consuming animal products (meat, fish, and milk). PAHs (Polycyclic Aromatic Hydrocarbons), i.e., auto exhaust, factory smoke stacks, petroleum tar products, and tobacco smoke, are metabolised but produce intermediate products that react with DNA to form adducts that cause cancer.

Electromagnetic Radiation Hazards

The increasing use of electromagnetic (EM) energy has become an integral part of our lives, resulting in possible health hazards associated with EM exposures. This is especially true when one lives near power lines or sleeps near the spot where the power enters the house. Sources of microwave (MW) and radio frequency (RF) radiation include radio and television broadcast antennas, mobile phones, computers, microwave ovens, industrial and medical devices. Recent US research by Drs. Lai and Singh found that cell phone radiation can cause DNA damage, which can be cumulative from repeated use of a mobile phone. Another study by one of Britain's top neurosurgeons, Dr. Vini Khurana, found that brain tumours are associated with heavy mobile phone use. Using mobile phones for more than ten years can double the risk of developing cancer. The specific absorption rate (SAR) needs to be monitored.

AUTISM

Autism spectrum disorder (ASD) is now a global epidemic, equally across nations, ethnicity, and social status. An estimated 1% of the global population has ASD—67 million people worldwide. The rate of autism among children worldwide is rising by as much as 800%. In the late 1970s, it was one in 2,500 children. Today it is one in 250 children in the world. Every 20 minutes, in the US, a child is diagnosed with autism. The US Department of Education states that autism cases in school children of 6–21 years of age increase yearly by about 25%.

ASD is a complex neurobiological disorder inhibiting communication and social relationships, with behavioural challenges, affecting four times as many boys as girls. It is likely caused by both genetic and environmental factors. Researchers at the University of Texas Health Science Center have found an association between environmental mercury emissions and the risk of developmental disorders including autism (November 2004). The study showed "a significant increase in the rates of special education students and autism rates associated with increases in environmentally released mercury."

This project examined participants in the *Childhood Autism Risks from Genetics and the Environment* (CHARGE) Study to evaluate household pesticide use during the prenatal and postnatal period. It was found that mothers who used pet shampoos were twice as likely to have children with ASD as those who did not. Pet shampoos often contain pyrethrins, which are designed to target the central nervous system in insects, rodents, and other species and can cause death of neurons and compromise the blood-brain barrier of humans in early life.

The most contentious issue of the autism debate is the link to routine childhood vaccines. This is especially so since Dr. Jon Poling's daughter, Hannah Poling, won her case on 9 November 2007 holding that her autism was triggered by nine childhood vaccinations administered when she was 19 months old. Over 5,000 autistic children are awaiting their day in vaccine court, a special US court to hear these cases.

Body Ecology Is Off

Donna Gates, founder and author of *The Body Ecology Diet*, describes how we are "destroying the delicate balance of the ecosystem that exists within our own bodies." Microorganisms inhabit our inner ecosystem. Friendly bacteria (probiotics) reside in our digestive track, defending us against unfriendly bacteria and pathogens which cause disease. They also strengthen our immune systems. Our body ecology is off when we take in chemicals through our food and environment, eat fast food, take medicines (especially antibiotics and hormones) frequently, and are under daily stress. This leads to the multiplication of unfriendly bacteria, causing headaches, flu, skin rashes, allergies, and other ailments.

Antibiotics kill both good and bad bacteria, allowing unfriendly organisms, like Candida albicans, to proliferate. This pathogenic yeast or fungus thrives on a high-sugar, acid-forming, low mineral diet. It produces a toxic waste that poisons and weakens the immune and endocrine systems. Candidiasis or Candida Related Complex (CRC) has many symptoms, including food allergies, digestive disorders, PMS, skin rashes, chronic constipation, headaches, vaginitis, chemical and environmental sensitivity, poor memory, mental fuzziness, and loss of sex drive. This is clearly a modern epidemic.

Probiotics and beneficial yeast aid digestion, help white blood cells fight disease, protect intestinal mucosa, prevent diarrhoea and constipation, and manufacture B vitamins, including B-12. Fermented foods are the missing link to restoring the ecology of our bodies.

It is also important to detoxify and cleanse the colon. Toxins accumulate first in the digestive track and colon. Waste on the walls of the colon breeds parasites, yeast, and viruses. When this happens, toxins, instead of nutrients, are absorbed into the body. Even a normal healthy adult can have 7 to 10 pounds of old faecal matter in the colon. Waste accumulated in the colon can back up into the digestive track, then into the liver and kidneys. The benefits of a healthy colon include higher energy and vitality, no more sugar cravings, a slower aging process, mental clarity, and inner calm. There are various ways to cleanse the colon, including enemas, colonics, herbs, and psyllium.

This cleansing process has helped many people overcome nagging and serious ailments, including immune disorders, Candida-related imbalances, weight issues, and even autism. Defeat Autism Now (DAN) doctors found that autistic children have a combination of conditions, including severe intestinal dysbiosis, systemic fungal and viral infections, mineral deficiencies, abnormal serotonin levels, and are loaded with toxic materials, such as pesticides, mercury, and other heavy metals. A healthy diet brings back the balance in the inner ecosystem, heals intestinal dysbiosis, corrects nutritional deficiencies, strengthens the adrenals, and conquers systemic infections.

Environmental Impact on Evolution

The environment is changing rapidly, largely due to the way our lifestyles have been evolving. Are our bodies and our descendants' bodies also changing under this environmental impact? In 1996, the book *Our Stolen Future: Are We Threatening Our Fertility, Intelligence, and Survival? A Scientific Detective Story* was published by Theo Colborn, Dianne Dumanoski, and John Peter Meyers. Here, two leading environmental scientists and an award-winning journalist present evidence of the ways chemical pollutants in the environment are disrupting human reproductive patterns and causing problems such as birth defects, sexual abnormalities, and reproductive failure.

Many studies, papers, and books have since been published, including *Endocrine Disruption: Biological Bases for Health Effects in Wildlife and Humans*, 2005, by David Norris and James Carr. This book highlights the biological effects of compounds (endocrine disruptors) that persist as pollutants in the environment and have caused developmental disorders and/or fertility problems in fish, amphibians, reptiles, birds, and humans.

Endocrine Disruptors

About 85,000 synthetic chemicals are registered for use in the United States, and 1,000–2,000 more are added every year. They pervade the air, water, food, homes, and our bodies. The National Infertility Association (NIA) has studied the impact of environmental factors, body weight, and exercise on fertility, and found that the average American has hundreds of manmade chemicals in his or her tissue (also in amniotic fluid and umbilical cord blood) at levels high enough to be of concern. Many of these chemicals are known to alter sexual development, undermine reproductive health, intelligence and behaviour, and decrease resistance to disease. They could also cross the placenta and reach the foetus.

Foetal and early life exposures affect the development up to reproductive maturity. Many of these impacts are irreversible, especially those involving abnormalities of the reproductive tract, impaired ability to respond to hormonal stimulation as an adult, and decreased sperm production or function in males. Children exposed to chemicals in the womb have higher aggression, lower IQ and reading ability, and simple motor skills compared to their less-exposed peers.

Synthetic oestrogen diethylstilbestrol, or DES, which was invented in 1938 to manage difficult pregnancies, is a good example. Research has revealed that not only did DES not help pregnancies, but it caused severe damage to foetuses in the womb. Damage to the children included subsequent development of rare cancers, deformed fallopian tubes, decreased sperm count, and increased risk of endometriosis, often detected after the victim had passed through puberty.

Humans carry PCBs and DDT in their body fat and breast milk, and pass these chemicals to their babies at levels which could endanger them. Prenatal and breast feeding exposure pose the greatest hazards as sensitive developmental processes, especially the brain and behaviour, are most vulnerable to endocrine disruption. In September 2002, Dutch scientists reported that boys exposed prenatally to higher levels of PCBs and dioxins were more likely to show demasculinised behaviours.

The NIA report presented the possibility that future generations might end up in "reproductive intensive care" if we don't pay attention to chemical research and regulation. It lamented that while pharmaceutical manufacturers are required by the US Food and Drug Administration to investigate product safety before marketing, most companies that produce personal care and home cleaning products that use toxic chemicals are not required to do so.

FASTING

> Our own physical body possesses a wisdom
> which we who inhabit the body lack.
>
> —*Henry Miller, novelist and painter*

With so much affecting our bodies in the stressful modern age we live in—unwise habits in diet and drinking, substance abuse, environmental pollution, and inorganic chemicals—, it is not surprising that we are dying prematurely. Unhealthy lifestyles kill millions every year. According to

Paul Bragg, author of many health books including *The Miracle of Fasting*, the best way to rejuvenate physically, mentally, and spiritually is by fasting; it is the key to internal purification.

When we fast, i.e., stop eating, the energy that was used for digestion and elimination is diverted to flushing toxins and poisons from our bodies. Paracelsus, a 15th-century physician who established the role of body chemistry in medicine, said, "Fasting is the greatest remedy—the physician within!"

Primitive man and animals have practised fasting as a method of healing since the beginning of time. The instinct of self-preservation takes away hunger, and the energy not needed for digestion is concentrated at the site of injury or illness to remove waste products, purifying and healing the body. It is also used for mental and spiritual growth.

Richard Anderson, ND, NMD, in his book titled *Cleanse and Purify Thyself*, says that as we clean out our homes and our cars, it is far more important to clean ourselves from the inside out. Water fasting is incredibly powerful in releasing sludge, mucus, and toxins. The body digests excess proteins, fats, sugars, pathogenic micro-organisms, weak and dying cells. Harmful negative emotions stored at cell level are removed, all senses and mental abilities are improved, and awareness and inner knowing are expanded.

On a global scale, a period of abstinence from consuming and producing waste materials would be reflected in a slower and calmer pace of life; it would decrease frenetic activities on our land, air, and waterways. The energy frittered away by rampant consumption would be harnessed to focus our attention on really important issues, such as helping us to heal as a species and form a sustainable relationship with our environment.

Pause: Wisdom From Our Bodies

"Our body is the barometer to our psychological, emotional, and spiritual health" is a statement that more and more people in modern medical communities are saying. However, this wisdom about the body has been practiced by spiritualists from different traditions for thousands of years. In an increasingly westernised world, especially our education and healthcare systems, we are indoctrinated to believe that our mind is superior to our body. Our body is a thing, a machine that we need to feed and exercise to stop it from breaking down. Occasionally, and more frequently with old age, this machine falls sick and we need to send it to the doctors. The repairs include surgery, radiation therapy, or drugs.

With an increase in population and demand, doctors are often left with no time to do more than just writing prescriptions. Instead of a personal relationship, modern medicine is becoming an impersonal transaction based on fixing machines. This disconnection or dis-ownership of our bodies blocks us from tapping into our body's inherent wisdom to heal and to be whole.

Mind-Body-Heart Connections

We are often conscious of simple connections between the body and the mind, as when we experience nervousness or excitement, when we break out rashes or become hyperventilated. However, when it comes to more complex emotions or illnesses we do not typically believe that there is a connection between our mind, body, and heart. Generally, people are not surprised that they blush when they are embarrassed, or that a scary thought causes their hearts to pump, or an unexpected bad news throws their entire body out of harmony. But they find it hard to believe that mental maladies like despair, loneliness, and anxiety can have damaging impacts on their lungs, heart, stomach, and immune systems.

Emotions are made up of energy, and this energy does not just vanish when it is suppressed or ignored. When we cannot, or do not express our emotional and psychological issues, that energy becomes solidified or trapped in our bodies. When this energy is reinforced through repeated negative experiences over time, it becomes bigger and stronger, and it may start to influence parts of the body to become uncooperative or send out unhealthy signals to other parts of the body. These bottled-up and growing feelings (energies) often manifest themselves as stress in the body. Metaphorically, it's like trying to squeeze toothpaste without taking the cap off. As you apply more and more pressure, it inevitably leads to the toothpaste breaking out of the casing, either from the bottom or the sides, whichever is the weakest point. Similarly, in our bodies, when we are under pressure, our emotional and mental energy starts to grow, and it has to find a way to express itself. It will find the weakest point, which can be our digestive, heart, respiratory, urinary, or immune systems.

Some of the bodily manifestations of high stress caused by psychological and emotional distress may be difficulties in breathing, loss of appetite, nausea, migraine, irritable bowel, insomnia, a racing heart, rashes, and backaches. These bodily changes may bring on or intensify excessive coffee or alcohol drinking, over- or under-eating, smoking, compulsive shopping, prolonged periods of indecisiveness and sleeping, tics and twitches, sudden mood swings, increased reliance on medication, impaired sexual performance, and much more. Over time, these psychological and emotional issues may result in chronic illnesses like stroke, pulmonary diseases, hypertension, heart diseases, diabetes, mental disorders, and cancers.

The connection between stress and physical problems was clearly illustrated on the BBC News in January 2008, when the headlines read: "work stress changes your body." According to the report this was based on, stress in the workplace is a major factor in the development of heart disease and diabetes. Lead researcher Tarani Chandola of University College London said: "Employees with chronic work stress have more than double the odds of the disease syndrome than those without work stress, after other risk factors are taken into account." In their research they established that stress upsets the part of the nervous system that controls the heart. The impulses from the vagus nerve of highly stressed people are poor. In addition, the neuro-endocrine system seems to be disturbed by stress, evidenced by the fact that anxious workers had higher levels of the stress hormone cortisol in the morning.

Dr. Larry Dossey, the famous doctor whose medical work is steeped in spiritual wisdom, made an interesting discovery in his research. Author of more than nine books, Dr. Dossey found that more heart attacks occur on Monday than on any other day of the week, and not only that, but most of them occur between eight to nine o'clock in the morning! His work has spurred medical schools in the United States to explore the role of religious practice and prayer in health. Many medical schools use Dr. Dossey's work as textbooks.

The Japanese have a term called *karōshi* which, when translated literally, means death from overwork. It is a sudden death at work, usually due to extreme hours of prolonged work. In 2007, the Japanese Ministry of Health published relevant statistics on karōshi: 189 workers died suddenly—primarily from strokes or heart attacks, about 208 fell severely ill, 921 workers contended they became mentally ill, and there were 201 cases of mentally troubled workers who killed themselves or attempted to do so. According to Dr. Richard Wokutch's research, there have been about 620 karōshi claims for death or disability submitted every year in Japan since 1987.

Another report featured in the *British Medical Journal* published in October 2002 and entitled "Work Stress and Risk of Cardiovascular Mortality: Prospective Cohort Study of Industrial Employees," showed similar findings. In a 10-year study they found that employees who reported high job strain and high "effort-reward imbalance" had a twofold higher risk of death from cardiovascular diseases than their colleagues scoring low in these dimensions.

Stress does not come just from work; it can come from many areas in life. It can often come from major life changes, such as moving house, getting married, or losing a loved one. In the 1960s, research was carried out by Dr. Thomas Holmes and psychiatrist Richard Rahe on the impact of major life changes. They interviewed more than 5,000 patients and compiled a list of stressful events, allocating a rating to each one. The more events with a high rating experienced over a short period of time, the greater the likelihood of physical illness. Top of the list are the death of a partner, divorce, critical illness, accident, moving house, and loss of a job.

Assessing Health through Muscle Testing

The human body possesses its own wisdom and it was in the field of kinesiology that many scientific discoveries have been made on the relationship between the different parts of our body and our emotional and psychological health. Kinesiology is the study of human movement, applying the sciences of physiology, anatomy, psychology, and neuroscience. Through a structured process of muscle testing, Dr. George Goodheart discovered in the 1960s that the strength or weakness of every muscle was connected to the health of a specific corresponding organ. He also found that the muscles of the body were associated to the Chinese acupuncture meridians. By treating the muscle in a variety of ways and making it strong again, he was able to improve the functioning of the organ as well. With this approach, which he called Applied Kinesiology, well-trained practitioners had a powerful tool, a system of feedback from the body itself. If a doctor gave a patient the proper treatment, their muscle would test strong and vice versa. According to Dr. Goodheart, at a subconscious level the body knows what is good and what is not.

Dr. John Diamond refined behavioural kinesiology from applied kinesiology in the late 1970s. A well-established psychiatrist in Australia, Britain, and the United States, Dr. Diamond discovered that certain muscles would strengthen or weaken in the presence of positive or negative emotional and intellectual states. Every disease, every bodily imbalance, and every muscle problem will have an emotional component which can be accurately determined by its mediation through the meridian system. His approach has offered new ways of healing in the world.

Through a structured muscle testing process, Dr. Diamond was able to clearly establish—with hundreds of test subjects—the following astounding findings:

- the muscles of the body tested weak when people: ate refined sugar, listened to certain pop songs, held pesticide grown vegetables to their chest, stared at a fluorescent light, and thought of an unpleasant situation;

- the muscles of the body tested strong when people: ate raw sugar (naturally processed), listened to certain classical songs, held organic vegetables to their chest, exposed themselves to natural sunlight, and thought of an empowering experience.

According to Dr. Diamond, besides being affected by stress and emotional states, which is obvious, the energy of our body is also strongly influenced by our physical and social environment, food, and posture. Through a series of specialised physical manipulations, expressions, exercises, verbal and vocal articulation, as well as nutritional supplements, Dr. Diamond is able to help his patients achieve balance and centeredness, which leads to physiological, psychological, and emotional health.

The proverbial saying that the "mind and body are one" is clearly demonstrated in Drs. Diamond and Goodheart's kinesiology. Most people accept the fact that they need to change their mindset and beliefs to improve their conditions in life, be it health, finances, relationships, or career. But in Drs. Diamond's and Goodheart's research and methodology, changing one's posture, facial expression, diet, and doing certain physical exercises can have a profound impact on one's overall health and stance in life.

Spirituality and the Body

Deep in the recess of our subconscious is the possible belief that it is wrong to pay attention to our body. Many people believe that spirituality means being free from the body and its impulses. We are not confident in dealing with its urges, failures, shortcomings, and pains. Thus, many perceive the human body to be the enemy of spirituality. "In Christian theology, spirituality is not freedom from the body, but freedom within the body," says Dennis Hollinger, Ph.D. He further adds, "Spiritual maturity comes not by negating the body, but by harnessing its capacities and impulses for the glory of God." In this light, though our fallen bodies have a propensity for sin and wrongdoing, we are called to use our hands, face, eyes, feet, stomach, etc. for love, compassion, and healing.

In Judaism, there is no absolute separation between the spiritual and the material, the body and the soul. "The human body expresses divine reality and is a key to divine knowledge," according to the Hebrew Bible. Before Adam ate from the Tree of Knowledge, his body shone like the sun and he was capable of living forever. So one of the goals of Jewish spirituality is to perfect the body and make it shine again. In Judaism, ideally, the soul must fully inhabit the body, and not be liberated from it. The Jewish esoteric text—the Zohar—states that the ultimate reward of a person who has succeeded in this world is that his soul and body will be reunited.

According to Islam, Allah appointed the human soul as His *Khalifah* (vice-gerent) in this world. God has invested it with a certain authority, and given it certain responsibilities and obligations, and He has bestowed it with the best and most suitable physical properties. The body is not a prison of the soul; it is a workshop and if the soul is to grow and develop, it is only through this workshop. Islam rejects the ascetic view of life, and puts forward processes for the spiritual development of man, not outside this world, but inside it. In other words, the real place for spiritual development is in the midst of life and not in solitary places of spiritual seclusion.

Sufi Wisdom on Healing

Sufism is a mystic approach to Islam whose sole aim is to be united with God. It can be achieved through a wide range of ideas and practices that emphasise the attainment of divine love and compassion of the heart. In Sufism, all of life is part of spiritual practice: family, work, and relationships provide as much opportunity for spiritual development as prayers and contemplation. Just like Islam, Sufism believes that illness comes from God, so God is the one who can heal it. Sufis are convinced that illnesses are the result of imbalances or disharmony within the body and/or the mind. In Sufi healing, the ideas of mental and physical illness are not typically separated from one another.

According to an article written by Sehnaz Kiymaz of the International Society for the History of Islamic Medicine, there are six stages the human soul can be in. Each stage has its own emotional and physical illnesses. Sufi healing is usually done through a *sheikh*—a person who mediates between God and the patient. The healing system incorporates prayers, energy work, massages, food, water, abstinence, and oils.

The six stages of the evolution of the soul are the ego, the heart, the soul, divine secrets, nearness, and union. The first and lowest stage is the stage of the ego. People at this stage are primarily driven by their desires and pleasures of their body. The emotional problems could be anxiety, fear, selfishness, weeping for no reason, depression, sexual perversions, etc. The possible physical troubles include alcoholism, drug abuse, criminal behaviour, heart attack, venereal diseases, and cancer. These challenges are the consequences of not exercising proper control over the appetites of one's ego.

The second stage is the stage of the heart. People at this stage show a basic goodness to themselves and others. This stage can be attained when people gain control over the appetites of their ego. The emotional problems involve inability to focus, fear of failure, certain types of hypocrisy, severe anger, arrogance, forgetfulness, etc. The physical troubles of this stage are of a cleansing nature, such as migraine, nausea, diarrhoea, skin eruptions, fever, gallbladder, kidney problems, etc. These diseases are trying to disperse the toxins that are leftovers from the stage of the ego.

The third stage is the stage of the soul. It is believed that it is difficult to achieve this stage without the help of a spiritual leader. To achieve it, one must attain higher levels of compassion, mercy, consideration, and self-discipline. At this stage, one would be able to see problems as ways to strengthen their faith in God. The emotional problems could include irreverent attitudes to life, the habit of degrading others, pride, self-deception, lack of concentration, etc. The possible physical challenges are nervous tremors, low energy, auto-intoxication, corrupted appetites, etc.

The fourth stage is the stage of divine secrets. Regular breathing exercises are usually necessary at this stage of soul development, in which people sometimes have difficulties in breathing—this is why having the guidance of a spiritual leader is important. People at this stage are disturbed by things from the non-manifest realms. The more one advances in stages, the more one is tested for their faith in God. Without proper training, one can easily lose their faith in God at this stage. They may experience "spiritual fever," the kind of fever that gets rid of toxic substances through body temperature and sweating.

The fifth stage is the one of nearness. People at this stage enjoy a "neighbourhood" close to heaven. They are able to live in this world and the one closer to heaven. At this stage they are almost free from all physical illnesses. However, they might experience heightened ecstasy, which could lead to total loss of interest or connection to this world. The testing still continues, in fact more intensely.

The sixth stage is the one of union with God. This is a stage given by God and it cannot be achieved through one's own efforts. This is the stage that all Sufis strive to reach. There are no more emotional and physical problems when one has achieved union with God. People at this stage see death as just a continuum of their existence, but in a different form.

From a Sufi perspective, pain is like a warning bell in our body which brings our attention to the fact that something is wrong and pushes us to ask for help and healing. It is therefore a key to the development of our soul, self-knowledge, and the proper functioning of the body. As mentioned earlier, "The body is the barometer to our psychological, emotional, and spiritual health," thus it is essential that we probe into the deeper meaning of our illnesses. In other words, through pain, our body is perhaps trying to teach us something about the way we live our lives, what we value and believe about ourselves and others, who we need to become, and possibly even about the meaning of life. We need to ask, "What is this illness and pain trying to tell me?" and, "What can I learn from this pain to help me become a more whole and wiser being?"

Healing Messages from Our Body

Using the body as a metaphor to understand the cause of our mental, emotional, and spiritual illnesses is highly controversial. Nevertheless, more and more psychologists and doctors around the world are subscribing to this wisdom and many have incorporated it into their practice. The idea of referring to a book for any illness, to understand the psychological and emotional issues, is well explained in Louise Hay's international best-selling book *You Can Heal Your Life*. It was sold in more than 40 million copies worldwide and translated into 27 languages. This book contains the principles of the author's beliefs about healing, as well as a list of common illnesses and their corresponding cause and necessary positive affirmations.

Born to a very poor mother, ill-treated by her violent stepfather, raped by a neighbour at age 5, Louise Hay dropped out of school and became pregnant at age 16, and after a 14-year marriage, her husband left her for another woman; she hit one of the most challenging points in her life when she was diagnosed with cervical cancer at age 51. With her history of being raped and long periods of sexual abuse, she said that "It was no wonder I manifested cancer in the vaginal area." Up to that point, she was already a Practitioner of Religious Science—a spiritual and metaphysical school of thought established by Ernest Holmes in the late 1920s. She also studied Transcendental Meditation at the Maharishi International University in Iowa.

She knew that the cancer was due to her longstanding deep resentment of her life and the people that wronged her. She was convinced that if she did not clear her mental patterns, the cancer would return, even if she had surgically removed it. She decided to forego conventional medical treatment, and went through a regime of forgiveness, working with a therapist, healthy nutrition, foot reflexology, and colonic cleansings. All these approaches to heal herself from cancer were very unconventional in the 1970s and many said that she was taking foolish risks. After six months of thorough physical and mental cleansing, she was medically declared cancer-free. From this experience, she is convinced that disease can be healed if we are willing to change the way we think, believe, and act.

It was clear that Louise Hay did not disown her body, but chose to tap into her inner wisdom. She knew that she was the cause of her own illness and therefore she was going to be the one to "un-cause" it. She did not simply let her body in the care of doctors, but made time to listen and connect to it, and took the difficult and less-readily-available steps to heal her body from the inside out; most importantly, she tuned into the message that her body was trying to convey, embraced it and made peace with it by releasing the emotional energy and limiting beliefs. She did not just work on her feelings and thoughts, but also aligned the appropriate actions and priorities in her life to fully support her healing.

Since her first book publication in 1976, many independent researchers have been widening and deepening the knowledge in this area. The late Annette Noontil, Australian author of *The Body Is the Barometer of the Soul* (1994), wrote a much more detailed book on using the Indian Chakra system. She systematically gave a detailed description of the symptoms of the illnesses related to each Chakra, as well as their corresponding meaning. Her main aim in life was to do everything within her power to help people understand that "everything we think has an effect on our bodies."

A more recent similar publication by another Australian author, Inna Segal, is entitled *The Secret Language of Your Body* (2010). Using the list that she has developed, she was able to help many of her clients gain understanding of the cause of their physical problems and achieve optimal health through visualisation and energy healing techniques. In her book, she included a list of emotions and what each emotion is trying to convey.

Inna Segal's personal story tells of her debilitating problem with her spine. At one point she was not able to walk, and all her doctors, including chiropractors, had given up on her. Her moment of epiphany came when she realised that she had totally relied on others to heal her and that the more she did that, the more she gave her inner power away, and the more her condition worsened.

She began taking charge of her own healing by communicating with her spine—literally! In a meditative state she saw her spine in bad shape and began to ask herself, "What are the thoughts, emotions, and experiences that contributed to my condition?" She then felt waves of emotions moving through her body and she consciously chose to release them. Within three weeks her spinal problem was completely gone. The key to her remarkable healing was understanding and appreciating the lessons that her emotions and body were trying to communicate to her.

Decoding the language of illnesses must first come from the fundamental understanding of the functions of different parts of our body.

In other words, the underlying psychological or emotional issues are usually related to the function of the part of our body that is not feeling well. Some of the important questions to ask ourselves are, "What does that part of my body do?" "What role does it play in the functioning of the whole?" "How does its function relate to what is happening in my life?" We also need to explore the function of an illness or discomfort. "Is it a muscle strain or bone break, an infection or a disease, a nerve problem, a digestive crisis, or a blood disorder?" Each one of these dis-eases has a different implication.

In *Your Body Speaks Your Mind* (1996), author Debbie Shapiro gives powerful insights into our body by helping us understand the nature and function of each major part and system in it. And in order to fully understand each part or system we must also appreciate the relationship it has with other parts and systems and the function it serves in the whole body. She goes on to detail the body parts, some common diseases, disorders, and issues for each part, and she outlines the possible psychological and emotional conflicts and disturbances. Her book cites many scientific research studies, quotes from medical and psychiatric professionals, and countless personal stories.

By combining the knowledge of Louise Hay, Annette Noontil, Inna Segal, and Debbie Shapiro, one can tap into a vast pool of self-help wisdom. And instead of popping a pill or handing over our bodies to the doctor, we can start taking charge of our health and personal and spiritual development with the messages from our bodies. Below is a short list of ailments and diseases with their corresponding messages.

Migraine	Dislike to be driven by others; it goes against your grain. Built-up unexpressed anger and frustration.
Influenza	Feeling mentally and emotionally overwhelmed. Buying into mass negativity and beliefs.
Stiff neck	Unbending bullheadedness. Not being able to see other people's viewpoint.
Shoulder pain	Carrying too many responsibilities. Not responding well to the change in direction in your life.
Back problems	Upper: Lack of support from others. Feeling unloved. Middle: Bending over backwards, helping others too much. Lower: Lack of financial support, fear of not being able to support yourself.
Stomach ulcer/ Gastritis	Reality is too corrosive and it's eroding your ability to cope. What is eating you up? / Long periods of uncertainty (worry). A growing sense of doom.

Constipation	Holding on to old ideas. Unwilling to let go. Clinging on to familiarity, safety, and security.
Stroke	Resistance to change. A rather-die-than-change type of attitude. Fear of what lies ahead.
Cancer	Lack of self care and love. Helping others in the detriment of helping yourself. Deep hurt. Long standing resentment. Deep grief eating away at the self.
Hypertension	What is getting you heated? Is the pressure coming from yourself or others? Letting your emotions and reactions to a situation rule you instead of letting go.
Diabetes	Lack of sweetness in your life. Not being able to give and receive love. Longing for what might have been.
Heart attack	Your heart is being attacked by unexpressed hurt, loss, or grief. Unwilling to forgive. Squeezing all the joy of your heart to pursue a goal.
Bronchitis / Asthma	Inflamed family environment; arguments and yelling. Need to get something off your chest. Feeling overwhelmed by a situation or person—feeling suffocated. Needing to cry, but holding back.

Sickness and ailments give us the opportunity to take a good look at our behaviours and life patterns, to uncover what our deeper intentions are trying to convey, intentions that come from our body-mind. By accepting the message of our symptoms, we begin our journey back to well-being. Committing to release old patterns and embracing new ones is the key to recovery, which will be hindered if we are not fully congruent, e.g., if we are holding on to the subconscious benefits that illnesses provide.

We really have to want to recover. And that comes with accepting the truth that our body-mind is trying to convey. A symptom communicates a problem that we have been rejecting, ignoring, or suppressing, and therefore, acceptance means being painfully honest to our inner wisdom. It is from this honest acceptance and commitment to taking the healthy steps and words of wisdom from our body-mind that we achieve physical wholeness and spiritual growth.

CHAPTER 8
WHAT'S HAPPENING TO OUR NEIGHBOURS?

TURNING BACK THE CLOCK

From a Singaporean perspective

When we were young, it was common to pay attention to what was going on in our neighbourhood. Two doors away there was a Malay family of seven that used to have pungent dinners with chilly and curry on the floor. They had a chubby orange cat that we could often play with in the afternoons. In another apartment we had our favourite aunty, who used to bring over rice dumplings once in a while for us to taste and we would often talk about the price of food and how we could conserve it. A couple of floors down there was a door with a strange picture on it: a dark-haired being with blue skin—a God in the Hindu religion. The adults were very warm and friendly, and their children had very big eyes. They often gave us plates of crunchy, spicy snacks that looked like deep fried noodles (muruku).

I also remember the times when my mum had to rush out of the house for some emergency and the next door aunty would come over with her children to spend time with us; sometimes our neighbour with the big family used to borrow rice from us and they always gave it back. When our iron or television broke down, we could always walk over to our neighbours to borrow their electrical appliances or watch our favourite cartoon programmes.

Those were the good old days when we knew help was not far away, and people were open to ask for it. The values of sharing and caring were not talked about, but lived; that was our way of life.

LIVING BETTER, BUT ISOLATED AND OPINIONATED

Today, most of us live in urban cities. Staying in flats, condominiums, and terraced houses has somehow created a sterile environment. We want our privacy and, perhaps in wanting that, the message we send out is "leave us alone." We may have no idea who lives next door to us. We form opinions about our neighbours based on the décor outside their houses, their plants, pets, etc. Even if we bump into each other on the veranda, in walkways, or lifts, the most we do is smile awkwardly. Some will look away, perhaps afraid of starting a conversation, or maybe scared of being "attacked." On the one hand, we have grown materialistically and have become more self-reliant, and on the other hand, we feel more and more lonely, isolated, and unable to reach out to others.

At the same time, because of a lack of interaction and understanding, we form opinions about our neighbours. Due to the differences in cultural and religious practices, we observe how "insensitive and odd" they are. We complain about their intolerable pets and inconsiderate karaoke volume. We judge their ostentatious lifestyles and snobbish attitudes. These judgements that we harbour and the contrasts we make between ourselves and others drive us to the desire of being alone, or wishing that others were more sensible, intelligent, considerate, and humble—like us. Do our community, societal, regional, and international relations mirror these day-to-day relationships that we have (or rather don't have) with our neighbours? How much do we know about what is going on in our neighbouring countries?

Malaysian Neighbour

From a national perspective, Malaysians feel insecure, confused, and afraid of the political in-fighting and dirty tactics used by politicians. Incidents range from the sodomy trial of a government minister to the gruesome death of a Mongolian woman linked to another senior minister, and the privacy-invasion postings of semi-nude photos of a female opposition party member on the Internet.

Even after 50 years of independence, the minority races (Chinese and Indians) still stick together for support, safety, and trust. The longstanding issue of the Bumiputra policy—that has existed since 1970—continues to undermine unity in the country. This policy favours Malays over other races, including preferential treatment in employment, education, scholarships, business, access to cheaper housing, and assisted savings. According to a study conducted in 2006 by a network of Malaysian social scientists of the Merdeka Centre for Opinion Research, 89% of the respondents hardly eat out with friends from other races, 42% do not consider themselves Malaysians first, and 70% would help their own ethnic group first.

Malaysia is much like the rest of the world, where development depends on cheaper labour. Richer than most of its neighbours, it has become irresistible for poor workers from the region. More than one thousand migrants arrive daily at the international airport and tens of thousands more arrive annually by boat, illegally, across the Strait of Malacca, or from Kalimantan into Sabah and Sarawak. These immigrants are aggressively competing for survival with Malaysians, causing a complex labyrinth of social, economic, and environmental problems.

With the global economic crisis affecting every country, many Malaysians live in fear, as robbing and stabbing on the street and in homes are on the rise. Adding on to the worries of personal safety, from 2007 to 2008, reported rape cases rose by 300%. In 2008, former Prime Minister Abdullah Badawi was himself stunned by the 45% rise in the National Crime Index from the time he took office.

INDONESIAN NEIGHBOUR

Coping with the rising prices of food and other daily necessities is a common challenge for most Indonesians. The global financial turbulence in November 2008 weakened their exports. Disruption in their balance of payments meant that a rising deficit began to accumulate, and the exchange rate underwent significant depreciation.

A growing concern in major cities is that people are feeling insecure at home and on the street. North Jakarta, for example, is known for violent attacks on motorists at busy intersections. Stories of daring ATM robberies and brutal murders make the front page. Armed robberies and vehicle theft are also common. Police blame it on the worsening economy.

Inculcating moral values in children is a challenge, as they live in a rapidly decaying society, influenced by the West, in which materialism, free sex, and drugs are commonplace. In 2008, the Indonesian Police seized large quantities of crystal methamphetamine and discovered several meth labs capable of producing huge amounts of this dangerous drug. In recent raids, they have also got hold of counterfeit prescription drugs worth millions of dollars.

Lack of education, human resource development, and good political role models are some of the major causes of the longstanding social and economic problems in Indonesia. A survey done in February 2009 revealed the police to be the country's most corrupt organisation. Though white collar crime numbers have lessened, corruption in parliament and among high-ranking officials is still happening.

The gap between the rich and the poor is distinctly huge and widening every year. The intensification of dissatisfaction and jealousy among the different income groups could likely be one of the driving forces behind terrorist activities and social and political unrest.

Furthermore, the nation was seriously impacted by a series of earthquakes in 2009, the one on 30 September hitting 7.6 on the Richter scale. It struck southern Sumatra and swallowed several villages. According to Indonesia's National Disaster Management Agency, 83,712 houses, 200 public buildings, and 285 schools were destroyed. An additional 100,000 buildings, twenty miles of road, and five bridges were badly damaged. Indonesia's tourism industry, an economic cornerstone, was battered, and 70% of hotel rooms in Padang, on Sumatra Island, were ruined. While more than 1,100 people died, thousands more were caught or buried under collapsed buildings.

THAI NEIGHBOUR

Thailand, a devout Buddhist country, is filled with political strife and corruption problems. There are unfair trade practices between government officers and people who pay for advantages. In other parts of the world, it is the duty of civil servants to serve the public. In Thailand, people often have to treat government officers like "gods"; they have to "pray" to them to obtain housing, telecommunications, electricity, and transport services.

The majority of Thais fear job retrenchment due to the political parties fighting and the economic crisis. The political conflict between the yellow, red, and white shirts has caused severe uncertainty in the business infrastructure and social welfare, as well as the closure of the airport in December 2008, which cost Thai Airways 20 million Bahts. This event led to the cancellation of hotel bookings and Thai exporters suffered tremendous losses because their orders could not be shipped out. The financial damage to the commercial and business sectors was colossal.

Foreign investors and employers are losing confidence in the country's political stability. Many multinational corporations are holding back long-term plans and are focusing on cutting short-term losses. All of this has created an atmosphere where Thai people feel uncertain about the prospects of their businesses, and they feel fearful—or at least uneasy—to walk in certain streets. In some parts of Bangkok, rioting, gang clashes, and demonstrations can break out very suddenly.

In October 2011, massive flood waters hit inner parts of Bangkok that are just 10 kilometres away from the heart of the capital. The widespread flooding started in the northern parts of Thailand in July 2011. It was the country's worst flooding in half a century, affecting more than a third of the country's provinces: 10 million homes were flooded and over 700 persons were killed nationwide. Fed by unusually heavy monsoon rains and a string of tropical storms, the flood destroyed large areas of arable land and forced 10,000 factories to close. The government has anticipated and set aside more than 100 billion Bahts to rebuild the damaged areas of the country.

FILIPINO NEIGHBOUR

For the majority, getting employment to sustain daily survival is still a big challenge in the Philippines. With a weak national currency, low wages, few job opportunities, wobbly political management, and the rising costs of basic commodities in cities, many Filipinos leave the country to seek a better life for themselves and their families. As of 2008, there were almost two million Filipinos working outside their country. The money sent back to the Philippines amounted to $15.9 billion in 2008, representing a whopping 13–15% of the country's GDP.

Like in many developing countries, there are insufficient government hospitals, and the few that exist are overcrowded. The cost of private hospitals is terrifying for 90% of the population. In major cities, rising gasoline prices and chronic traffic jams are serious daily problems for most Filipinos. Traffic and related environmental problems affect more than 10 million residents in Manila. Every day, thousands of commuters from all sectors are stranded for hours, and when they do catch transport, they cling on to already full buses and jeepneys—means of public transportation—as if clinging on to life itself.

One of the oldest and biggest national problems is resolving government corruption. In 2007, the Philippines was ranked among the world's most corrupt countries by the World Bank and the Berlin-based Transparency International. The disparity between the rich and the poor is obvious: the rich are very powerful, influential, and able to get away with atrocities. Skilled professionals can still opt to leave, but the majority of the poor are compelled to stay on and watch corruption leach away the funds needed to deliver them basic social services.

In September 2009 devastating floods spawned by the fierce tropical storm Ketsanawas, the latest in a series of natural disasters, afflicted this nation stuck in both the Pacific typhoon belt and its volcanic Ring of Fire. Storms, landslides, or earthquakes—sometimes all three—are almost monthly occurrences, displacing as many as eight million people a year. The Ketsana typhoon battered the country with the heaviest rainfall in 40 years. Cities were submerged in 10 to 20 feet of water and 25 provinces were affected. Total damage was estimated at $100 million. Entertainment plazas, banks, stores, and building agencies were soaked in flood and mud.

Chinese Neighbour

Around 17% of the total manufacturing output in the world comes from China. While it has become the world's largest manufacturer with the fastest growing economy, China's population is still struggling to converse with the world, with less than 5% of the Chinese speaking English. China has more than 26 million rural people living in poverty and nearly 20 million urban people living on the government's minimum allowance.

With the largest population in the world, they face the problem of offering everyone equal opportunities in terms of jobs, education, and healthcare.

Economic progress at breakneck speed has caused severe environmental pollution. People are getting sick and will continue to get even sicker from air, water, and soil pollution. In the process of rapid industrialisation and urbanisation, the loss of farmland has brought a serious problem to society. Some 40 million farmers have lost their land.

The price of housing in large cities is very high. Even with a graduate's pay, most residents cannot afford city living. Another problem that people in China face is the escalating cost in healthcare and basic living necessities.

The gap between the rich and the poor is immense. The fact that the rich are living luxuriously, while the poor are struggling with basic living, causes societal instability and high crime rates. Many people are involved in unethical practices (white collar crimes). To better their lives, people dedicate time to build important relationships—*guanxi*—with government officials and key influential businessmen.

The widening crevice between the two classes influences people's outlook and priorities in life and their social attitudes. People are valued according to the amount of money they earn or materials they possess. With this mindset, they place a lesser value on spiritual practices and moral development.

Indian Neighbour

The contrast between the wealthy and those in need is starkly and distinctively clear in India. According to the World Bank there are approximately 456 million Indians (42% of the total Indian population) now living under the global poverty line of $1.25 per day (PPP). This means that a third of the global poor now reside in India, and yet, India ranked number six in the world for having the most billionaires. Most major cities have beggars on the roads, at bus stops, and outside hotels and temples. In many parts of the country one will witness modern affluence and abject poverty side-by-side.

India lacks infrastructure. Any new business venture will face a nightmare, as governmental formalities are cumbersome, with too many protocols. Public projects (e.g., roads, drainage, power supply, etc.) are run by the government and take a long time to get started and then finished. This problem is linked to the fact that there is a lack of cooperation between political parties and well-educated government officials to carry out critical projects properly. In other words, without improved governance, delivery systems, and effective implementation, India will continue to have problems in educating its citizens, building infrastructures, improving health facilities, and increasing agricultural productivity.

With issues of severe land degradation, water and air pollution, and dangerous waste management practices, raising the national health status and alleviating poverty and starvation remains a herculean task. According to Dr. Rashmi Mayur, an Indian urban planner and futurist, if India's environment continues to decline, the country will become one of the largest wastelands on earth. Overuse of pesticides is destroying India's soil and groundwater, affecting millions of people with sharp rises of cancer cases in young adults, crippling birth defects, lowered sperm count, and so much more. One in four Indians has no access to water, and for those who do, the water is dangerously polluted.

Most Indians criticise the government for not taking India to greater heights, despite the fact that the country is filled with many natural resources. In such a densely populated place, the dissatisfaction and agony of the poor will grow stronger if they can't see a way to improve their condition.

Singapore's Challenges

According to the data released in May 2009 by the Ministry of Trade and Industry (MTI), Singapore's economy contracted by 14.6% in the first quarter of 2009. Compared to the same period in 2008, the economy fell by 16.4%. Taking into account all factors, MTI was looking at an economic forecast of -9.0% to -6.0% for 2009. Across all income groups, Singaporeans are vigilant on managing the rising cost of living. The lower income groups are concerned about retrenchment or cheaper labour. For Singaporeans, such statistics and experiences are disconcerting and humbling. In the last decade, despite wars around the world, disease, climate change, economic upheavals, and political turmoil, Singapore has been able to maintain a positive annual GDP. That is, until 2008 and 2009.

Singapore's list of economic and social issues includes:

a) wanting the government to channel resources to focus on the disadvantaged minorities (children with learning disabilities, the aged poor, single-parent families, etc.), moving toward holistic living rather than running Singapore like a mega-corporation;
b) stress and dissatisfaction about life, wanting more work-life balance, and, on the other hand, worrying less about the rising cost of living; an increasing number of people in their late 30s and 40s wanting to spend additional time with their families;

c) wanting the government to involve the nation in creating new innovations or areas of specialisation to secure the country's growth in the globalised world, as Singaporeans are seriously feeling the challenge of competing in it;
d) teenage/youth Internet and gaming addictions and the growing number of young people facing challenges in raising their own families independently; what was accepted as part and parcel of daily life 25 years ago is now perceived as stressful or depressive for young adult families;
e) the rise of teenage pregnancies and inappropriate sexual relationships; this includes teenage sex among the same age group and with older age groups; this problem is intimately connected to the easy access to Internet and social networks, and the absence of guidance from parents who are busy working.

Shifting from Independence to Mutualism

In a planet where our natural resources are being depleted and many countries are not able to produce sufficient food and water to sustain their own populations, it is no longer wise or ethical to shut our eyes and hearts to our neighbours and the world. We can learn a lesson from nature about this. Mutualism is a biological interaction between two organisms, where each individual derives a reciprocal fitness benefit. Instead of following this pattern, our modern globalised economic systems have given unprecedented wealth to a few and have marginalised billions. These systems have globalised trade, finance, manufacturing, and communication, but they have also created national and regional unemployment, widening income gaps and escalating the degradation of local and global environments.

In 1960, the per capita gross domestic product of the richest 20 countries was 18 times that of the poorest 20. In 1995, this gap increased to 37 times. By 2008, this void has perilously widened to 63 times! While the richest 20% of the world are becoming richer still, the poorest 20% (1.5 billion people) are directly or indirectly pressed into abject poverty, barely surviving on less than $1 a day. They live in uninhabitable urban slums, shack houses, and rural hinterlands.

Perhaps it is about time the world considered mutualism: a social organisation based on common ownership, effort, and control, regulated by sentiments of mutual help and brotherhood. Unlike inter-nation competition, where one nation benefits at the expense of the other, mutualism increases the survivorship of the citizens of the world.

Cross Cultural Intelligence

Mutualism would take place if citizens of the world genuinely wanted to understand and appreciate different cultures. In 2003, the term "cultural quotient" (CQ) was coined by Professors Christopher Earley (United States) and Soon Ang (Singapore) in their book *Cultural Intelligence: Individual Interactions across Cultures*. As defined by Earley and Soon Ang, cultural intelligence is "the capability to function effectively across national, ethnic, and organisational cultures." Rooted in rigorous empirical work that spans 25 countries, their cultural intelligence model provides a practical 4-step cycle:

- **CQ drive**: deals with one's motivation; the tangible and intangible benefits of doing well in culturally diverse situations;
- **CQ knowledge**: defines the cultural information needed to do well, such as understanding the cultural systems, norms and values;
- **CQ strategy**: how to crystallise a plan for implementation, which includes developing awareness and anticipation and monitoring feedback;
- **CQ action**: tailoring actions—verbal and non-verbal—and specific words and phrases to specific cultural contexts.

This model is not about getting people to master all the norms of the various cultures encountered. Instead, it helps a person develop an overall repertoire and perspective that results in more harmonious and productive cross-cultural interactions. In a practical sense, cultural intelligence provides the know-how for people to become more benevolent in how they view others who see the world differently from themselves. Most of all, if we seriously want to build a better world with others, we must change the way we see our fellow citizens around the world. Rather than pretending to be respectful, we must genuinely strive to understand and value people from different cultural backgrounds.

Wealth and Resource Distribution

To gain further clarity and understanding of the magnitude of this problem, let us imagine that the global village is made up of 1,000 people, of which 440 are men and 560 are women. In terms of race compositions, 576 are Asians, 320 are Europeans, Americans, Arabs, and Australians, and 104 are Africans. In the context of the average earning power, 149 enjoy $78 a day, 445 live on $16 a day, 406 survive on $5 a day, and 227 struggle to make it on $1 a day. The 200 better-offs consume 86% of everything that is on the market of this village, while the poorer 800 fight to live on the remaining 14% (source: *The Chaos Point*, 2006, by Ervin László).

The 1.5 billion poorest people on earth are actually destroying the environment on which they depend. They overwork their land, rivers, and lakes, and they have less access to drinkable water. Poverty and lack of education leads to high birth rates. Population increase creates more poverty and an augmenting number of poor people destroy more of the environment. The rich overuse the planet's resources and the poor misuse them. If we look at this unequal wealth distribution from another angle, the net worth of 500 billionaires, in 2006, was equivalent to the net worth of half the planet's population! Yet these 500 individuals represent only 0.0000076% of the world's population. In other words, if the wealth of these 500 persons was equally redistributed, we would bring tears of joy to the lives of 3,250,000,000 human beings.

Violent Consequences

The world's population, which has exceeded 7 billion in 2012, is still increasing, with ever more people moving into cities and living closer to each other. These conditions are highly volatile; they spur resentment, dissatisfaction, the desire to make things equal, and ultimately lead to violence manifested in different forms of conflict.

A quarter of the world's conflicts in recent years have involved a struggle for natural resources, according to the Worldwatch Institute. The wars in the Democratic Republic of Congo, Rwanda, Zimbabwe, Uganda, and Colombia are clear examples. People are fighting over deposits of gold, diamonds, and a mineral ore called *coltan*. Not long ago, wars were fought over oil, with the Iraqi invasion of Kuwait in 1991.

In the near future, as demand for water will hit the limits of rapidly shrinking supplies, potential conflicts are brewing between nations that share trans-boundary freshwater reserves. More than 50 countries on 5 continents might soon be caught up in water disputes, unless they move quickly to establish agreements on how to share reservoirs, rivers, and underground water aquifers. The regions severely affected are in the Middle East, northern Africa, India, Pakistan, Nepal, and China.

According to a Canadian study done in 2005 (Project Ploughshares), a third of the world's population is involved in some form of war or conflict. A total of 27 countries were fighting in 32 armed conflicts, which affected 2.33 billion people in the world. A conflict is recorded in the Ploughshares list when more than 1,000 people are killed in it. For people living in conflict, food, social, and economic securities are disrupted. The daily reality of these people is that they are driven off their land, deprived of food and water, their family members are killed, maimed, or incapacitated, and the psychological scarring on children is immeasurable. The land and the environment take 25 years or more to heal, and in some places—hundreds of years.

Whether violence is due to the wounded ego of an individual, the violation of the respect of a group of people, whether it is based on personal vengeance, a holy war in justification of one's faith, or on the fight for the right to live, to gain access to food, water, and valuable natural resources, the possibility of violence remains a constant threat.

Breaking the Socio-Economic Double-Bind

Relieving and eradicating social and economic problems requires socio-economic development, and that is not possible without better communication, information, and education. It is, therefore, a debilitating chicken-or-the-egg type of dilemma which produces under-development that, in turn, builds frustration. From a humanitarian standpoint, it would be fairly easy to do all this, but when we get down to smaller levels—be it the level of individuals, groups, or organisations—the solution suddenly becomes too big to be handled. Faulty and ineffective behaviours are thus created and development is diluted or impeded. This cycle with no beginning and no end can only be broken if we foster a new consciousness—a collective one—a large, unified "we," which all of us can connect to.

It was Einstein who said that the same level of thinking that creates a problem cannot solve it. In other words, in order to solve a problem we must move to a higher level. This higher level of thinking may just be in a realm of reality that is completely foreign to us and often treated—by billions—with fear, ridicule, and scepticism.

This higher level is loosely referred to by scientists as a field of intelligent energy with a "we consciousness." It constantly surrounds us and we are inevitably part of it. We can say that this field of intelligent energy contains and connects all of our thoughts, emotions, and experiences and it is communicating to us every day.

The chicken-or-the-egg cycle of our socio-economic problems can be broken if we foster and align to this new collective consciousness—this large, unified "we", which all of us can connect to.

Tuning into a Morphic Field

All of us have the capacity to resonate with our collective thoughts and emotions, to add on to, and to amplify their richness, as suggested in the works of biochemist Rupert Sheldrake (Cambridge, UK) in the early

1980s. Known as the "morphic field," this pool of collective memory contains an inherent memory of, for example, animal and human behaviour, social and cultural systems, and mental activity. Each unit of a collective morphic field tunes in to the collective information for guidance, through morphic resonance—the vibration between the morphic field and morphic units. The stronger and more stable the field, the easier it is to tune in to it. As each unit develops, it further enhances its field, leading to the habituation or proliferation of all the units in that field. So the existence of such a field makes the creation of a new, similar form easier. Morphic fields of mental activity (the mind) extend beyond our brains, through intention and attention, just as magnetic fields extend beyond magnets.

Deep within each of us—bypassing the layers of envy, pride, righteousness, hurt, grudge, and hate—we all have the power to resonate with love, peace, and faith, to forgive, heal, and uplift ourselves and others. The more we tap into this field of love, peace, and faith, the more we amplify it, and the more we are all able to trust, support, and work with each other through this field.

Here is what the representatives of the Club of Budapest (an eclectic association, consisting of Nobel Laureates the 14th Dalai Lama, Mikhail Gorbachev, Sir Joseph Rotblat, and Archbishop Desmond Tutu, along with many other luminaries including Dr. Ervin László, Edgar Mitchell, Masami Saionji, and Karan Singh, all dedicated to the proposition that only by changing ourselves can we change the world) have to say:

> Unless people's spirit and consciousness evolve to the planetary dimension, the processes that stress both society and nature will intensify and create a shock wave that could jeopardize the entire transition toward a peaceful and cooperative global society.

Clearly, the message reinforces the importance of human beings tapping into and expanding our collective morphic field, and resonating with the common deepest positive values that we share as one giant collective consciousness.

Solidifying an Infrastructure of Collation & Distribution

Along with this collective intentional energy, we would need to put in place an effective collation and distribution network. In order to reduce the problem of chronic poverty around the world, a robust social and economic infrastructure must be set up, and this includes hospitals, water and sanitation systems, food supplies, and schools. These require large injections of donor finance.

The UN Human Development Report puts forth good practices for donors and recipients. Broadly speaking, on the donors' side there must be a decentralisation of decision-making, alignment of projects to a country's needs, and full accountability taken by the members involved. On the recipients' side there must be institutional reforms to promote transparency, widespread participation in development issues, and increased levels of non-governmental oversight to deepen accountability. To put it simply, donors are to make sure that the money goes where it is needed, and recipients—that it gets spent on the right things, by giving a voice to people in the respective communities.

PAUSE: RAINBOW FOR LIGHTER BEINGS

This afternoon, there was a sudden rain that ended at seven o'clock and we still have bright, visible sunlight. As I look out the window, I can see a beautiful rainbow. I haven't seen one so clearly for decades and I am excited about what messages it brings. In ancient traditions, the rainbow is often seen as a giant bridge between the physical and non-physical worlds. In Hawaii, Polynesia, Japan, and among Native American tribes, the rainbow is the path souls take on their way to heaven, and has been called "a bridge to higher worlds." In order for us to walk on the rainbow toward a higher world, we must transform our old beliefs, values, and self-concept. What an appropriate and powerful message from nature!

Bridging Science to Spirituality

Dr. Edgar Mitchell, a former Apollo 14 astronaut, trained as an engineer and scientist, was most comfortable in the world of rationality and physical precision. Yet the Institute of Noetic Sciences (IONS), which he founded in 1976, is a world-renowned non-profit organisation that builds bridges between science and spirit, researches subtle energies of healing, and probes into the realm of love, forgiveness, and gratitude.

In his research on morphic fields, he studied quantum physics and the principles of non-locality, zero-point energy, chaos theory, and the power of human perception and intention to influence and shape reality. His studies in advanced physics included the scholarly and visionary works of Rupert Sheldrake, Larry Dossey, Karl Pribram, Sir Roger Penrose, and many more. His research points to the proposition that human minds have the ability to tap into a field of intelligence in the Universe (that connects everything) and influence it through thoughts, intentions, and beliefs. Mitchell eventually called it "Nature's Mind."

IONS cited numerous examples of such studies, a notable one having been carried out by Dr. Elisabeth Targ, Director of the Complementary Research Institute at California Pacific Medical Center. In 1996, she performed a stringent double-blind study of 40 advanced AIDS patients. Half the patients received healing for ten weeks, the other half did not. The healers worked from a distance, receiving only an envelope for each patient, containing the patient's first name and health statistics. Neither the patients nor the doctors knew who was receiving healing. After six months, the researchers confirmed that the patients who had received healing had significantly fewer new illnesses (83% fewer), fewer visits to doctors (71% fewer), and fewer days in the hospital (85% fewer). The results of this study point to the existence of an all-encompassing field of life-supportive energy that connects all parts of the field like a hologram. After the study, Dr. Targ declared about it that:

"It says something about the nature and power of our own intentionality. It suggests that we can always be helping each other, that we're more connected than we realize."

In another study, Dr. Robert Jahn, Dean of the School of Engineering and Applied Science at Princeton University, tried to determine whether a person's thoughts could influence electronic devices. He built a machine known as a *random event generator* (REG)—the electronic and mechanical equivalent of flipping a coin. In one experiment, a toy frog sat on top of a motorised robot, with its movements directed by an REG. A human study subject tried to get the robot to move in a particular direction, without touching it. Through countless experiments, researchers collected three billion bits of binary information. Upon careful analysis, the subtle ability to influence chance was unmistakable. The probability that the trials' results were chance was one in a billion. Interestingly, one of the findings was that the people who had the greatest ability to influence the machine said they had "resonance," "bonding," or "fell in love" with it. In Lynne McTaggart's international bestseller, *The Intention Experiment*, she authoritatively asserts that human emotions and intentions have the power to influence not only other humans and machines, but also plants, animals, and the environment.

NATURE'S MIND AND THE FIVE MIRRORS

The experiments that have been expounded here tell us that nature or Nature's Mind has the power to relate to us and to respond to our intentions. Gregg Braden's discovery and development that he called the "Divine Matrix" expands on this notion. A former aerospace computer systems designer, Braden sought to understand the purpose of human existence and this brought him to different ancient places of worship in Egypt, Peru, Tibet, and various Native American sites, where he studied ancient texts and traditions. In his journey, he amalgamated a vast body of knowledge (often paradigm-shifting) that united science and spirituality.

Braden cited scientific study after study in his research, showing that modern science is beginning to prove that there is a field of intelligence that permeates all creation, labelled differently by many luminaries: called the collective unconscious by Carl Jung, the quantum hologram by Karl Pribram (Nobel Laureate), and the database of consciousness by Dr. David Hawkins.

This ever-present field is continually communicating with us through what Braden calls the Five Divine Mirrors. These Mirrors serve as reminders, messengers, reflectors, and actualisers of our deepest emotions, beliefs, and intentions. According to Braden, our most valued beliefs about ourselves, others, and our world are mirrored in our closest relationships.

Braden extrapolated the wisdom of the Mirrors from the ancient Coptic, Gnostic, and Essene texts that were discovered as part of the Nag Hammadi library in 1945, some of which date back 1800 years (2nd century BCE). These texts are currently conserved at the Coptic Museum in Cairo. They contain one text in particular that made the headlines: the Gospel according to Thomas, which was originally called *The Secret Words of Jesus* and some believe was written by Thomas. Through these Mirrors we discover the hidden relationships between the seemingly normal, meaningless events that occur in our lives every day and what those events are saying to us about our beliefs, fears, and judgements at a much deeper level.

The Mirrors are everywhere, all the time; they can be in our business, health, family, career, and romantic relationships, as reminders, messengers, and opportunities for us to become better people and build a better world.

The Five Divine Mirrors show and teach us the following:

(1) perception is projection: the flaws that we criticise in others could be reflections of the unresolved flaws we possess, that we need to see in ourselves; once this mirror is recognised, the flaws can be healed immediately;

(2) what we judge: when we experience highly charged emotions of offense or violation, they might be reflections of what we judge in others through our values formed from past experiences; instead of intensifying that judgement within us by getting even with others, this Mirror offers us the opportunity to re-examine our values and release negative emotions;

(3) what we have lost: when we are intensely drawn or attracted to the qualities or lifestyle of a certain individual or group, it could be a reflection of certain desirable inner qualities that we used to possess but have lost through repeated experiences of being jaded, suppressed, and/or denied in life; for example, each time we trust someone enough to love or nurture them and that trust is violated, we lose a little of ourselves to that experience. Over time, this becomes a void that we unconsciously strive to complete within us. When we find our "missing pieces" in others, we find ourselves powerfully and irresistibly drawn to them. The purpose of this Mirror is to awaken the missing parts in us that we find attractive in others. In doing so, we reclaim those parts that are essential in making us whole;

(4) dark night of the soul: this is when we experience total desolation, despair, and powerlessness after taking action on a long-awaited opportunity presented by a person, group, or situation. This difficult and seemingly unending dark, hopeless, and lonely experience is actually the manifestation of our greatest fear that we have been subconsciously evading all our lives. This Mirror often puts us into a situation where we can't turn back to the life we had before, and at the same time we are totally lost in how we can change the suffering we are currently in. This Mirror can be seen as a blessing in disguise, one which helps us redefine our identity. The purpose of the dark night of the soul is for us to heal and master our own great fears and doubts;

(5) self-acceptance: in the pursuit of an ideal, we put ourselves into life-threatening situations (sickness, financial, familial, or societal destruction); this ideal is subconsciously learnt from comparing ourselves with others. Here, we might be presented with the mirror of our ego, whereby we try to be superior to others or to match our own ideals. The way we feel about ourselves—our performances, appearance, and achievements—is mirrored back to us by the reality we have created. Essentially, this mirror is asking us to be compassionate to ourselves and to accept the fact that we are a constant work-in-progress.

All in all, the relationship Mirrors help us to pause, take stock, and tune in to each moment of our lives, to empower ourselves and others. According to Braden's studies, once a pattern changes in one relationship, all other relationships that are based on the same pattern benefit from that change. All the Mirrors ask us to be compassionate, loving, trusting, and fully accountable for our thoughts, emotions, and actions. It is through this collective inner-work, through connecting to our environment, and openly relating to each other in order to share, learn, and collaborate that we can make ourselves lighter to walk up the rainbow (as mentioned earlier) to a higher, more harmonious, and sustainable new world.

Common Threads in Spirituality

Probably two of the most highly valued human qualities emphasised in almost every spiritual tradition are love and compassion. By looking at the quotes or scriptures in these traditions, we come to see the importance of love and compassion in all aspects of our lives, but especially in human interactions—in family, romantic, professional, or social contexts:

Love your neighbour as yourself.
—*Leviticus 19:18, Mark 12:31*

One who looks with equality upon all,
in the image of his own self, whether in pleasure or in pain,
such a person is considered a perfect yogi.
—*the Bhagavad Gita, 6:32*

Treat well those who are good.
Also treat well those who are not good.
Thus is goodness attained.
—*the Tao Te Ching, 49:3-5*

Mettā (loving kindness) embraces all beings without exception.
The culmination of mettā is
the identification of oneself with all beings (sabbattatā).
—*Abhidhammattha Sangaha (Buddhist teaching), 2.45*

The current Dalai Lama (Tenzin Gyatso) is one of the greatest exemplars of love and compassion of our time. Exiled from his country since 1959 when the Chinese Communists invaded Tibet, the Dalai Lama has nonetheless become an emissary and emblem of love for the whole world. Often, when he is asked what his religion is, he does not go into a long discourse of what Tibetan Buddhism is; instead, he simply says, "My religion is love." Despite the fact that the Chinese government hates him and has brutally killed and tortured many of his countrymen, he refers to them as "my friends, the enemy."

Compassion Promotes Health

"If you want to be happy," reads the first part of the Dalai Lama's famous advice, "practice compassion." Then comes the second part: "If you want others to be happy, practice compassion." For many, this advice may sound dull, stale, and trite, but new research shows that it is not only the receiver who enjoys the benefits of compassion, but also the giver. And it is not just at the emotional or mental levels, but also at the physical level. Research and studies on affective neuroscience are blossoming at universities like Emory, Harvard, North Carolina, Maastricht, Florence, Tel Aviv, Bonn, and Imperial College London, to name but a few.

Larry Gallagher's article in *Ode Magazine* (July 2011) talks about researchers of Emory University (2009) who randomly assigned 61 students to either a 6-week series of classes on compassionate meditation or a health discussion group. Among the meditators, there was a distinct difference between those who spent the greatest number of hours in daily practice and those who spent only a few hours. When they were subjected to a stress test, all of the heavy meditators demonstrated a clear reduction in interleukin-6, a protein marker for inflammatory response regulated by the immune system, after the 6-week training. This inflammatory protein marker links to studies in social neuroscience which state that people with strong, positive social connections have lower inflammatory responses—a stress reaction linked to cancer, depression, and arthritis—compared to people who are socially isolated or in conflict.

Studies have shown that meditation has positive impact on the vagus nerve, which is a cranial nerve that runs between the brain, heart, and gut. The vagus nerve helps the brain regulate the heart rate and respiration, contributes to an efficient regulation of glucose, a lower incidence of heart disease and diabetes, and reduces inflammations. In their book *Ultraprevention*, Drs. Liponis and Hyman clearly state that inflammation is one of the key determinants of disease and aging. And by activating the vagus nerve through meditation—as meditation masters have done for centuries—we can control inflammation. Through positively influencing the vagus nerve via mediation, Drs. Liponis and Hyman believe that we have the capability to reverse aging and chronic diseases. Their findings are fully supported by other scientists in several scientific studies in this area.

Barbara Fredrickson, an award-winning professor of psychology at the University of North Carolina and author of a book titled *Positivity* (2009), has shown that in six weeks of "loving kindness" meditation, subjects raised their vagal function, reaping all the positive effects that come along with it. This Buddhist technique involves the meditator extending positive feelings toward friends, colleagues, neighbours, and all beings, even enemies. In her entry in the *Journal of Personality and Social Psychology* (2008), Fredrickson's empirical study on 202 employees of Compuware Corporation in Michigan, clearly demonstrated that employees who practiced loving-kindness meditation for 7 weeks had increased daily experiences of positive emotions which led to greater access of personal resources such as heightened mindfulness, purpose in life, social care and support, and decreased illness symptoms.

Compassion Is Learnable

Empathic responses of Tibetan monks with 10,000 hours of practice were contrasted against those of new initiates of compassion meditation at the University of Wisconsin (2008). All test subjects were put into a brain scanner and exposed to sounds of human joy and distress, as well as neutral sounds. The researchers found significantly more synapses firing in the monk's insulas—tissues in the brain that regulate emotions and control functions—and the anterior cingulate cortex—which regulates blood pressure, heart rate, and rational cognitive functions—during distressing sounds compared to the newbies. This clearly suggests that compassion can be developed as a skill.

When empathic caregivers are exposed to the suffering of others day after day, they often experience burnout. In this instance, what Buddhist meditators do is that they add a powerful feeling of unconditional love and compassion, and according to Mathieu Ricard—a molecular biologist who became a monk—they are able to dissolve the distressing aspects of empathy and turn them into compassionate courage to do whatever they can to soothe the suffering of others. These techniques of life-supportive meditation have been handed down through old Buddhist scriptures and learnt by Monks who have applied them into their daily lives for generations.

In the journal *Clinical Psychological Review*, Stefan G. Hofmann clearly states that compassion meditation reduces stress-induced distress and immune responses. Neuro-imaging suggests that compassion meditation enhances activation of brain areas that are involved in emotional processing and empathy. According to Hofmann, combined with cognitive-behavioural therapy, compassion meditation can provide powerful strategies for targeting a variety of different psychological problems, such as depression, social anxiety, marital conflict, and anger management.

Tania Singer, a leading compassion researcher in Europe, is actively exploring the use of brain imaging and biofeedback to teach subjects to activate parts of the brain associated with compassion. In her laboratory at the Max Planck Institute, she is trying to train compassion in western societies in such a way that it can be integrated into busy and stressful daily lives. Singer and many other neuro-scientists strongly believe that we can train ourselves to be compassionate.

Harry Palmer, author of the book *Resurfacing* (1997), states that compassion can be practiced anywhere, with friends, colleagues, family members, neighbours, and even strangers. He suggests practicing the five steps below in a quiet manner, by focusing our attention on the same person at every step, and saying these sentences to ourselves.

- Step 1: "Just like me, this person is looking for happiness in his or her life."
- Step 2: "Just like me, this person is trying to avoid suffering in his or her life."
- Step 3: "Just like me, this person is experiencing sorrow, loneliness, and despair."
- Step 4: "Just like me, this person is trying to satisfy his or her own needs."
- Step 5: "Just like me, this person is learning about life."

Many people who have tried the above steps were able to notice a deeper connection with the person they were focusing on or interacting with. It is almost as if the layers of judgement and fear dissolved and feelings of openness and familiar bond emerged. Rather than experiencing separation, suspicion, and difference, the simple compassion practice of Harry Palmer promotes feelings of similarity, understanding, and acceptance.

Separation versus Holism

One of the core problems with being human is the centuries of paradigm conditioning encouraging us to believe that we are separate and detached from our environment and other living things. Quantum physicists now recognise that the Universe is not a collection of separate things moving around in empty space. Rather than a cluster of individual, self-contained atoms and molecules, objects and living beings are now more appropriately understood as lively and protean processes, in which parts of one thing and parts of another constantly exchange places. This discovery is not restricted to physics; new parallel discoveries in biology, social sciences, and psychology are changing our view of the relationship between living things and their environment.

Leading-edge biologists, sociologists, and psychologists have all found evidence that between the tiniest particles of our being, between our body and our environment, and between ourselves and all people, there is an inseparable connection. The world fundamentally works not through the activity of individual things, but it operates in the exchange and relationship between them. This dynamic exchange and relationship between things holds the key to the life of every organism, from subatomic particles to large-scale societies; this is also the key to our viable future.

Such startling discoveries in the last century have been explained and embraced by spiritual traditions for several millennia. The boundaries between science and spirituality are quickly disappearing. Combining spiritual approaches with scientific endeavours is fast becoming a common practice among leading researchers and educators in the 21st century.

> An object does not have any "intrinsic" properties (for instance, wave and particle) belonging to itself alone; instead it shares all its properties mutually and indivisibly with the systems with which it interacts.
>
> —*David Bohm*

> There are not many but only one. He who sees variety and not the unity wanders from death to death.
>
> —*the Upanishads*

> Things derive their being and nature by mutual dependence and are nothing in themselves.
>
> —*Siddha Nagarjuna*

The frontiers of science and the mystical experience of spiritual traditions seem to be telling us that nature's most fundamental impulse is not a struggle for domination but a continual strive for wholeness. These discoveries hold vast implications about how we choose to define ourselves and more importantly, they tell us that we need to live our lives differently if we want to see a viable future ahead of us; and this desired future is not likely to happen if we continue on a path of competition and individuality: we need cooperation and mutual support instead.

Having a compassion meditation practice daily can help us to dissolve barriers and connect with others and our world, as well as promote giving help to others unconditionally. Scientists like Mathieu Ricard, Drs. James H. Austin, and Andrew Newberg have discovered through brain imaging that practitioners of meditation lose their boundary of self and others. This innate impulse to connect, share, and help through compassionate meditation is the foundation on which we can develop our true sense of community and mutual support. This all-encompassing emotional and mental state is ideal for us to explore our common purpose and unite in it, to find and support each other in achieving a common goal.

SUPERORDINATE GOALS

One of the powerful and promising ways to form a community goal is through developing superordinate goals. This is a process whereby two or more groups of people are committed to achieving specific goals and each member is actively contributing to them. Superordinate goals are designed to be so large that more than one group is required to achieve them. Through a process of developing team structures and leadership, identifying and experiencing the problems of group conflicts, and appreciating the dynamics of conflict resolution, Muzafer Sherif, the developer of superordinate goals, has provided a powerful framework for social scientists, to help them achieve impossible objectives among disparate groups of people.

Using the process of developing superordinate goals, Dr. Don Beck, co-author of *Spiral Dynamics*, has helped South Africa unify from apartheid. Dr. Beck utilised South Africa's entry into the rugby world cup as a focal point for building national spirit. At that time, the black South Africans fiercely hated the national rugby team because all the players were white and closely associated to the apartheid.

He suggested that the rugby team take on a common identity using green-golden team shirts and a team song with Zulu drums to invoke the spirit of the nation. By working intimately with the rugby coach and through a series of well-planned experiential activities with the rugby players, he was able to get the team to gain a powerful sense of brotherhood and see the larger role they played in inspiring the country to unite as one nation. They eventually won the world cup in 1995 and Nelson Mandela congratulated them by wearing the team's green-golden shirt to signify unity and the healing of the racial wounds of the nation.

"Developing a superordinate goal is one of the best ways to achieve peace in areas of political conflict," declares to Dr. Beck. Frequently, he meets with both conflicting sides and helps them create a positive vision of the future. Both sides clearly understand that the superordinate goals can only be achieved when they work synergistically together, using their common resources to create a win-win solution for all.

By studying wars and conflicts, Dr. Riane Eisler found two opposing models for moulding and organising relationships, which she called the "domination and partnership models." Looking at some of the most brutal societies such as Hitler's Germany, Stalin's USSR, Khomeini's Iran, and Idi Amin's Uganda, she was able to identify common characteristics that define a domination model: authoritarianism, male dominance, acceptance of social violence, and a belief that it is honourable and moral to kill or enslave others.

On the other hand, partnership cultures are based on linking instead of ranking and they uphold democratic social and family structures. They promote gender equality, practice mutual aid, establish social policies that support caregiving and environmentally sound manufacturing processes. The nations that practice the partnership model regularly rank at the top of the UN national quality-of-life charts.

In the words of Dr. Riane Eisler, "When you look around our world, it sometimes seems like we need to change everything. Actually it comes down to one thing: relationships. As we shift our relationships from the domination to the partnership model for our families, communities, and world, as we relate more in partnership to ourselves, others, and our natural habitat, we have better lives and a better world." Dr. Eisler's message reinforces all the major themes that we have discussed this far—instead of being driven by competition, domination, and individuality, it is imperative that we practice cooperation, partnership, and expand the "we-field" in our daily lives.

With the information and evidence presented so far it seems that we are in our most resourceful state when we and our neighbours are cooperative and supportive. Perhaps the most powerful way to re-establish the bonds within our neighbourhood and societies is to expand the definition of who we are. In our truest essence, we are not separate from others, but intimately connected; we are governed by the laws of holism. By creating an all-encompassing identity, we are able to embrace more groups of people and greatly strengthen our resources.

In view of the multi-systems problems that we are facing today, it is far beyond the capability of one country, or even of a group of countries to take action; it is the responsibility of the entire world, and the world is made up of you, me, and every living being on this planet. We need to see beyond our tiny selves; we need to transcend beyond our nationality, race, or religion. We have to see ourselves as planetary citizens, working toward the superordinate goal of creating a viable future. The sooner we start, the more time we have to make the appropriate changes. The initial step is to connect to our environment and other people—our family members, neighbours, and community first.

CHAPTER 9
WHAT'S HAPPENING TO OUR WORK AND MONEY?

PLEASE GIVE US OUR DAILY BREAD

Since the beginning of time, man has expended considerable effort attempting to meet his daily survival needs or, if you will, to get his daily bread; and for the longest time, this was literally the case: bread was actually what people used to work for. Later on, work started to involve employment in different industries, and it was a means of earning one's livelihood, one's bread and butter. Now, *bread* has become slang for *money*. The work we do and the money we earn have become measures of our success. The achievement of economic success has largely overshadowed other possible interpretations of success. A person's value is largely pegged to his financial accomplishment. A company's value is assessed by cash flow, profits, and assets. A country's success is also viewed as its economic output, measured by the gross domestic product (GDP).

A widespread focus across races, religions, or nationalities is the daily "bread": how to make it, how much is made, how much can be spent or borrowed, etc. It is this mindset that has led to the kind of world and lives we lead today.

> The moral flabbiness born of the bitch-goddess SUCCESS.
> That, with the squalid [lacking moral standards] interpretation
> put on the word success, is our national disease.
>
> —*William James to H.G. Wells, 11 September 1906*

Work is central to a person's well-being, providing income and paving the way for broader social and economic advancement; it is the main route out of poverty. Decent work sums up aspirations for productive work with a fair income, workplace security, and social protection for families, prospects for personal development and social integration, and equality of opportunity and treatment.

The International Labour Organization (ILO), founded in 1919, pursues a vision that universal peace can be established only if it is based on the decent treatment of working people. We will ask you to hold the ILO's vision in mind as we explore what's happening to our work and money in the world today.

MARCH OF URBANISATION AND INDUSTRIALISATION

The proportion of the world's population living in urban areas has greatly increased from just 13% in 1900 to 50% in 2007. Considering the trend, this number will probably rise to 70% by 2050.

More and more people will be living in megacities with populations of over 10 million; and these are major global risk areas, vulnerable to supply crises, social disorganisation, political conflicts, and natural disasters. They pose risks of infrastructural, socio-economic, and ecological overload.

People are drawn to cities for economic opportunities: jobs, money, services, variety in lifestyle, and social mobility. With the industrial age, machines, electricity, computerisation, and technological innovations, the nature of work has been changing, massively affecting social structures and lifestyles. Extended families spanning many generations have been replaced by nuclear families—separated, mobile, and broken.

Pollution, stress, over-crowding, high prices, fast food, consumerism, substance abuse, and life in the virtual world are some of the less desirable facets of urban living. Slum dwellers increase by 25 million every year through internal and transnational migration into cities, many with insufficient housing and sanitation, and having little or no access to education, healthcare, or the urban economy.

MORE EDUCATED, BUT INCOME GAP WIDENS

Worldwide illiteracy (no schooling at all) has been substantially reduced from 1970 to 2000. However, it is still much higher in developing countries compared to developed countries. Worldwide, the main share of the labour force has primary or secondary education, indicating low- or medium-level skills. The growing wage gap between high- and low-skilled workers shows the demand for high-skilled tertiary educated workers, who are in shorter supply.

In October 2008, the International Institute for Labour Studies released a report revealing that the gap between richer and poorer households has greatly increased since the 1990s, reflecting the impact of financial globalisation.

The report stated that excessive income inequalities could lead to higher crime rates, lower life-expectancy, malnutrition in poor countries, and greater likelihood of children being taken out of school to work. In Eastern and Western Africa, more than half the population subsists on less than $1 a day, working hard to generate an income.

Work and Life in the Information Age

> Information Technology and business are becoming inextricably interwoven. I don't think anybody can talk meaningfully about one without talking about the other.
>
> —*Bill Gates*

With the advent of the personal computer, information technology, communications technology, the mobile phone, and the way people and businesses work have changed dramatically. Working hours and places have also become much more flexible and people's professional and social lives have changed. The digital content market is growing tremendously.

More People Work from Home

The ability to work from home or a satellite location, known as "teleworking" or "telecommuting," is increasing rapidly. This allows people to save the time they used to spend commuting to and from work and it saves space on roads, thus reducing our carbon footprint. It permits more frequent family interaction and flexible working hours, which is especially useful for work-at-home mums.

The availability of broadband offers new businesses the chance to be set up in the owner's home or in a remote or isolated area, which results in an increase in the number of home offices and self-employed people. For employers, office and overhead costs are reduced, and the possibility of teleworking may entice existing staff to keep their jobs and new staff to join companies, even though team work may be more difficult.

Cybercrimes, the New Menace & Threat

One of the downsides of this technological advancement is that computer crimes or cybercrimes have become more rampant, with viruses and malware, identity theft, credit card and Internet auction frauds, embezzlement, phishing scams (even work-at-home scams), slander, child pornography, cyber stalking, and even cyber-terrorism. The Internet's basic architecture, initially not designed to grow in all directions, has revealed gaping flaws in its security. There is no global authority that is in charge of the 30,000 computing networks worldwide, with currently 1.6 billion users.

This lack of security could affect the future of e-commerce, as people and businesses reconsider transacting over the web. Michael Fraser, director of the Communications Law Centre at the University of Technology, Sydney, said, "The way it's trending now, the Web could be so full of rubbish that people won't trust it. That could destroy the potential of the whole knowledge economy, which so many developed economies are counting on for the competitive advantage."

FINANCIAL CRISIS STRIKES

The financial crisis of 2007–2009 has been the most serious crisis since the Great Depression. It was triggered by a peak in the US housing market in 2005–2006 with loan incentives and easy initial terms, followed by high default rates on sub-prime and adjustable rate mortgages and increased foreclosures. *Sub-prime* refers to the credit quality of borrowers who have a greater risk of loan default than prime borrowers. Sub-prime mortgages made up less than 10% of new mortgages until 2004, when they rose to nearly 20%.

Besides mortgage loans, auto and credit-card loans were easy to obtain, and people were under high debt loads. Mortgage-backed securities (MBS) increased, enabling institutions and investors worldwide to invest in the US housing market. Financial institutions, such as investment banks and hedge funds (known as the "shadow banking system"), provided credit that was not subject to the regulation imposed on commercial banks. They used financial innovation (with the limited regulation of banks and financial services) to synthesise more complex and highly lucrative financial products or loans, known as "derivatives." They provided loans without sufficient financial reserves to absorb loan defaults or MBS losses. Warren Buffett, ranked by Forbes as the richest person in the world in 2008, referred to derivatives as "financial weapons of mass destruction" in early 2003.

Globally, trillions of US dollars were lost. The top five US investment banks reported over $4.1 trillion in debt for fiscal year 2007. Lehman Brothers went bankrupt in September 2008, having incurred losses of billions of dollars in the US mortgage market, the largest bankruptcy in the US corporate history. Lehman's collapse had huge repercussions on the psyche of investors. Share prices plunged around the world.

AIG could not support its credit insurance commitments and was taken over by the US Government in September 2008, with over $180 billion in US Government support. By September 2009, 94 US banks had failed. The US Government and the Federal Reserve committed $13.9 trillion, of which $6.8 trillion had been used by June 2009. The troubled asset relief programme of $700 billion includes an automotive industry bailout of $80.1 billion. The US unemployment rate rose to 9.5% by June

2009, the highest since 1983 and the average hours per work week fell to 33, the lowest since 1964.

Worldwide, governments had to bail out key financial institutions and other organisations and implement economic stimulus programmes. By September 2009, $10.8 trillion had been poured globally into the financial system by central banks and governments. The United State and the UK were severely affected, as they already had large public sector budget deficits, and they implemented huge fiscal stimulus plans. The bailout in the United States was 25.8% of its GDP, while the bailout in the UK was 94.4% of its GDP. This is equivalent to $10,000 per person in the United States and $50,000 per person in the UK.

The International Monetary Fund (IMF), the lender of last resort to the world economy, dipped into its $200 billion bailout fund to help struggling nations. In February 2009, the IMF aimed to double its lendable resources to $500 billion.

Central Europe has been called "ground zero of the global recession" by Stratfor, a global intelligence company. Before the crisis, central Europe had been the top destination for foreign capital since 2002, mainly from Sweden, Italy, Austria, and Greece. Burdened with $870 billion in external debt (77% of the region's combined GDP), a large proportion in foreign currency, central European governments were scrambling to prevent their domestic currencies from depreciating to pre-empt a deluge of defaults.

A study published in *Evolution and Human Behaviour* in September 2008 related that testosterone may have played a role in the downfall of investment firms, mortgage financers, and insurers. Financial markets are dominated by capitalistic male egos and research studies indicate that men with higher testosterone levels make riskier financial investments. Women tend to be more risk-averse regarding financial gambles and tend to trade less.

Bankers' bonuses have been blamed for encouraging risk-taking, when banker compensation should reward long-term success rather than short-term risk-taking. An overhaul of the financial system was met with opposition in the banking industry. Finally, on 15 July 2010, a sweeping financial reform was passed by the US Senate, aimed at avoiding a replay of the 2008 crisis. The reform would protect consumers and reduce risks that banks took.

The interdependence and fragility of the financial system has shown us, without a doubt, that we are one world and whatever happens in one area or aspect can severely affect the rest. We are lined up like dominoes ready to topple.

Mental Health Woes

Mental health is defined as a state of well-being where a person can cope with the normal stresses of life, can work productively and fruitfully, and is able to make a contribution to the community. The Organisation for Economic Cooperation and Development (OECD) says that there is evidence that work-related mental health has worsened, especially for those with low skills.

A study by professor Jean-Pierre Brun, *Work-Related Stress: Scientific Evidence-Base of Risk Factors, Prevention and Cost*, found that increases in shift work, night shifts, weekend work, part-time work, overtime, working very fast, and general intensification of work resulted in more work stress. This gets translated into sick leave, absenteeism, lateness, job dissatisfaction, job turnover, lower productivity, and higher insurance costs.

The annual cost for mental health problems in the United States in 2002 was 1.3% of the GDP, and the stress cost was 2.6% of the GDP. The conclusion was that "the health of people is a corporate business decision criterion." The OECD says that mental health problems account for a quarter of all disability benefits paid across the European Union.

The WHO estimated that in 2002, 154 million people worldwide suffered from depression, 25 million from schizophrenia, 91 million from alcohol use disorders, 15 million from drug use disorders, 50 million from epilepsy, and 24 million from Alzheimer and other forms of dementia. These disorders have staggering economic and social costs.

In 2005, a WHO study found that 326 million people suffered from migraine, and about 877,000 people die by suicide every year. *The Global Burden of Disease* study conducted by the WHO, the World Bank, and Harvard University, revealed that mental illness and suicide accounted for over 15% of the burden of disease in developed countries, more than that of all cancers.

According to the US National Institute of Mental Health, one in four US adults aged 18 and above suffers from a diagnosable mental disorder in a given year. However, about 6% suffer from a serious mental illness. The cost of treating mental disorders rose sharply in the United States between 1996 and 2006, from $35 billion to $58 billion. The number of people with mental conditions has gone up much more rapidly compared to the number of people with heart disease, cancer, trauma-linked disorders, and asthma. The number of Americans seeking treatment for depression, bipolar disorder, and other mental illnesses almost doubled, from 19 million to 36 million. Antidepressant use also doubled, from 13.3 million people in 1996 to 27 million in 2005.

The WHO predicts that within 20 years more people will be affected by depression than any other health problem. In 2009, 450 million people were directly affected by mental disorders or disabilities, most of them living in developing countries. Yet these countries spend less than 2% of their national budgets on mental healthcare. About 800,000 people commit suicide every year, 86% from low- and middle-income countries. The highest suicide rates are among men in eastern European countries. For every person who completes a suicide, 20 or more may attempt to end their lives.

As the global financial crisis hit, Fox News reported in October 2008 that "suicides from financial crisis cause concern." AsiaOne Health highlighted on 28 October 2008 that suicides and mental illness increased in Korea with the stock market meltdown. In Australia, the Mental Health Council said that there was a 40% increase in the number of Medicare claims for mental health consultations in March and April 2009. The number of people getting psychological services rose from 54,000 in March 2008 to 83,000 in March 2009. On 8 September 2009, Uganda People News reported Asha-Rose Migiro, Deputy Secretary General of the UN, citing that the current global credit crunch is increasing violence against women in different parts of the world. She said that countries should ensure the protection of civilians against sexual and gender-based violence.

Pause: the Importance of Money

Work and money are on practically everyone's mind a large portion of the time. What makes us view "our daily bread" as so very important? Abraham Maslow, the American psychologist who is considered the founder of humanistic psychology, conceptualised a Hierarchy of Human Needs. The five levels of basic needs are:

I. physiological needs: breathing, water, food, sleep, shelter, clothing, sex, homeostasis (ability to regulate the body's inner environment in response to the outside environment);
II. safety needs: personal and financial security, health and well-being, a safety net against accidents/illness and their adverse impacts;
III. social needs: friendship, affection, love, intimacy, a supportive and communicative family, belonging, social connections;
IV. esteem: respect, desire to be accepted and valued by others, success, status, recognition, prestige, attention, self-esteem, self-respect, competence, mastery, self-confidence, independence, freedom;

V. self-actualisation: the motivation to realise one's maximum potential and possibilities, harmony, and understanding.

These five needs are predetermined in order of importance, where lower-level needs have to be met first. Maslow studied people he felt were self-actualised, including Lao Tzu, the father of Taoism. The basis of Taoism is that people do not obtain personal meaning or pleasure by seeking material possessions. Taoism also has a different perspective on affluence, whereby the greater the biodiversity in the environment, the greater the community's affluence.

Near the end of his life, Maslow found the sixth need: the need for self-transcendence, which is the need to be aware of and living in the realm of *being*, to have what he called unitive consciousness, and peak experiences with illuminative insights. Maslow estimated that only 2% of the population achieves this self-transcendence level which involves empathy with and responsibility for others' misfortunes, humility, creativity, intelligence, and divergent thinking. These descriptions match Bill O'Hanlon's framework on integrating spirituality and brief therapy. As the developer of Solution-Oriented Therapy, O'Hanlon demonstrated to psychotherapists that by using spiritual resources one can help others to change with greater impact and ease. The elements of O'Hanlon's spiritual approach to psychotherapy are:

- connection—moving beyond one little isolated ego and connecting with something bigger, within or outside of oneself;
- compassion—softening toward oneself or others by being *with* rather than *against* oneself, others, or the world;
- contribution—being of unselfish service to others or the world.

Bill O'Hanlon's framework may be a practical approach to help more people move into the realm of self-transcendence. A world filled with more self-transcended beings would certainly bring forth a new age of cooperative, supportive, and harmonious living on our planet.

Stuck in Materialism

> There is no calamity greater than lavish desires.
> There is no greater guilt than discontentment.
> And there is no greater disaster than greed.
> —*Lao Tzu*

Living in a material reality and striving to meet physiological needs, especially for food, clothing, and shelter, it is easy for people to get caught up in materialism and to need more money. This need for money then gets fused with our financial security needs. It gets further connected to our social needs, "paying" for friendship, love, and social connections by buying gifts and entertainment. By the time the need for esteem is reached, money is already deeply lodged into our sense of value, success, and recognition.

We look up to people with wealth, fame, and good looks who are seen as financially successful and having a luxurious lifestyle. We constantly compare ourselves with others. Thus, we put in more and more effort and hard work to study, get a "good" job, earn more, have more, look better, and be more valued (perhaps envied) by others. So, self-actualisation is mistaken by many with maximising their financial status, looks, lifestyle, connections, fame, and power. This is not what Maslow's self-actualisation was about. Lao Tzu, a self-actualised sage, said: "Manifest plainness, embrace simplicity, reduce selfishness, have few desires."

Having—Consumerism and Receiving

It is precisely this craving for more and more—materialism—that has brought us to our current level of rampant consumerism. "I want to have this, I want to have that and the other..." Anyone with children will know that this constantly wanting to receive is at a juvenile level of maturity.

We are now very much a throw-away society. Our food and drink purchases come in containers, bottles, cans, plates, utensils, and plastic bags, which get thrown away (another problem in itself). Our electrical and electronic gadgets, like phones, MP3- and CD-players, TVs, computers, printers, ovens, washing machines, etc., get discarded when new and better models come onto the market. When we started writing this book, we installed an air-conditioner that cost $600. Within a couple of weeks into the writing, we realised that using the air-conditioner was not in line with the message of this book, so we uninstalled it, but could only get around $15 for it, as people do not want a (barely) used air-conditioner.

Doing—Working Hard and Earning

As we mature, we act as if there is a need to work hard in order to earn more and then have more. Thus, we get onto the treadmill of the rat race, doing things at a faster and faster pace, running around, becoming *human doings* instead of *beings*. This "earning to have..." takes a heavy toll on our physical, emotional, mental, and spiritual well-being. True meaning or purpose in life easily gets lost in building our material façade.

We see many people who are successful in their careers but feel there is a gap in their lives, so they start looking for a meaning and purpose to it all. Many of them make dramatic changes in their choices of career and lifestyle. Frequently, this involves becoming self-employed, starting a business, balancing family and work, pursuing courses and studies, or contributing back to society in various ways.

Being—Self-Transcendence and Giving

The Merriam-Webster dictionary defines *being* as "the quality or state of having existence." As mentioned previously, only 2% of the population is living in the realm of *being*, therefore interactions with others are more focused on giving. In relation to the spirit of giving, psychologist, author, futurist, and motivational speaker, Leland Kaiser, Ph.D., says:

> Self-transcendence is growing into your unfulfilled potential.
> [It] is moving beyond the orbit of your ego into your soul.
> [It] is gaining a new concept of self that is much expanded
> and includes more of the Universe.
> You are part of much more than you know or imagine.
> The Universe is interconnected, [it] is non local.
> You are everywhere. Only in the third dimension do you
> occupy such a small place in the scheme of things.

He believes that four great waves of social transformation have swept over the planet: (i) "the hunter and gatherer wave," (ii) "the agricultural wave," (iii) "the industrial wave," and (iv) "the information wave." Each wave transformed humankind and prepared the way for the next one. A fifth wave of consciousness is building and will envelop the human race. The (v) "consciousness wave" emphasises holistic human development, the experience of the spiritual dimension of love through personal/global/galactic interconnectedness, compassion, empathy and giving, an inner expansion of dreams, visions, intentions, ideals, commitments, and inclusiveness. There is a focus on ecological and healthy living, a sustainable economy, and responsible investing.

Spirituality and Work

While we may be going through the fifth wave of social transformation according to many leading evolutionary theorists, the idea of mixing spirituality and work is not popular among us. Though a large percentage of the population believes in some form of spirituality, the notion of keeping one's spiritual beliefs out of one's work is an expected norm. We realise that the workplace is a very competitive place which requires us to set aside some of our values and virtues, such as compassion, fairness, and honesty. We believe that those values are fine for private life, but the business environment is such that we must play by a different set of rules in order to succeed.

It's been reported that Ray Kroc, the founder of the great McDonald's empire, who claimed to be a Christian, said: "My priorities are God first, family second, and McDonald's hamburgers third. And when I go to work on Monday morning, that order reverses." Though we can't be certain about what he meant by this, it can be taken as he had two sets of priorities, one for the workplace and the other for his private life.

Recent studies suggest that the practice of "compartmentalisation" is a crucial component in the acceptance and continuation of corruption within an organisation. For instance, Kenneth Lay, the former CEO of Enron, who was indicted for multiple counts of fraud, was a self-proclaimed Christian. In his term at Enron he worked diligently in several charitable causes in the Houston area and was well-known for his philanthropic endeavours. Another example, Bernie Ebbers, the CEO of WorldCom, is also a self-proclaimed Christian who taught Bible studies in his church in Mississippi during his term with the company. Clearly, compartmentalising our core values and beliefs is not the path that leads to our self-transcendence or spiritual development.

Many spiritual traditions, namely Christianity, Hinduism, Islam, Judaism, and Taoism have clearly stated that work is one of the pathways to God. These spiritual traditions do not see a dichotomy between a spiritual and a working life. Dallas Willard, Ph.D., a professor of philosophy at the University of Southern California, says that the primary place of our spiritual formation is not in our churches (or temples), or small gatherings of 15 minutes of reading the Holy Scriptures; it is the time we spend at work, at school, and at home, mopping, washing, ironing, and cooking.

Doing Good Work in the Workplace

In the GoodWork Project led by Howard Gardner et al., conducted over 10 years and 1,200 interviews, it was found that most workers aspire to do "good work," to take pride in doing something that matters, that serves society, that enhances the lives of others, and that is conducted in an ethical manner. While good work is generally the goal of most working people, powerful barriers exist. Internal and external conditions can make good work extremely challenging. Pressure to keep costs low, profits high, and to fulfil several life roles (parent, spouse, friend, co-worker, boss, etc.) can make cutting corners tempting. In addition, in countries with high levels of corruption or governmental interference, good work is extremely hard to practice.

One of the many useful findings this project has produced is the uncovering of the effects of spirituality on the quality and enjoyment of work. A number of studies from the project have pointed out the impact of religion on imparting values that workers apply to their daily working lives. In addition, they studied contemplation and meditation and established how quasi-spiritual practices shape professional identity.

Building on the GoodWork Project, Seth Wax, a Harvard graduate and former researcher on the project, did a research called *Spirituality at Work* (2005), which focused on uncovering unifying spiritual patterns across diverse work practices and age cohorts. In each interview, he posed questions such as, "Do spiritual or religious beliefs guide you in your work?" and "What are some core beliefs or values that guide you in your work?" He presented four models of how modern workers experience personalised spirituality at work:

I. sensing the presence of an impersonal higher force during a work performance;
II. intuiting the mystery of the unknown through making discoveries;
III. envisioning a personal God who is active in all personal and occupational aspects of the believers' life;
IV. acknowledging a higher force as beyond one's immediate experience, yet supporting one's occupational activities.

In the first model, subjects interviewed described their performance experience in terms of "universal energies," "vibrations," "elemental forces," and "God presence." The descriptions in this model were found exclusively among professionals in the performing arts. These descriptions point to the phenomenon of flow, an experience in which individuals feel like being carried away by a current, with everything moving smoothly, without effort. The flow is associated with the divine presence, it supports the performers, but does not assume agency in worldly affairs.

The second model findings are drawn from professionals in scientific research. Though many scientists have rejected religion and spirituality, a surprising 40% of scientists maintain a belief in a personal God. Many of those interviewed claimed they may be spiritual, but they are definitely not religious. It is also noted that scientists do not seem to maintain a sense of personal agency in the spiritual realm.

The third model findings are drawn from people in business, social entrepreneurs, non-profit workers, and philanthropy professionals. These professionals possess a God-centred spirituality, i.e., they accept their role as simple cast members in a universal divine plan, but they feel they can interface with and experience God directly through prayer and active dialogue. Many of the professionals interviewed see themselves as performing God's good works on earth. While they show no evidence of taking credit for successes, neither do they relinquish responsibility for negative outcomes.

The fourth model findings are drawn from the same categories of professionals in the third model. These professionals report a general spiritual or metaphysical feeling in their job. They draw support from the higher power in their work and attempt to live in accordance with it. These individuals view spirituality as a source of encouragement, but their worldview does not sense an immediate and personal presence of the divine. They consider themselves to be directly responsible for their work, and the higher power in which they believe serves more to comfort them than to specifically guide their actions.

The models and findings presented demonstrate the different ways in which spirituality can be infused into one's daily life. While believing in the spiritual may not automatically bring about good work in the workplace, it certainly has the greatest potential of influencing a person's belief in his or her function in the world and how best they can actualise their spiritual beliefs at work.

Uniting Spirituality & Work

The friction between spirituality and work is beautifully mitigated in an ancient wisdom from India. Yoga, in its original spiritual context, has little to do with exercise. In Sanskrit, *yoga* means "to yoke" or "unite," and the goal of yoga is actually to unite oneself with God. The practice of yoga is to facilitate this union. People pursuing this spiritual union can be broadly classified into four psychological types: the emotional, intellectual, meditative, and physical. There are, thus, four forms of Yoga designed to facilitate each type. It is important to understand that the four types are not mutually exclusive. In fact, each one blends into the next, and each one balances and strengthens the others. Practicing this knowledge may bring about a total transformation of one's emotional, mental, spiritual, and physical experience, which strengthens one's ability and desire to do one's best.

The Path of Emotion

Bhakti Yoga is the path of devotion, the method of unifying with God (Brāhman) through love and the loving memory of God. Through prayer and worship, practitioners surrender to God, channelling and transmuting their emotions into unconditional love of devotion. According to Hindu philosophy, when one loves another, one is really responding unconsciously to the divinity within them. It is the divine self within each person that one is emotionally connected to. Our love for others becomes unselfish and motiveless when we are able to see the divinity in them. Chanting and singing the praises of God form a large part of Bhakti Yoga.

The Path of Intellect

Jñāna Yoga is the path of knowledge, particularly the knowledge of self (*Ātman*) and the universal spirit (*Brāhman*). The Jñāna practitioner uses the power of his/her mind to discriminate between the real and the unreal, as well as the permanent and impermanent. Jñāna practitioners are monists—there is only one sole reality: the Brāhman. There is no need to look outside ourselves for divinity: we ourselves are divine. In Jñāna Yoga, many mental exercises are practiced every day to help one realise the unity of the self with the universal spirit.

The Path of Meditation

Next is the *Rāja Yoga*—the practice of body and mind control through meditation. It is the process of turning mental and physical energy into spiritual energy. Rāja Yoga is essential for all the other paths because all types of Yoga involve some form of meditation. Rāja Yoga incorporates moral virtues which must be infused into one's thoughts, words, and deeds. It also includes specific postures, breathing techniques, and concentration exercises. All of the steps prescribed in Rāja Yoga are to help the practitioner reach the final step, known as *Samādhi,* the state where the practitioner is one with self (*Ātman*) and God (*Brāhman*).

The Path of Action

Last but not least, *Karma Yoga* is the path of dedicated work and involves renouncing the rewards of one's actions. In other words, performing work to one's best ability with spiritual awareness, without attachment to the outcome. In Hinduism, *karma* refers to the law of action: every action directed toward the external world and others reacts on the doer. What we experience today is the result of our karma—both good and bad—created by our previous actions. When we disengage our ego from the work process and offer the results to a higher power, we stop our chain of cause and effect. In this way, our work takes us in the direction of God, rather than away from God.

Karma Yoga—Path to God through Work

In modern society, one has to work in order to live. Almost all of us work with expectations: we work hard to get respect and appreciation from our colleagues and promotions from our boss, we study hard to get good grades, in the hope to have a bright future, we cook a good meal with the expectation that we will receive praises. This cause-and-effect thinking is very much wired into our neurology.

From a spiritual standpoint, all these expectations and anticipations are Trojan horses that will bring us misery sooner or later. And our expectations and desires are unending and unappeasable. We live from one disappointment to another because our motivation gratifies and enlarges our ego. The process of practising Karma Yoga (renouncing the results) helps us break out of this vicious cycle. At the same time, it empowers us to achieve extraordinary results in whatever we want to do.

Very often, creators of exquisite results—whether in the arts, business, or science—see themselves as an instrument of the divine. The performers can see, hear, and feel themselves doing the tasks, but they know they are not actually orchestrating them; the performance has a life of its own. Frequently, the miraculous happens when we are able to get the self out of the way, so that the creative divine energy can flow into our performance. The diminishment of one's ego is necessary for the divine presence to emerge. The realisation and acceptance of this divine intelligence is one of the critical milestones of Karma Yoga.

Below is a summation of the key principles and benefits of this type of yoga:

- *Right Attitude While Doing*
  In the Bhagavad Gita (revered Hindu scripture) it states, "See Me in all things, and all things within Me." This means that the divine exists in the hearts of all beings and creatures. So we must strive to do our work with love, compassion, and integrity, as if we were in front of God. It is not what we do that counts, but our attitude while doing it. At the same time, we must see our actions as being part of something greater than ourselves and see ourselves as an instrument for the divine power to manifest. The point of all yoga practices is to align and synergise our entire life instead of compartmentalising our days into *work* and *personal* segments.
- *Focus on the Process, Surrender the Results*
  The Chinese Sage Chuang Tsu taught this: "When you are betting for tiles in an archery contest, you shoot with skill. When you are betting for fancy buckles, you worry about your aim. And when you are betting for gold, you are a nervous wreck." Our skill is the same in all these scenarios, but it is our attachment to the perceived value of the outcome that creates the different results. By surrendering the outcome to a higher wisdom, one can then direct all their thoughts and actions on the

process of the work. Many Karma Yogis confess that by doing so, their success is more likely to happen. People practicing Karma Yoga may appear to others to be goal-oriented, but inside they are peaceful because they are unattached to the outcome.

- *Assume Full Ownership—Put in Your Best*
  According to Gandhi, one of the most well-known Karma Yogis, "renunciation of fruit in no way means indifference to the result… He who is fully equipped, is without desire for the result, and is wholly engrossed in the due fulfilment of the task before him, is said to have renounced the fruits of his actions. Do not fear criticism and failure, fully commit to the work that you undertake." Many people pray for a particular result in their lives, but they do not align their actions to their prayers. So, work in a manner that honours your prayer. Treat your work as an active form of prayer.
- *See the Work as a Teacher*
  It is by interacting with others that our engrained conflicts, blockages, and habits can be identified and recognised, and in time, be fully healed with the practice of the other types of Yoga (Bhakti, Jñāna, and Rāja). To attain greater wisdom in life, yoga says that we have to introspect and understand how we function on a deep inner level. Seen in this manner, each job is a teacher of some sort. We can learn different knowledge and skills by doing different kinds of work. So no job is superior or inferior, they all offer us different knowledge and skills, and therefore we do them with diligence and devotion.

STUDIES ON SPIRITUALITY AT THE WORKPLACE

Spirituality in the workplace is a hot topic today. Several major organisations are advocating meditation in the workplace, sponsoring retreats and training programmes, and providing employees appropriate spaces in the workplace to meditate for 10–20 minutes a day. Companies like Apple, Prentice Hall Publishing, Google, Nike, Deutsche Bank, Procter & Gamble not only witness an improvement in their employees' health and temperaments, but also find them much more able to think out their problems and raise bottom-line profits. CEO of Procter & Gamble, Alan G. Lafley, puts it aptly, "You can't out-work a problem; you have to out-meditate it."

According to Wisnieski, Askar, and Syed, authors of *Toward a Theory of Spirituality in the Workplace* (2004), the stronger the spiritual factor of a worker's personality, the more tolerant he is of work failure and work stress, and the more he exhibits altruistic and organisational citizenship behaviours. They also demonstrate more trust in people and greater tolerance of human diversity.

Chris Sangster, in his article called "Spirituality in the Workplace" (2003), states that spiritual workers are those who think cooperatively and altruistically, who listen more than they speak, believe in a higher driving force and purpose beyond mankind, work open-mindedly with a wide range of people, encourage and empower others selflessly, and demonstrate integrity and trust.

In a controlled experiment conducted in a manufacturing company called Birla Celluloise, in Gujarat (2007), Adhia, Nagendra, and Mahadevan obtained clear evidence that by adopting a yoga way of life, managers have positively impacted four out of five of the organisational performance indicators. The five indicators evaluated were: job satisfaction, job involvement, goal orientation, affective organisational commitment, and organisational citizenship behaviour.

Managers were given the choice of joining this experiment after having been explained its purpose and modality. Those who opted to do it were divided into 2 equal groups of 42 managers: group 1 was called the "Yoga Group" and group 2 was called the "Physical Exercise Group," which was the control group. The Yoga Group was given 30 hours of yoga practise and 25 hours of theory lectures on the philosophy of yoga. The experiment was completed in 6 weeks. The control group was given the same number of training hours, consisting of normal physical workout and lectures on success factors in life. This control group eliminated distortion of results from the tendency of some people to work harder and perform better when they are participants in an experiment (the Hawthorne effect).

The distinct positive results obtained in this research confirmed the power of the yoga philosophy, which is to help a person gain a broader view in life and greater awareness of his actions. It also enhances one's commitment to his job and organisation, and therefore raises one's willingness to go beyond their call of duty.

All this is a result of the embracing of the concept of Karma Yoga, implying the integration of behavioural and introspective approaches to personal growth. It provides a framework and process by which one can disengage from the unhappy personality one has created for himself and the negative role one has adopted. It smoothly fits into a training program for changing habits, thought patterns, and self-concepts.

Altruism Is Good for Business

Discussing the hot topic of uniting spirituality with work naturally brings up the notion of altruism. Loosely defined, altruism is a genuine concern for the welfare of others without a calculated consideration of one's own possible benefits in the context. It is a traditional virtue in many cultures and a core aspect of various religious traditions. The notion of altruism is slowly spreading in the business world as more and more companies are incorporating the principle of "do good, do well" into their mission and strategic planning. Though the ultimate intention of incorporating this principle is not entirely an uncalculated consideration for most companies, it is proving to be a powerful approach to attracting new customers and retaining old ones, as well as recruiting and retaining quality employees. This principle of "doing good and doing well" is also making a positive impact on the environment, safeguarding animal welfare, promoting fair trade, and protecting human and workers' rights.

When Pfizer partnered with Grameen Health to launch their "Do Good Do Well" programme (2008) in order to give the working poor in emerging economies unprecedented access to medicine and healthcare via a self-sustaining business model, the director of Global Access, Ms. Ponni Subbiah, was flooded with emails from Pfizer employees expressing passionate support. Many employees applied for jobs or asked if they could volunteer; they were even willing to work on weekends and evenings. "We all want to feel that we can have an impact on the world. That's why I like Global Access," said Ms. Subbiah. "The fact that we are going to increase access to our medicines in a part of the world where people are needy [...] that's very gratifying for me," she continued. It is also gratifying for many Pfizer employees and life-giving for thousands of poor people in emerging economies. And it is good business for Pfizer.

In 2010, the Pepsi Refresh Project awarded more than $20 million to 1,000 individuals and businesses that promoted a new idea that had a positive impact on their community, state, or nation. All this is still done on Pepsi's website and on Facebook, and consumers decide who they want to receive the money, by voting online. In its second year (2012), Pepsi Refresh was heralded by *Forbes* magazine as the fifth best social-media campaign. This project would not have been possible without a team of young idealists from *GOOD Magazine*, a publication that believes in "living well and doing good." Kirk Souber, creative director of GOOD Corps (agency arm of GOOD) said that over the last 5-to-6 years, more and more people have looked at brands, asking, "Is this brand part of the solution or is this brand part of the problem?" In current time, many consumers are actually following brands in the same manner they follow their friends on Facebook and Twitter; "The values revolution is in many ways being amplified by the digital revolution," said Sebastian Buck, a co-leader of GOOD Corps.

Save the Ta-tas, a company founded by Julia Fiske in 2004, donates at least 25% of their proceeds to support the fight against breast cancer by selling cancer awareness T-shirts, gifts, and accessories. The money goes directly to the cause and Save the Ta-tas is currently providing salary support for breast cancer researchers. Up until 2012, Fiske's company has donated $814,000. As she chose to leverage her brand with cause marketing, her profits have been increasing every year.

Another "do good do well" company is TOMS Shoes, a for-profit company based in Santa Monica, which also operates a non-profit subsidiary called Friends of TOMS. When TOMS sells a pair of shoes, Friends of TOMS donates a pair of shoes to a person in need. The founder, Blake Mycoskie, conceived the idea in 2006 after returning from a vacation in Argentina; he saw many children without shoes in poor villages and suffering from podoconiosis—a foot disease. Since then, he has given shoes to children in more than 20 countries, including Argentina, Ethiopia, Guatemala, Haiti, and South Africa. Until 2010, he has given away 1 million pairs of shoes. His product has been sold widely in the United States and internationally, with partners like Neiman Marcus, Nordstrom, and Whole Foods Market, featuring styles made from recycled materials.

So, does promoting a worthy cause impact business profitability? If someone is still in doubt, then perhaps the compilation of facts and statistics below will squarely show that a company's reputation around corporate social and environmental responsibility is paramount to consumer's purchasing decisions.

- 80% of consumers are likely to switch brands similar in price and quality, to one that supports a cause (Cone 2010 Cause Evolution Study).
- 54% of shoppers say they consider elements of sustainability—sourcing, manufacturing, packaging, disposability, etc.—as they select products (GMA/Deloitte Green Shopper Study, 2009).
- Mums and Millennials—people born between approximately 1980 and 2000—are two big consumer groups strongly influenced by cause marketing. Mums want to buy products that support a cause and Millennials are willing to pay a premium price to buy "green" products (Cone 2010 Cause Evolution Study).

- 76% of consumers say they would purchase products from a socially and environmentally responsible company (2009 Tiller LLC Social Action Survey).
- 59% of consumers aged 18-45 are willing to pay more for green products (Capstrat-Public Policy Polling, 2009).
- 88% of women say they like brands that "allow me to do something good" (Women, Power & Money: The Shift to the Female-Driven Economy, Fleishman-Hillard/Harrison Group, 2010)
- Nearly 1 in 5 (19%) young adults (aged 18–29) say they are willing to pay "significantly more" for green goods (Capstrat-Public Policy Polling, 2009).
- 76% of Millennials want brands to be ecologically conscious (Generate Insight, millennial demographic research, 2009).
- 46% of Americans say that companies should support an issue in the communities where they do business (Cone 2010 Cause Evolution Study).

The worldview and consciousness of more and more people around the world are changing; we are slowly awakening to the notion that doing good is as important as doing well. Virtues like integrity, truly caring for people, animals, and the environment, treating others the way we like to be treated, being compassionate and loving, and being honest in our interactions, are all qualities that make us feel good and truly alive. It's time for small business owners and big corporations, as well as people who are going to work every day to think and act in life-supportive ways. As it has been clearly shown, our ideal to be and do good, need not be in conflict with our desire to do well in our workplace. Our spiritual life need not be compartmentalised separately from our work life. In fact, work is one of the spiritual pathways to God.

CHAPTER 10
WHAT'S HAPPENING TO OUR CHILDREN?

I believe the children are our future.
Teach them well and let them lead the way.
Show them all the beauty they possess inside.
Give them a sense of pride to make it easier.
Let the children's laughter remind us how we used to be.
—George Benson, "Greatest Love of All" *(1977)*

One generation plants the trees, another gets the shade.
—*Chinese proverb*

What kind of world are we leaving to our children? What values and beliefs do we have which will come to shape our young ones? What are our expectations for them? Given the ways they are being brought up, how do they spend their days? What are their aspirations?

According to Hillary R. Clinton, who wrote a book in 1996 titled *It Takes a Village and Other Lessons Children Teach Us*:

> I chose that old African proverb to title my book because it offers a timeless reminder that children will thrive only if their families thrive and if the whole of society cares enough to provide for them. For a child, the global village must remain personal. Parenting tasks cannot be carried out in cyberspace. Children require the presence of caring adults who are dedicated to children's growth, nurturing, and well-being.

Are we these caring adults?

CHANGES IN POPULATION AND FAMILY STRUCTURE

After the Second World War, the world's population more than doubled from 2.52 billion in 1950 to over 7 billion in 2012. One-third (2.21 billion) were under 18 years of age, with only 1 in 10 coming from developed regions and nearly 2 in 10 from the least developed regions. The rate of population increase was the highest in Africa, followed by Latin America and Asia. America and Europe, especially, had the lowest increase rate, with much lower birth rates. According to the UN Population Division Policy Brief in March 2009, most of the least developed countries still have high fertility, with more than five children per woman. There is a high unmet need for family planning, low levels of modern contraceptive use, and childbearing starts early in adolescence, increasing the morbidity and mortality risks of mothers and infants. Children born into large families have fewer opportunities to receive schooling, and women with less schooling tend to have higher fertility (more children).

In 2007, the life expectancy at birth in developed regions was 76 years, but only 49 years in sub-Saharan Africa. While 9 out of 1,000 children died before reaching the age of 5 in developed regions, in sub-Saharan Africa, 167 out of 1,000 died. In developed regions, 11 mothers died out of 100,000 live births, compared with 900 in sub-Saharan Africa. According to UNICEF's *The State of the World's Children 2009* report, 1,500 women die every day while giving birth, which comes to 500,000 mothers every year. This shows the wide disparity in a child's life in different regions of the world.

Overall, 21% of the population in developed regions was below 18 years of age, rising to 36% in developing regions, and to 47% in the least developed regions.

Children of Divorce

Divorce numbers have been soaring around the world. In the United States, 49% of marriages end in divorce; in Sweden, 64%; in France, 43%; in Israel, 26%; and in Italy, 12%. The median duration of first marriages ending in divorce is slightly less than 8 years.

Many children in the world are affected by their parents' divorce and their emotional trauma carries forward into their physical well-being, schoolwork, social behaviour, and future relationships. Divorce statistics in the United States show that:

- over one quarter of children are living in single-parent or step/blended families; two fifths do not live with their fathers;
- divorced women with children are four times more likely than married women to have an income below the poverty line;
- rates of child abuse are eight to ten times higher in single-parent or step/blended families, and children are thirty-three times more likely to be so seriously abused that they require medical attention;
- children of divorce are two to three times as likely to drop out of school, require psychiatric treatment, and commit suicide as adolescents; 71% of all high school dropouts and 63% of youth suicides are from fatherless homes;
- children raised in single-parent homes are less likely to marry and more likely to divorce;

- teenage girls from single-parent homes are two to three times as likely to give birth to an out of wedlock child;
- children of divorce are 12 times more apt to be incarcerated; among long-term prison inmates, 70% grew up without fathers, as did 60% of rapists, and 75% of the adolescents charged with murder.

Daily Life in Developed Regions

Too many parents make life hard for their children
by trying, too zealously, to make it easy for them.
—attributed to Johann Wolfgang von Goethe, writer and polymath

Do not handicap your children by making their lives easy.
—Robert A. Heinlein, writer

Of the 2.2 billion young people below 18 years in 2008, only 0.2 billion (or 9.2%) lived in developed regions. In developed and many developing areas, both parents or single parents are working (oftentimes long hours). With fewer children and greater affluence, children are typically treated as precious and sometimes pampered with gifts and instant gratification. Sometimes parents do this to make up for not being around, thus assuaging guilt. High aspirations for each child, together with the competitive educational environment, may result in many hours of tuition, extracurricular activities, and enrichment programmes, such as music, dance, art, drama, martial arts, sports, and learning/mind/brain development. In many homes, especially those with domestic help, children do not need to lift a finger as cleaning, washing, ironing, meal preparation, tidying are all done for them.

Nancy Gibbs, in her article "The Growing Backlash against Overparenting," published at Time.com on 20 November 2009, describes parents hovering over their kids and obsessing over their safety and success, whom she calls "helicopter parents." There are even helicopter grandparents. She presents fear of danger and fear of failure as a kind of parenting fungus, invisible and insidious. Some parents ghost-write their kids' homework, airbrush their lab reports, lobby for them to be in certain classes, and rescue them when they forget to bring their books or lunch boxes. Overprotectiveness and overinvestment have raised kids who may seem ready to break at the tiniest stress. The over-parenting or hyper-parenting backlash she calls "slow parenting," "simplicity parenting," or "free-range parenting."

Kim John Payne, author of *Simplicity Parenting*, points out that the average child studied has 150 toys and suggests a reduction of 75%, keeping those which leave the most to children's imagination and build calm into their schedules, so that they can actually enjoy their toys.

The Net Generation: Internet and Computer Gaming

The Internet Generation, iGeneration, or Generation Y, born between around 1977 and the mid-1990s, is accustomed to digital technologies. Don Tapscott in his book *Grown Up Digital: How the Net Generation is Changing Your World* describes them as doing five things at once: texting friends, downloading music, uploading videos, watching a movie on a two-inch screen, and socialising on Facebook or MySpace. He presents them as living partially in this world and partially in a virtual world, as spending hours each day interacting with digital devices—computers, Internet, handphones, game consoles, audio and video players, etc.

IBM released the personal computer in August 1981. By the end of the 1980s, video games dominated the toy market. In the 1990s, handphones and the Internet provided a swift means for social interaction. The digital revolution allows for constant connectivity to the networked world, but such progress is a double-edged sword; Internet and computer gaming addiction has become widespread, with associated withdrawal symptoms. A survey in September 2009 by the market research company NPD Group found that 82% of US children aged 2 to 17 years played video games, meaning some 55.7 million US children are "gamers." An article in the *New Scientist* on 16 November 2005, by Alison Motluk describes how "gaming fanatics show hallmarks of drug addiction." While playing to relieve stress from studies, violent video games can increase aggressive thoughts, feelings, and behaviour. Besides problems with diving school grades, behaviour deterioration and disorders, such as Attention Deficit Disorder (ADD), Attention Deficit Hyperactive Disorder (ADHD), and other cognitive and development problems have surfaced.

A two-year longitudinal study, *Pathological Video Game Use Among Youths*, published in the medical journal *Pediatrics* online on 17 January 2011 by American, Hong Kong, and Singapore researchers, studied over 3,000 children and youths in Singapore and found that the average amount of time spent playing was 21 hours per week. Pathological gaming (addiction to video games) was reported to be similar in various countries, with around 9% being hardcore gamers who play over 31 hours per week and who usually also become increasingly interested in games with violence. The higher the frequency of gaming, the greater the risk of poor social skills and impulsivity. While children do play computer games as a coping mechanism for stress in their lives, pathological gaming, in fact, increases depression, anxiety, social phobias, and poorer school performance.

Health issues also surface, including epileptic fits triggered by flashing lights in games and deep vein thrombosis where the flow of blood is restricted in veins. On 1 June 2007, *The Straits Times*, Singapore, reported the death of a university undergraduate, an ex-avid gamer diagnosed with deep vein thrombosis. Four years earlier, at age 16, he had had his first attack and near-death experience after he had spent long hours during the previous weeks glued to his computer, playing games online. At that time, computer gaming was his life, and he was very shy by nature. A second near-death experience occurred the following year. He did not survive the third experience. On 19 June 2008, BBC's *Press Pack Reports* described how one boy from Somerset, Huw, aged 11, was campaigning to make games safer for kids after he was hospitalised for an epileptic fit the year before. He wanted to make sure that all game companies screen their games to ensure they are safe for people who are photosensitive, as flashing lights in games trigger epileptic fits.

Cyber Bullying and Online Predators

Cyber bullying is also of concern, as some researchers say it causes longer lasting psychological damage than traditional bullying. Victims of cyber abuse have little means to get back at cyber bullies, who are protected by freedom of speech and privacy laws. The BBC reported on 12 November 2009 that a study of 10,000 teenagers by the English Government found that 47% of 14 year-olds were bullied, through name calling and cyber bullying. Another BBC report, on 16 November 2009, highlighted that more and more children in primary schools are being affected by cyber bullying, according to a survey by the Anti-Bullying Alliance, with one in five having been bullied by phone or online.

Over a quarter of the girls in the UK, some even as young as 10, suffer because of cyber bullying. Kirsty Perkins from Wales, aged 14 in 2006, made a suicide attempt after having been bullied on a social networking website. According to Dan Simmons in his November 2006 BBC News article, "Cyber bullying rises in South Korea," South Korea is one of the most connected places on the planet, where social networking is as popular as meeting up for a drink. The growing phenomenon of cyber violence sees online mobs demonising and harassing those they disagree with, affecting their social statuses, jobs, and even mental health. All the police stations in Korea now have a cyber-terror unit to help deal with this problem. Videos of bullying are also being uploaded for public viewing.

A lot of 10- and 11-year-olds also use social networking sites (even though sites like Facebook and MySpace are meant for persons at least 13 years old). Some scientists worry that pre-adolescent use of these sites, which have been linked to Internet addiction among adults, could damage children's relationships and infantilise their brains.

Online chatting has led to online sexual predators, including paedophiles who are trawling cyberspace. They do not use their real identities, which makes it difficult to track them down. When Singapore's *The Straits Times* polled 50 teenagers between 11 and 19 years old in December 2008, 18 of them said they had been sexually propositioned online, 10 agreed to a meeting, and 4 ended up having sex with the strangers they met online. Girls aged 13 to 17 were most at risk. Those who posted pictures of themselves on social networking profiles were particularly vulnerable, allowing predators to take their pick.

Fast Food Kids and Obesity

According to the WHO, 43 million pre-school children worldwide are either obese or overweight and this is usually because they eat and overeat fast food. Jeanie Lerche Davis reported results from a study of 6,000 children in the WedMD Health News, on 5 January 2004, that on a typical day:

- as much as 30% of children eat fast food;
- fast food is the main food source for 29% to 38% of children;
- fast-food eaters get 15% more calories than other kids.

The fats, sugar, and salt in fast food appeal to a child's primordial tastes, triggering more eating later in the day, as fast food does not contain much fibre and kids do not feel full, nourished, or satisfied afterwards.

A study in 2007 of 63 pre-schoolers, aged 3 to 5 years, by the John Hopkins Bloomberg School of Health in Baltimore, United States, found that 76% preferred fries in branded packaging and 60% preferred branded chicken nuggets, compared with the same items in unbranded, plain wrapping. The study also found that kids in homes with more TVs were more likely to prefer a branded meal, revealing the strong influence of commercials. The $10 billion spent every year to market food and beverage products to US children could easily be used to market healthy food choices instead, helping to reverse the obesity epidemic among children.

Health.com, on 26 November 2008, reported that the National Bureau of Economic Research calculated that a ban on fast-food advertising during children's programmes would reduce the number of overweight children aged 3-11 by 18%, and overweight adolescents aged 12-18 by 14%. While not practical in the United States, there are outright bans on fast-food ads in countries such as Sweden and Norway.

In Singapore, males on reaching 18 years old are conscripted into the military for 2 years. Because they aren't fit enough, more youths are engaging fitness trainers to help them get in shape for National Service. This generation of adolescents is, to a large extent, on a steady diet of computer gaming and fast food.

Diabetes and Cancer

Diabetes is now one of the most common chronic childhood diseases, on reported on 14 November, World Diabetes Day, 2008. Around 500,000 children around the world under 15 have type 1 diabetes (an autoimmune disease), while type 2 diabetes, once considered an adult health problem, is growing at a high rate among children. Between 1982 and 1994, the incidence of type 2 diabetes in children multiplied by 10. In 1996, a third of all new diabetes cases in children were type 2. This paralleled the rising epidemic in childhood obesity in the United States and worldwide, especially Europe and Japan. The main risk factors for type 2 diabetes in children are the fact that they are overweight, physically inactive, and have a family history of the disease. Obesity contributes to both insulin resistance and cardiovascular problems.

In the United States, the percentage of children aged 6-17 years who were overweight was 5.7% in 1976-1980, rising to 16.5% in 2005-2006. Among adolescents in the United States, 29% have type 2. The National Institute of Health recommends eliminating sweetened beverages, such as soft drinks, entirely, and limiting television and video game viewing to no more than one hour per day. The *China Daily* reported on 15 November 2007 a survey of 17,311 children aged 8-18 by the Beijing Children's Hospital, which found that more than 1 in 5 children were classified as obese and over 2% suffered from type 2 diabetes (the numbers are rising sharply every year).

Malignant tumours have also become a major killer of children in China. Professor Pang Da, president of Heilongjiang's Anti-Cancer Association, said on 3 February 2009 that the current child obesity problem in China will lead to a large-scale outbreak of cancer in the future. Sun Aiming, former vice chairman of China's Red Cross Society, added that air pollution also contributed greatly to the high rate of blood cancer in children, as carcinogenic substances in the atmosphere have a greater impact on children than on adults. Also, the formaldehyde used in home decoration materials can cause problems in children's hematopoietic system and can even cause leukaemia.

In developed countries, cancer is the second most common cause of death among children, surpassed only by accidents. More than 16 out of every 100,000 children and teens in the United States were diagnosed with cancer, and nearly 3 out of 100,000 died. The most common cancers in children are leukaemia, brain, and central nervous system cancers. There are approximately 90,000 childhood cancer deaths annually around the world, representing an important global public health problem.

The Union for International Cancer Control has highlighted the UN's call for action to protect children against the marketing of unhealthy food, especially television advertising that influences their food preferences. A lot of these are non-core food products, which are low in nutritional value and high in saturated fats, trans-fatty acids, free sugars, or salt. As adults and parents, why do we choose to harm our children with the food we buy for them?

Schooling and Activities

> Too often we give our children answers to remember
> rather than problems to solve.
> —*Roger Lewin, British prize-winning science writer*

School Stress and Violence

A 2007 American Academy of Pediatrics report highlighted that too much work and too little play is manifesting a generation of students with increased signs of depression, anxiety, perfectionism, and stress. A great deal of the pressure stems from the college admissions race. In the United States, in 2007, 11.9% of females aged 12-17 had at least one major depressive episode compared with 4.6% of males aged 12-17. Around 40% of them received treatment for depression.

School violence has become a serious problem in recent decades, especially when guns or knives are involved. A survey by the Japanese Ministry of Education found that students in public schools were involved in a record 52,756 cases of violence in 2007, up 8,000 from the previous year. In 2007, in the UK, the police were called to deal with violence in schools in more than 7,000 recorded cases. A US nationwide survey in 2007 among high school students found that 5.9% carried a weapon (gun or knife) on school property; this percentage was three times higher among males than females. Over the previous year, 7.8% of students reported having been threatened or injured with a weapon on school property. The Virginia Tech shooting on 16 April 2007, when a student killed 32 people and then himself, and the Columbine High School shooting rampage by two students, killing 13 people and wounding 23 others before killing themselves on 20 April 1999 attest to the horror of violence in schools.

Suicides

Suicide is one of the leading causes of death among teenagers and young adults. It indicates severe emotional distress, unhappiness, and/or mental illness. At least 100,000 adolescents die by suicide every year. It is estimated that there are 10 to 20 times as many suicide attempts as suicide deaths. Risk factors include suicidal thoughts, psychiatric disorders (such as depression, impulsive aggressive behaviour, bipolar and anxiety disorders), drug and/or alcohol abuse, previous suicide attempts, and access to firearms.

Japan has one of the world's highest suicide rates, compounded by the cultural attitude that suicide is not criminal and can even be considered a morally responsible action. In Taiwan, a survey of 4,475 students aged 15-22, revealed that about 60% have considered suicide, with 23% still considering it. A total of 34% had no idea what to do in life besides getting into good universities. The survey director, Ms. Huang Ching-Hsuan, was reported as saying in *The Straits Times* on 19 November 2009 that Taiwan's youth lacked public role models and faced weakened family support, with fewer children per household, while both parents worked. She added, "Children are often sheltered to a point where they can't handle any setbacks, even the death of a pet."

Sexual Experiences and Pregnancies

The Taiwan Fund for Children and Families—a child and family welfare organisation—reported a four-fold increase in the number of juvenile (under 18 years) sex offenders over the past 5 years, rising from 153 in 2005 to 751 in 2008 and to 730 in the first 9 months of 2009. The number of sex offenders who were younger than 12, as well as female minor offenders, had both doubled in the past 5 years. Most of the troubled youngsters come from broken or dysfunctional families, displaying more violent, aggressive, and impulsive behaviour

According to the Institute for Population and Social Research at Mahidol University, Thailand, as reported on 26 October 2008 in the *Bangkok Post*, the age of rape and sexual assault victims ranged from as young as 7 months to as old as 105, while the perpetrators can be as young as 7. Over the past 5 years, 68% of the victims have been under 18 years old. The Pavena Foundation for Women and Children assisted 33,719 victims of sexual violence in Thailand between 2000 and 2007.

Vaginal, anal, and oral intercourse place young people at risk of HIV infections, sexually transmitted diseases (STDs), pregnancies, and (unsafe)

abortions. The WHO 2006 report *Promoting and Safeguarding the Sexual and Reproductive Health of Adolescents* found that:

- worldwide, 10 million young people live with HIV/AIDS, many in Africa and Asia; around 2.5 million new HIV infections occur every year among people aged 15–24, with a higher rate among young females than males;
- over 100 million cases of curable STDs are contracted each year by people under 25;
- pregnancy related problems are a leading cause of death for females aged 15-19, and these include complications from unsafe abortions.

Children in Poverty

On the other side of the world, in less developed and the least developed regions, children are struggling a great deal. Over 200 million young people, or 18% of all youth, live on less than $1 a day, and 515 million on less than $2 a day.

UNICEF Child Soldiers Global Report 2008 estimated that 300,000 child soldiers under 18 years are fighting in conflicts around the world, participating as armed combatants, spies, messengers, minesweepers, sex slaves, and suicide bombers. Many are abducted or recruited by force, while others join to escape poverty or to avenge violence against family members.

On 20 November 2009, the global community celebrated the 20th anniversary of the adoption by the United Nations General Assembly of the Convention on the Rights of the Child. The Convention offers a vision of a world fit for children. Some of the findings are as follows:

- globally, the annual number of deaths of children under 5 has dropped from 12.5 million in 1990 to 8.8 million in 2008; 4 million new-borns worldwide are dying in the first month of life; every year, 4 million deaths of children under 5 are due to diarrhoea, malaria, or pneumonia;
- vaccines save millions of lives and have helped reduce global measles deaths by 74% since 2000; however, 2 million infants are not protected from diseases by routine immunisation;
- HIV treatment for children under 15 has risen dramatically, most significantly in sub-Saharan Africa; unfortunately, 2 million children under 15 worldwide are living with HIV; 15 million children have lost one or both parents to AIDS;

- the number of children not enrolled in primary school declined from 115 million in 2002 to 101 million in 2007; more girls are missing schooling than boys;
- 1 billion children are deprived of one or more services essential to survival and development;
- 500 million to 1.5 billion children have been affected by violence; 18 million children are living with the effects of displacement;
- 150 million children aged 5-14 are engaged in child labour; from 2000 onwards, 1.2 million children were trafficked every year;
- more than 64 million women aged 20-24 in developing regions reported being married before age 18; 14 million young women give birth between 15-19 years of age.

YOUTH'S NEW TRENDS AND CONNECTIONS

The Youth Global Consultation 2005–2007 by the International Federation of Red Cross and Red Crescent Societies found that, contrary to popular stereotypes, youths in general do not regard themselves as alienated, rebellious, or antagonistic toward their families and adults. Despite the changing structure, the family remains the first social institution where generations meet and interact. The majority expressed positive views about themselves and their life situations. A minority had serious problems, such as substance abuse, teenage pregnancy, gangs, and crime.

Youths were concerned about the sustainable future and needed to increase their involvement in decision-making processes that related to the environment. Globalisation has led to deep changes in youth culture, consumerism, and in different manifestations of global youth citizenship and activism.

In the midst of alarming statistics about the challenges facing our children, there are also some shining examples of caring and resourceful youths.

1) Severn Cullis-Suzuki, Canadian, born in 1979

At the age of 9 she founded the Environmental Children's Organization, a group of children dedicated to learning and teaching other youngsters about environmental issues. When she was 12, they raised money to attend the United Nations Conference on Environment and Development, in June 1992, travelling 5,000 miles to Rio de Janeiro. Severn presented environmental issues from a youth perspective and was applauded by the delegates. This video, popularly known as "The Girl

Who Silenced the World for 5 Minutes" can be viewed at this time at http://www.youtube.com/watch?v=uZsDliXzyAY&feature=related.

In 1993, Severn received the UN Environment Programme's Global 500 Award. Since then, she has spoken worldwide about the necessity of defining our values, acting with the future in mind, and about individual responsibility. She is passionate about social and ecological issues, climate change and intergenerational injustice. She has appeared in many programmes and has hosted *Suzuki's Nature Quest* on the Discovery Channel. She has published two children's books: *Tell the World* and *The Day You Will Change the World.* She promotes a culture of diversity, sustainability, and joy.

2) William Kamkwamba, Malawian, born in 1987

William was forced to quit school at age 14, when his family could no longer afford the $80 a year fee. The teenager dreamt of bringing electricity and running water to his village, Masitala. He kept up his education by using a local library. When he saw a windmill in a tattered textbook, he thought he could build one to make electricity and pump water, as a defence against hunger. In 2002, following one of Malawi's worst droughts, his family was on the brink of starvation, when he finally got his 12-watt prototype, made from spare parts and scrap, to work. It began pumping power into his family's mud brick compound. He has since built a solar-powered water pump that supplies the first drinking water in his village, and two other windmills. His story *Moving Windmills* can be viewed at http://www.youtube.com/watch?v=arD374MFk4w.

He intends to bring power to all Malawians, only 2% of whom have electricity. He plans to manufacture low-cost, distributed, clean energy and clean water systems, including wind and solar power, to serve the rural poor. His story is captured by reporter Bryan Mealer in the 2009 book *The Boy Who Harnessed the Wind*, chosen by *Amazon.com* as one of the Top 10 Best Books of the Year in November 2009, and selected for the Publishers Weekly list of best books. Mealer describes William as representing Africa's new "cheetah generation:" technology-hungry, energetic young people who are taking control of their own destiny.

3) Babar Ali, Indian, born in 1993

At age 16, Babar Ali is the world's youngest headmaster. He lives in Murshidabad, in West Bengal, and is the first member of his family to get a proper education. Rising early to travel over 10 kilometres to school, with his parents paying $40 per year for it, he arrives back home at 4 p.m. to start school in his home backyard! His afternoon school has 800 village students, all from poor families, who come after finishing their day's work. Books and food are given free, funded by donations. Ali started his school when he was only 9 years old, and he now has 10 volunteer teachers (including himself), all school or college students. His school has been recognised by local authorities for helping to increase literacy, and Ali has been given awards for his outstanding work. Helping others learn amidst abject poverty, sharing the knowledge he learns in school—this is his vision of how to help his country build a better future. His story, available at http://news.bbc.co.uk/2/hi/south_asia/8299780.stm, was reported and his school videoed in BBC's *Hunger to Learn across the World* series on 12 October 2009.

In the book *Grown up Digital*, author Don Tapscott surveyed more than 11,000 young people in a $4 million private research study. He found a bright community with revolutionary new ways of thinking, interacting, working, and socialising. Instead of passively watching TV, they actively participate in the distribution of information and entertainment. Instead of consuming content on the Web, they constantly create or change online content. Instead of just viewing videos on YouTube, they post videos of themselves, levelling the playing field for talent and fame. They instinctively turn to the web to communicate, to learn, or find information, to understand, to shop for goods, tickets, or experiences, to get the news, watch movies, download music, set up a business, and so on. Tapscott found that young people are more tolerant of racial diversity, work collaboratively, are innovative, and have global reach and social activism. The bottom line is that the Net Generation is the future and we need to adapt to it—today.

PAUSE: CHILDREN ARE OUR FUTURE

Before you were conceived I wanted you.
Before you were born I loved you.
Before you were here an hour I would die for you.
This is the miracle of life.

—*Maureen Hawkins, poet*

Looking at the main and oldest religions in the world, they all point to children being blessings or gifts from God or the divine. Parents are guided to nurture their children well, to prepare them for the future. Christianity, the largest group with around 2.2 billion people, accounts for 1 out of 3 persons in the world. The Hebrew Bible, also called the Old Testament, declared the blessing of having children, "Behold! The heritage of God is children; the fruit of the womb is a reward" (Psalms 127:3). It tells the father that his children will be like the "shoots of olive trees surrounding

[the] table" (Psalm 128:3). Tender saplings would not grow up into fruit-bearing trees without careful cultivation, without being given the right nourishment, soil, and moisture. Likewise, successful child rearing requires work and care. Children need a healthy environment to grow to maturity, physically, emotionally, and spiritually. In the New Testament parents are further warned, "You fathers, do not embitter your children, or they will become discouraged" (Colossians 3:21). Parents are asked to look for the good in their children, even though it is easy to find fault, to reduce the risk of their growing up downhearted, convinced that nothing they do will ever be good enough.

Islam is the next largest group, with 1.6 billion people, or almost 1 in 4 persons in the world. According to the Quranic text, progeny is a gift from the Almighty Allah to His faithful servants. In Islam, fertility is highly prized and children are a gift of God to bring "joy to our eyes" (Surah 25, Al-Furqān: 74). If God intends for a child to be born, she/he will be born. It is the duty of the parents to bring up children who are the blessing of family life, in a manner beneficial to their life on earth and life after death. Parents should take care of the material and spiritual needs of their children, love and respect them, encourage them to do good, and prevent them from indulging in evil deeds. Islam permits minor punishments to correct wrong deeds without causing injury to their pride.

Hinduism accounts for around 1 billion people, almost 1 in 5 persons in the world. Hindus love their children dearly, believing that they are gifts from gods and products of their previous karma. Many presume that their children were related to them in their past lives or were their close friends. Hindus are very possessive about their children and spend a great deal of time and energy bringing them up. They are sentimentally and emotionally attached to them and experience a great warmth and intimacy in their relationships. It is the obligatory religious duty of sons to look after their aged parents and provide them with decent means of living. If a child strays and brings a bad name to the family, parents blame it upon themselves and their previous karma.

Buddhism accounts for almost half a billion people, or about 1 in 10 persons in the world. While it is not laid down anywhere that Buddhists must produce children or regulate the number of children that they produce, they think that the basis of all human society is the intricate relationship between parent and child. Parents should care for the child as the earth cares for all its plants and creatures. It is the duty of the parent to guide the child on the proper path. Showering parental love, care, and attention does not mean pandering to all the demands of the child. Too much pampering would spoil them.

According to Dr. K. Sri Dhammananda in *A Happy Married Life—A Buddhist View*, present-day parents are in a rush for material advancement and spend a great deal of time working. Children are left to care-givers or paid servants, and are denied tender, motherly love and care. Mothers, feeling guilty about their lack of attention, try to placate children by giving in to all sorts of demands from them, thus spoiling them. Parents who have no time for their children should not complain when these same children have no time for them when they are old.

Professor Juliet Schor of Harvard University, author of *The Overworked American*, emphasised that the culture of long hours has spread like cancer in the Westernised world, leading to more divorces, delayed marriages, and child bearing, dysfunctional parent-child relationships, delinquent children, and other social problems.

It is not what the parents profess, but what they really are and do that the child absorbs unconsciously. As the child becomes an adolescent, parents should practise *karuna* or compassion toward them. Adolescence is a very difficult time for children. Having loving kindness and compassion, and not hate, directed at them, they will become better people and radiate love and compassion toward others.

Judaism, one of the older religions, accounts for around 15 million people, less than 1 in 100 persons in the world. Among the Jewish people, children are viewed as a Divine trust and guarantors of the future. Judaism teaches that all human beings are created in the image of God, human life being therefore sacrosanct. The sacredness of human life is applied to infants as soon as they are born. Jewish law specifies the rights of children, which are the primary obligation of the natural parents, but which, in the latter's absence, incapacity or failure, become the responsibility of the community. The abuse of children is prohibited even to parents and teachers with good intentions. Judaism recognises that the well-being of a society as a whole is determined by how its children are treated. In the Talmud, there is a story of a man who planted a carob tree, which is known to bear fruit only after 70 years. When asked whether he thought he would live to eat from the tree, the man replied, "I am doing as my ancestors did. Just as they planted a carob tree for their children, I am planting for my children" (Babylonian Talmud, Ta'anit, 23).

Have we been treating children as blessings? Have we nurtured them well? What kind of world are we leaving to them?

Our Values and Theirs

Fundamental value differences exist between those of different generations, their mindsets, and how they see the world based on their experiences.

Dr. Morris Massey, an organisation-development expert, worked on generational values and said that our behaviours are driven by our value system and value programming. In these very terms, Massey presents generalizations defining values schematically by generation, giving a number of overall contrasts in the psychology of individuals as groups based on birth years, which we may use in viewing the different age groups in today's world (at least in western-influenced countries):

- *The Silent Generation*, sometimes called "Traditionalists"; they are the children of the Great Depression, born in the early 1920s-early 1940s; they were too young to join the service during the Second World War. They tend to be grave, fatalistic, conventional, cautious, withdrawn, and expecting disappointment, but desiring faith. Women desire both a career and a family. They grew up with financial and global insecurity and were poised to take advantage of economic opportunities, harnessing scientific and technological advances in the second half of the 20th century. Many are hardworking people who focus on getting things done and advancing their careers.
- The *Baby Boom Generation*, born between the Second World War and the late 1950s—a time of marked increases in birth rates; they are associated with privilege, growing up in a time of affluence. They genuinely expected the world to improve with time and had an assumption of lifelong prosperity and entitlement. In 1993, *Time* magazine reported that about 42% of baby boomers were dropouts from formal religion, a third had never strayed from church, and a quarter were returning to religious practice. Growing up, they had personal transistor radios, allowing them to listen to rock and roll, The Beatles or The Motown Sound. In their teenage and college years, they were characteristically part of the counter-culture of the 1960s, organising anti-war protests, doing social and drug experiments, enjoying sexual freedom, and participating in civil rights movements. They were committed to values such as gender- and racial-equality and environmental stewardship. They are free spirited, social-cause oriented, and individualistic. Commonly, due to denial of aging and a failure of long term planning, they

are deemed to be leaving an undue economic burden on their children for their retirement and care.

- *Generation X* or *baby bust* generation, born during declining birth rates times, from the 1960s to the 1970s; they were brought up watching TV, had home computers and videogames, were not taught to believe in God as much, didn't respect parents or authority, had pre-marital sex, were pragmatic and perceptive, savvy but amoral, more focused on money than on art. They are associated with dot-com businesses, early MTV, grunge, hip hop, and punk and rock bands. While holding the highest educational levels by age group, a US study, *Economic Mobility: Is the American Dream Alive and Well?*, by the Census/BLS released on 25 May 2007, found that in real dollars, this generation's men (in their 30s in 2004) made 12% less than their fathers did at the same age in 1974. A survey reported by Anne Fisher in the article "What Do Gen Xers Want?" in *Fortune* magazine on 20 January 2006, shows that the top three things they want in a job are: positive relationships with colleagues, interesting work, and continuous opportunities for learning. They are not looking for power, prestige, or recognition, and they are not as preoccupied with salary as Baby Boomers. They are not big on loyalty for loyalty's sake.
- *Generation Y*, *Generation Next*, *Millennials*, *Net Generation*, *N-Gen*, *Internet Generation*, or *iGeneration*, born between the late 1970s and the mid-1990s; they have had easy communication through technology and are more peer-oriented. Self-expression and social acceptance are important and they are achieved through blogs, social networking sites, online gaming, virtual worlds, and online communities. They are culturally liberal, more civic-minded, and reject the counterculture of the Baby Boomers and Generation X. They have a trend of living with their parents for longer, are frequently in touch with their parents and speak with them about a wide range of topics. With the recession in the late 2000s, many face youth unemployment. Those in employment seek feedback, involvement in decision-making, and desire to shape their jobs to fit their lives, rather than adapt their lives to the workplace.
- *Generation Z*, born between the mid-1990s and the end of the 2000s; they are highly connected and technologically advanced, earning the nickname of "digital natives."

With these schematic differences in mind, we need to recognise that our children are quite distinct from us. They are much more advanced in terms of being global citizens, with networks of instant communication through the likes of Facebook and Twitter. If we nurtured them properly, we would realise that they are more able to rally support from people around the world and can be great instruments of change.

What are Children Mirroring Back to Us?

If you must hold yourself up to your children
as an object lesson, hold yourself up
as a warning and not as an example.
—George Bernard Shaw, playwright

Grown-ups have collectively created the world children are living in now. They are also collectively creating the future generations. What do we see around us?

- Divorced, single-parent, unmarried, and step-parent families; emotional instability and child abuse.
- Working parents, latch-key children, day care, lack of parental contact and concern, or over-parenting and disempowerment.
- Fast food, microwaved and instant meals, food additives, soda drinks; obesity, cancer, diabetes, HIV, allergies, asthma, learning disorders, and a host of other ailments.
- Computers, Internet, virtual worlds, telecommunications, gaming (addiction), handheld gadgets and objects of instant gratification, online predators and cyber bullying.
- School stress, extra-curricular activities, school violence, school drop outs, substance abuse, crime, suicides.
- Early sexual activity, multiple partners, STDs, teen pregnancies, abortions.

What do we see in poorer parts of the world?

- High infant mortality, young mothers, high birth rates.
- Lower accessibility to food, water, sanitation, healthcare, education; child labour.
- War, violence; child soldiers.
- Abject poverty and starvation.

What is the message here for us? What are children and their circumstances saying about us?

Charity Begins At Home

The issues many of us have today about relationships can be turned around when we make our family and children the centre of our lives. Our sense of who we are and what we value is formed in our childhood—it shapes how we treat and relate to our family, relatives, and friends. The way in which we are brought up at home moulds our values about success and happiness and therefore influences our choices and priorities later in life.

Our relationships with our children can be improved even though we need to work hard to ensure a good living and even though we can't afford to be constantly by their side. Our efforts need to be directed at finding a balance between a taxing career and having enough time to care for and educate our little ones. When we lose sight of our children's activities, friendships, and behaviours, they may fall into bad company and/or get exposed to negative habits and addictions.

It is important that we encourage children to make their own decisions, to take their own actions, and that we let them learn from their own mistakes; it is not in their best interest that we control most aspects of their lives and do everything for them. By doing so, we only deprive them of the strength they will need when they grow up. As we can see in our society today, even or especially teenagers are highly stressed when they have to do things on their own.

Two other very important aspects we must take into account when expecting or raising children are our diet and lifestyle habits. Without being aware, we can pass on our ill-health habits to them at very early ages, or we may even pass them on through our genes before conception. Diet may just be the easiest thing we can control and the way we live our lives can progressively be changed when we pay attention to what is and isn't good for our children's development. For example, over-consumption of things may very easily create the belief that having material things will make us happy. Like us, our children start to compare themselves with their friends and feel dissatisfied with their lot in life. A constant desire to excel, to be better, and have more than others, is putting a lot of pressure on our children when it comes to performing well in school, getting into a good institution, achieving accolades.

As we have seen up to this point, if we want to make sure that future generations live in a world where basic needs are met for everyone, it is time for us to reconsider how we are living our lives and find out what we can do differently.

Green Parenting

The concept of green parenting is spreading rapidly among parents who want to instill in their children values that enable life-supportive actions. It is about raising children with a conscience toward conservation, sustainability, and the environment.

One of the recommendations of experts in green parenting is to keep children in direct contact with all that nature has to offer, as much as possible; this is one of the best ways to teach them to be responsible stewards of the planet. Today, children can't wander around forests and meadows the way they did a generation or two ago; it may require advance planning, but they should be given a chance to hike on a trail, skip stones in a creek, or hunt for bugs and worms. It's cheap (or free) and it doesn't require travelling great distances.

Green parents understand that children's perception of reality is based on what they are in contact with every day. And by spending a great deal of time in shopping malls, children develop values and beliefs in line with money, possessions, and outward beauty. By exposing them to nature while they are young, children will understand ecology and its importance and will live their lives in ways that do not deplete the planet. They will not want to destroy their playground, and with proper education, they will see the importance of safeguarding the playground of the planet.

Children's daily activities determine the way they define success, needs, and wants, and eventually their mindset and worldviews, so we need to make sure that what they do every day is aligned to what we want them to value when they grow up.

A seemingly insignificant action that green parenting experts focus on which has a great impact on the environment is the way we shop. As we buy a lot of groceries, we can make our trips to the supermarket greener by bringing reusable bags; and one way to make this simple action stick with our children is to put them in charge of decorating the bags; thus, they will be happy to remind us to bring these along and we will make a difference by lessening the amount of plastic we use.

There are many more things that green parents can do with their children, but given the richness of the subject, we are not going to further expand on it here; ideas and tips are available online, in magazines, and entire books dedicated to this topic.

Better Parent-Child Relationships

Leaders in child psychology today provide empirically tested insights for managing children's behaviour and bettering parent-child relationships. One of their recommendations is for parents to make one-on-one time for their children; to avoid doing this only after life's obligations are met—

housework, paying bills, etc.—, parents are advised to spend at least one hour a week paying attention to and expressing positive thoughts and feelings toward each child. During this time, they are to avoid teaching, inquiring, sharing alternative perspectives, or offering corrections.

Dr. David Palmiter, child psychologist, says many families he has recommended the strategy to over the years have told him that adding an hour of special time in addition to the quality time they spend with their children—such as attending various events together—has significantly improved the parent-child relationship. Furthermore, a study published in January 2013 in the *Proceedings of the National Academy of Sciences* shows that, particularly among younger children, a parent's demonstration of love, shown through nurturing behaviour and expressions of support can improve a child's brain development and lead to a significantly larger hippocampus, a brain component that plays an essential role in cognition.

Another recommendation from experts is for parents to embrace labelled praise—specific feedback that tells the child exactly what he or she did that the parent liked. Simply put, giving attention to undesired behaviour increases undesired behaviour, while giving attention to good behaviour increases good behaviour, says Alan Kazdin, Ph.D., a Yale University psychology professor.

A critical issue in our modern society is stress; and what adds on to the gravity of the issue is that it doesn't only negatively affect us, but our children as well. According to a 2010 Stress in America survey, most children reported feeling sad, worried, or frustrated about their parents' stress. Another study published in the *Child Development* journal of The Society for Research in Child Development found that parents' stress imprints on children's genes and the effects last a very long time. Psychologists say that this is why modeling good stress management can make a very positive difference in children's behaviour, as well as how they themselves cope with stress.

On this matter, psychologist David Palmiter, Ph.D., recommends that parents make time for their own well-being as well, through exercise, hobbies, maintaining their friendships, and connecting with their partners. That could imply committing to spending regular time at the gym or making date night a priority. When parents lead a balanced, stress-free life, children can experience the same.

Conscious Parenting

A relatively modern form of parenting that a growing number of parents are adopting nowadays is called conscious parenting. It emphasises growth, truth, authenticity, and presence, and its advocates believe in providing children with a physically, emotionally, mentally, and spiritually safe environment with lots of love and connection. Every parent can practice these skills and achieve various levels of mastery; however, the genuine intention toward the journey is the most important aspect.

According to Erika McDaniel, clinical nutritionist, writer, and speaker, being a conscious parent implies, among other things:

- taking pause before acting, no matter how little or how big the action, in order to be thoughtful with one's choices;
- understanding that one's actions have repercussions beyond the immediate moment and oneself, and consider those when making decisions for oneself and one's family;
- having the courage to speak from the heart, and follow one's intuition;
- focusing one's energies on what one can do to help oneself, rather than waiting for others to do it;
- modeling our values to our children and our communities—one does one's best by walking the walk instead of just talking the talk.

Conscious parenting and green parenting are both life-supportive: by adopting the actions and strategies they imply, today's parents have a good chance not only to have better relationships with their children and see them grow up harmoniously, but also to teach them the importance of being eco-literate in today's rapidly evolving planet.

Children to Pay for Our Carbon-Happy Generations

According to Lord David Puttnam, ambassador for UNICEF UK, as quoted on the BBC on 23 November 2009, climate change poses a huge barrier to a fulfilling future for the generations ahead. With three or four carbon-happy generations before them, he asks what price children will have to pay. The Intergovernmental Panel on Climate Change (IPCC) stresses the grave consequences of continued high CO_2 emissions, including increased child poverty, inequity, and death (up to 160,000 child deaths annually in sub-Saharan Africa and south Asia are anticipated).

How long more will we carry on cutting down forests, eroding the topsoil, growing cash crops and crops to feed livestock, using land, water, and resources to farm livestock, wasting precious water, burning fossil fuels, over-fishing, wasting energy, over-consuming resources for industrialisation, emitting CO_2, polluting the land, air, waterways, and the sea, all in the name of economic success? Aren't we learning anything from the recent years' melting ice caps, floods, typhoons, tsunamis, droughts, and forest fires?

Climate change is not just an environmental issue; it is a human rights issue. It is about the way our children are going to live in the future. Will the world be green, murky, or grey? What have we taught our children by the way we live? Have we prepared them well? Are they going to give life to or deplete life from the planet? The answer lies in our actions—today!

CHAPTER 11
WHAT'S HAPPENING TO OUR LIVES?

A PERFECT LIFE?

For many people, a perfect life would involve having a well-paying job that they enjoy doing, with work hours that they like. They would be surrounded by meaningful, fun, and loving relationships with their spouse, colleagues, friends, and neighbours. Their children would be intelligent and sensible; they would just know how to make full use of their time and plan ahead. They would be in good health with all the energy to do whatever they want. Their home would be filled with stimulating things to do and they would travel twice a year to new exotic places. At the end of each day, they would sleep well, knowing three years of living expenses are stashed in their banks and their cosy home is secure in a peaceful and clean neighbourhood. This would be a typical city dweller's dream.

The perfect life for a poor sub-Saharan African, Afghan, Nepalese, Cambodian, Bhutanese, and Bangladeshi would be very different. According to the World Bank, in 2008, the poorest of the poor representing 1.4 billion people, or one-fifth of the world's population, was living on less than $1.25 a day, and 3.14 billion people, or almost half of the world's population—on less than $2.60 a day. Finding one wholesome meal a day and having access to safe drinking water is still largely a life and death battle for two billion people every day. In another analysis done by the World Bank in 2005, the Gross Domestic Product of 41 heavily indebted poor countries (567 million people) was less than the wealth of the world's seven richest people! With more than 80% of the world's population living in countries where income differentials are widening, there are volatile levels of anguish and discontentment.

The poorest of the poor need help. They need education to create products/services to generate trade and build infrastructure, access to clean water and food, basic housing, sanitation, and healthcare. Providing long-term financial aid is not sustainable; they need strong and dedicated administrators and systems implementers, and this is going to take time.

From a global resource standpoint, the focus, therefore, should be on the world's wealthiest nations, approximately 20% of the world's population (1.3 billion), who are consuming 77% of the planet's resources; the remaining 23% are left to be shared by 5.5 billion people. The wealthy are primarily living the affluent modern city life at great expense to others.

Earn More Money, Work Harder

In Chapter 9, core issues and challenges surrounding work and money have been discussed extensively.

In a nutshell, more and more people are living in urban areas. Environmental pollution, over-crowding, soaring prices, increased competition, high consumption lifestyles, stress, reliance on pharmaceutical drugs, living in the virtual world, and a fast-paced changing world are some of the debilitating effects of modernisation and globalisation.

The financial crisis of 2007-2009 has resulted in trillions of US dollars being wiped out globally. Banks, corporations, and even countries (such as Iceland) went bankrupt overnight. Other countries, like Mexico, Japan, Germany, and the UK experienced sharp declines in their GDP, ranging between 8-22%. Worldwide, governments had to bail out key financial institutions. The world has been over-spending on future probable earnings that are built upon weak banking foundations. This, coupled with short-term, focused, unsound risk-management practices, means the world is poised for a total financial collapse.

On 25 February 2009, Reuters reported that suicides were on a sharp rise in Asia. With the economic crisis, many people lost their jobs and their savings to the stock market. According to mental health professionals, work is very important to Asians, as they don't have good social security, and losing their jobs is associated with losing face. Among developed Asian countries, South Korea, Japan, and Hong Kong ranked the highest in suicides—mainly committed by males.

No Time for Health

Our hectic lifestyle is reflected in the condition of our health and without it we cannot enjoy what we have earned. The fast-food way of living has created not only health problems, but also a myriad of social, environmental, and climatological issues.

Thoroughly discussed in Chapter 6, the concept of getting quick and tasty food without looking at its nutritional value and how it was produced or cooked has created a global epidemic of health issues. With a more-is-good kind of mindset, millions of people around the world are working themselves to death so that they can consume and have more in life; the stress they deal with at work creates a wide range of lifestyle diseases and ailments.

Relationship Trends

Cohabitation

Studies carried out in the United States and Europe indicate that a high percentage of people are cohabitating; 30-50% of children are born to unmarried couples. Denmark, Norway, and Sweden have the highest incidence of unmarried couples living together (50%), followed by the United States and the UK (37-42%), and then Australia and New Zealand (18-22%). Though Asians still view cohabitation as a taboo or religious violation, this practice is growing and spreading steadily in Japan, India, China, the Philippines, and other parts of Asia. In short, more people are saying "yes" to sharing homes and "no" to marriage. Though there may be many factors to consider, the three most common reasons cited were: (i) wanting to share the cost of living, (ii) wanting to prepare for married life before actually getting married, and (iii) wanting to be in a close relationship without marital commitments.

In spite of this, research has shown that couples who cohabitate prior to marriage have less stable marriages and are more likely to divorce than couples who do not cohabitate prior to marriage. Children born to cohabitating parents are more likely to experience instability in their living arrangement, which undermines their development. The family structure that helps children the most is a family headed by two biological parents in a low-conflict marriage. Children born in cohabitating relationships or single-parent families, and children in stepfamilies, face higher risks of poor academic and behavioural outcomes than do children in intact families headed by two biological parents.

Singlehood

People now have the tendency to marry later and divorce more frequently than before, leading to growing numbers of single adults. The feminist movement has given women more freedom to make choices in their love lives, including the choice to remain single. More women in their late thirties to early forties have no children. According to the UN statistics and indicators on women and men (2000), informal unions are common and more people live alone in developed regions. Common reasons for the rise in singlehood are that people put a high value on independence, have no compelling desire to have children, and are afraid that marriage will lower their health and attractiveness; many fear divorce.

Divorces

Divorce rates around the world have increased dramatically after the Second World War. Studies by the Americans for Divorce Reform in 2002 revealed that the total number of marriages that ended in divorce in

Sweden, the United States, Belarus, and Finland accounted for 50-55% of all marriages. In most parts of Europe and Russia, rates were ranging from 37-47%, while in Singapore, Malaysia, Japan, and Israel—from 10-27%. Divorced families put more pressure on the economy, as they create a need for more homes and electricity. With divorce rates continuing to climb globally, affected children who later become adults are more pessimistic about marriage, and this increases the trend in people wanting to cohabitate, choosing to be single, limiting their emotional investment and, therefore, "keeping their options open."

INFOTAINMENT

When it comes to 21st-century information-management and entertainment, nothing is more pervasive than the Internet. It started with four American universities in 1969 that had the intention to share their knowledge. The ARPAnet network was opened to the public in 1983, which led to the creation of the World Wide Web by the Particle Physics Laboratory (CERN) in Switzerland in the early 1990s. From then on, it grew to a global network of 1.6 billion users (rapidly rising annually) and has revolutionised the world.

Pornography

The total pornography revenue of China, South Korea, Japan, and the United States is $86.4 billion. Global statistics show that every second, $3,076 are spent on pornography and 28,258 Internet users are watching pornography. Research revealed many systemic effects that are undermining an already-vulnerable culture of marriage and family, indicating that Internet pornography decreases marital intimacy, increases infidelity, devalues monogamy, marriage, and child rearing and forms addictive sexual behaviours.

The impact on adolescents includes the formation of a belief that superior sexual satisfaction is attainable without having affection for one's partner (thereby reinforcing the commoditisation of sex). There is an increased risk of exposure to unhealthy information about human sexuality long before a minor is able to contextualise it, and of overestimation of the prevalence of less common practices such as orgies, gangbangs, bestiality, or sadomasochistic activity.

Online Gaming

According to a US market research report by DFC Intelligence, the top 17 online game development companies made $28.5 billion in 2005. Susan Greenfield of the Royal Institution of UK said that consistent computer use could infantilise the brain, by affecting the way it naturally learns; when a child falls from a tree, he/she will quickly learn not to repeat that mistake, whereas someone who does the same thing in a computer game will just keep playing. Other negative effects include attention problems and stifled empathy and imagination—which correlate with the working of the pre-frontal cortex of the human brain. Scientific studies show that when the pre-frontal cortex is damaged or under-developed, it makes a person less active, willing to take more risks and become more reckless. Exposure to violent video games is linked with the increased teenage aggressive behaviour—shooting in schools and public places, confrontational behaviours with parents and teachers—and the decline in school achievements that we are witnessing today.

Human Evolution Affected by Internet

The Internet has evolved so much that one can say almost everything is hooked on to it; figuratively speaking, there is even lesser need for people to get up from their chairs due to the exponential growth of products and services available online. With the greatest success of the 20th century comes the price of its impact on human development and interaction.

Studies done at Stanford and Carnegie Mellon University in 2000 indicated that increased use of the Internet leads to social isolation and depression. People who spend more time on the Internet, spend less time socialising with peers, communicate less with their families, and feel lonelier. Sociologists McPherson, Lynn, and Brashears of Duke University and the University of Arizona found that over the past 20 years, the number of Americans who have no one at all to confide in has more than doubled, reaching 25%.

With the trend of personalised services in almost all areas of business and the all-round interfacing of the Internet into our lives, technology has created a generation of "dis-armed and dis-legged" teenagers and young adults. A significant portion of this cyber-generation is characterised by difficulties in relating with others, unresponsive eye and facial expressions, poor body and limbs coordination, and a high percentage of short-sightedness. The adults of this generation find it hard to cope with day-to-day demands.

In 1999, an internationally acclaimed science fiction action movie, *The Matrix*, described a future in which the reality perceived by humans is actually a Matrix—a computer simulated reality created by sentient machines in order to pacify and subdue the human population, so that their body heat and electricity can be used as an energy source. In reality, the Matrix has arrived. While the planet is dying, hundreds of millions of people are hooked on to the Internet for stimulation, distraction, shopping, making money, and for love and friendship fulfilment.

On the other hand, the Internet has greatly helped individuals and groups from all corners of the world share information, appreciate cultural differences, understand global and national issues, and exchange and implement ideas. People are able to express their thoughts freely and in this diverse interaction, creativity is mushrooming all around the globe. According to Peter Russell, a renowned scientist and author of *From Science to God*, humanity has reached a crossroads on its evolutionary path. The Internet is linking humanity into one global community, one global brain. Combined with an explosion of spiritual awakening around the world, it is creating a collective consciousness that could save humanity from itself.

However, just like everything else, the goodness or evilness of the Internet depends on the users and their intentions. At present, humanity has vast amounts of knowledge, but still very little wisdom. Without developing wisdom, we will head toward catastrophe. It can be said that we are currently facing our final evolutionary exam: are we (the human species) fit to survive? Will we wake up, grow up in time, and combine our tremendous powers for the good of all of us and of the generations to come?

CONSUMERISM

Human beings need to consume resources to survive. However, consumption has evolved as people have ingeniously found ways to help make their lives more comfortable and/or use their resources more efficiently. Over time, as powerful groups of people (governments and business corporations) grow, the desire to control consumption systems naturally comes to pass.

According to Richard Robbins, author of *Global Problems and the Culture of Capitalism*, the consumer revolution of the late 19th century was caused largely by a crisis in production. New technologies resulted in the production of more goods, but there were not enough people to buy them. Since production is such an essential part of the culture of capitalism, society quickly adapted to the crisis by convincing people to buy things, by altering basic institutions, and even generating new ideologies of pleasure.

Advertising became a major industry by imbuing products and services with the power to transform the consumer into a more desirable and valuable person. In 1880, the US spent the equivalent of $30 million on advertising; by 1910 this exploded to $600 million. Today, the figure has increased exponentially to $120 billion.

Increasing Buying Frequency

Victor Lebow, a successful US economist who worked with large corporations in the United States and Europe in the 1950s, said:

> Our enormously productive economy demands that we make consumption our way of life, that we convert the buying and use of goods into rituals, that we seek our spiritual satisfaction, our ego, satisfaction, in consumption. We need things consumed, burned up, worn out, replaced and discarded at an ever-increasing rate.

As it turns out, by looking at history, the later part of the 20th century mirrors Lebow's assertion. Two important strategies were used by businesses to power unparalleled economic growth: planned obsolescence and perceived obsolescence.

Planned obsolescence is an intentional deliberation about how long a product should last at the design stage, before manufacturing. The shorter the lifespan of the product, the more frequently the consumer needs to buy again. It also helps the producer choose the least expensive components to satisfy the product lifetime projections. It instils in the buyer the desire to own something newer and better a little sooner than is necessary.

Perceived obsolescence is about introducing new designs to redirect the consumer's focus away from their existing products and/or individualising the product to differentiate it from the competition. Examples of such product categories include automobiles, with a strict yearly schedule of new models, the almost entirely fashion-driven clothing industry, and the mobile phone industry, with constant minor feature enhancements and re-styling. These changes are designed to make owners of the old model feel out of date. People want to stay current, be in-line with others, and also demonstrate that they have the power to spend. In short, people want to be valued, and the way they show it is through buying and owning things.

Conspicuous Consumption

Coined by Thorstein Veblen, a Norwegian-American sociologist and economist in 1899, *conspicuous consumption* is a term used to describe the lavish spending on goods and services acquired mainly for the purpose of displaying income or wealth. In the mind of a conspicuous consumer, such display serves as a means of attaining or maintaining social status. A BBC documentary aired in September 2001, examined the psychology of shoppers in countries like the UK, the United States, and Japan. Psychologists uncovered the following: (i) consumption defines who we are, (ii) people now essentially "buy" a lifestyle, (iii) brands help turn perceptions into reality, thus encouraging purchases based on fashion and peer/social pressures to fit in, (iv) to deal with social and consumerism pressures, more and more people resort to compensatory consumption—consuming more to feel better.

Another BBC documentary aired in April 2003 looked at the issues of whether increased wealth and consumption led to more content and satisfied individuals. In the documentary, Professor Andrew Oswald of Warwick University explained that as people get wealthier, they tend to compare themselves more with others, which leads to increased anxiety and dissatisfaction. They also compare themselves with more successful or higher status people. Oswald concluded that it is hard to make society happier as people get richer, because human beings constantly make comparisons. In another decade-long research, Professor Juliet Schor stated that people who care about how much they make, about their possessions, about their economic and social status are often depressed and have lower self-esteem.

Ecological Buying

"Many of the things we buy are not ecological; the way they are grown or extracted, produced, transported, distributed, and discarded has environmental and human consequences," said Daniel Goleman, renowned psychologist and bestselling author of *Ecological Intelligence* (2009). In this book, Goleman reveals the hidden environmental consequences of what we make and buy, and how, with that knowledge, we can drive the critical changes necessary to save the planet and ourselves.

We wear organic cotton shirts, but we don't know that the dye used to make them may put factory workers at risk of cancer. We dive into coral reefs, not realising that there is an ingredient in our sunscreen that feeds a virus that kills the reef. Biofuels cut down greenhouse gases and provide income to poor countries; however recent news revealed that 133 workers in a Brazilian ethanol company were suffering from hunger and cold, living in over-crowded rooms and terribly unsanitary conditions. Through his research, Goleman explains why shoppers are clueless about the hidden negative impacts of the goods and services people make and consume.

But the balance of power is about to shift from seller to buyer, as a new generation of technologies informs people of the ecological facts about the products at the point of purchase. One such example is

GoodGuide, an ethical company that provides the world's largest and most reliable source of information on the impact of products used at home. In the near future, a shopper will be able to type the bar code of a product into a text message on their mobile and send it. Within seconds, an image will appear that will rate the product in terms of its environmental, health, and social impact. Goleman calls it *radical transparency*, with customers making well-informed decisions. It will push companies to re-think and reform their businesses, ushering in, he hopes, a new age of competitive advantage.

Macro View on Human Priorities

Taking a macro view of the global spending would help us see the mismanagement of our priorities. The following figures were taken from the United Nations' 1998 Human Development Report, entitled *The State of Human Development*, or, in other words, what we have been spending on:

Global Priority	$ Billions
Cosmetics in the United States	8
Ice cream in Europe	11
Perfumes in Europe and the United States	12
Business entertainment in Japan	35
Cigarettes in Europe	50
Alcoholic drinks in Europe	105
Narcotics in the world	400
Military spending in the world	780

Looking at the above, many of the items we consume or spend on are *wants* as opposed to *needs*. Let us contrast this against another list of true needs (below are estimated costs to achieve universal access to basic social services):

Global Priority	$ Billions
Basic education for all	6
Water sanitation for all	9
Reproductive health for all women	12
Basic health and nutrition	13

If this paradigm continues—not distributing from high-income to low-income consumers, not shifting from polluting to cleaner goods and production methods, not promoting goods that empower poor producers, not shifting priority from conspicuous consumption to meeting true basic needs—our future, our children, our way of life will change dramatically very soon. For billions of people, young and old, this long overdue massive change could be sudden. It can come in the form of violent social upheavals, financial and economic collapses, outbreak of wars, and regional food shortages due to environmental deterioration.

Peaceful Neighbourhood?

In any human society, low crime rates allow citizens to live with peace of mind and go about their daily lives happily and productively. A secure and friendly world would mean a society where people respected authorities, appreciated the elderly, and felt totally safe with their personal belongings. It would also mean a world where people were comfortable in approaching and helping each other and were calm about leaving their doors unlocked, where families and communities were stable and supportive, human lives and children were cherished, and family values upheld. In reality, how many of us live in this kind of world?

The International Crime Victim Survey (ICVS)

In 1989, a group of European Criminologists led by Jan van Dijk, Pat Mayhew, and Martin Killias started an international victimisation study. The United Nations got involved in the project in 1992. To date, more than 82 countries have participated in it. Based on a study in 2003 to determine trends in global crime, there was a steady increase from 2,300 incidents per 100,000 people in 1980 to over 3,000 in the year 2000. All over the world, crime problems have worsened (30% increase) over the past two decades. Though crime rates in North America and the European Union have fallen in recent years, they are still, by far, the highest in the world.

Rape and Sexual Assault

In some countries, approximately one in five women have experienced rape or attempted rape in their lifetime. According to the WHO, the negative impact on women comprises (a) sexually transmitted infections including HIV, unintended pregnancies, and subsequent unsafe abortions and injuries to the reproductive tract, (b) post-traumatic stress disorder, depression, and/or suicide attempts, (c) experiencing social stigma and facing rejection by husbands, families, and communities.

Total Recorded Rapes	Rate per 100,000	
Country	1995	2002
Belgium	12.65	23.57
Canada*	95.34	77.64
Chile	4.73	8.99
Costa Rica	7.02	16.06

Denmark	8.42	13.12
Finland	8.73	10.60
Germany	7.56	10.44
Hungary	4.08	5.89
Italy	1.65	4.41
Japan	1.20	1.85
The Netherlands	9.14	11.16
New Zealand	24.42	26.88
Peru	17.65	22.31
South Africa*	121.44	115.61
Sweden	15.43	24.47
Switzerland	4.27	6.64
The United States*	37.09	32.99

Source: 6th and 8th UN Survey of Crime Trends and Operations of Criminal Justice Systems, 1995 and 2002, total recorded rapes

As one can clearly see from the above, in areas other than Canada, South Africa, and the United States, rape cases are on the rise globally and we can safely say that the actual number of cases is much higher because women who are victims of sexual violence are often reluctant to report the crime to the police, their family, or to others.

Terrorism

In 2009 there was a suicide bombing of the Marriott and Ritz-Carlton hotels in Jakarta, Indonesia. Experts linked the bombings, which killed 9 and wounded 50 people, to Jemaah Islamiyah, an Indonesian Islamist militant group connected to the Al Qaeda terrorist regime. Though in recent years the country has achieved a level of stability not seen throughout most of its turbulent history, the attack has succeeded in showing to the world that Indonesia is still an insecure country prone to violence.

Suicide attacks are associated with the Al Qaeda ideology. They have been adopted by the Tamil Tigers, by militant groups in Iraq, the Taliban and the Jemaah Islamiyah. It is believed that extreme religious education flourishes in Pakistan, where inadequate funding for state education has allowed unregulated religious education to take hold. About 1.5 million children attend such schools, which are open to foreign visitors and a number of terrorists belonging to the Jemaah Islamiyah group have been identified as alumni of these schools.

According to the 2007 report of the US National Counterterrorism Center (NCTC), there were 22,685 deaths caused by 14,449 terrorist attacks around the world. The most terrorist-ridden region in the world is the Middle-East, followed by south Asia and Africa. Below are highlights of some of the major terrorist attacks around the world in the past eight years:

Date	Details	Killed	Injured
Sep 2001	Crashing of hijacked planes into buildings in New York City, the Pentagon, and Pennsylvania	2,993	8,900
Oct 2002	Car bombing outside nightclub in Bali, Indonesia	202	350
Aug 2003	Car bombing outside mosque in Najaf, Iraq	125	500
Mar 2004	Bombing of four trains in Madrid, Spain	191	1,876
Sep 2004	Hostage-taking at school in Beslan, Russia	366	747
Sep 2005	Multiple suicide bombings and shooting attacks in Baghdad, Iraq	182	679
Jul 2006	Multiple bombings on commuter trains in Mumbai, India	200	714
Aug 2007	Multiple bombings in Al-Qataniyah and Al-Adnaniyah, Iraq	520	1,500
Oct 2007	Bombing of motorcade in Karachi, Pakistan	140	540
Nov 2008	Multiple shooting and grenade attacks in Mumbai, India	174	370

Global terrorism threatens to undo a generation of multi-lateral endeavours for human development, inspired by principles of social justice and human rights.

Most countries around the world set aside substantial amounts of funds to counter terrorism; for example, the budget allocated in the United States for this matter was $142 billion, a figure which dwarfs the shortfall in annual funding required to meet the Millennium Development Goals ($40 billion), which would assist almost 1 billion people in extreme poverty. The huge budget did not substantially stem terrorism, although terrorist attacks did decrease somewhat in 2008 and 2009. The NCTC 2009 report counted 11,000 terrorist attacks in 83 countries in 2009, resulting in 15,000 deaths and 58,000 victims.

The daily images of violence, poverty, and disaster have become such a normal part of the mass media news that many people have become desensitised. Until we become uncomfortable enough, little gets done. However, even if we aren't consciously aware of being stressed by the constant bombardment of war, terrorism, violence, environmental and climate disasters, economic and financial upheavals, and the various sexual perversions, our bodies are responding anyway, secreting the stress hormones of adrenalin and cortisol.

One of the most telling indicators that levels of stress have exceeded our ability to cope with it is the mushrooming sales of antidepressants and prescription drugs: from $79 billion in 1997, it has escalated to $189 billion in 2004 (source: Kaiser Family Foundation, American Institute for Research).

Pause: Bird Nest—The Next Generation

Another unmistakable sign from the divine matrix came to me in the midst of writing this chapter: a bird nest with two hatchlings in it, right outside my bedroom window. A bird nest symbolises that parents have found or prepared a good place for their young. The hatchlings symbolise the promise of our future, the assurance that our species/legacy will continue. They are fragile, need to be looked after with love by the parents, and their growth is supported by the foundation of the nest.

To prepare our young for a new world to come, we must lead them by example, living the values that will support the kind of world that we want to create. In other words, if we continue to live our lives according to our old values, we will pass on these same ones to our children, with a way of life that is not sustainable.

Supporting this point is the study of epigenetics, the science of what regulates our genes—what turns them on and off. Dr. Michael Skinner, a geneticist at Washington State University, discovered an epigenetic effect in rats that lasts for four generations. This science is proving that we have the responsibility for the integrity of our genome. Everything we do, everything we eat can affect our gene expression and that of future generations. Therefore, we have the potential to influence human evolution.

In service of this quest, the skills of anticipating, adapting, and experimenting to change are vital. We are in the cusp of an unprecedented shift in human history. As adults today we are faced with a colossal challenge because we are part of a world which we have built on old values, yet we must embrace new and seemingly conflicting ones, rapidly change our way of living, and learn new skills to bring forth a new, ecological world.

LESSONS FROM LOST CIVILISATIONS

The earth's past offers innumerable lessons, though many of them have been lost through time and war. There have been a number of archaeological finds related to the Harappan, Incan, Mayan, Akkadian, Minoan, Angkor, and several other lost civilisations that have baffled scientists for decades. Yet, most of them have a common, unifying thread: their disappearance was linked to some form of cataclysm.

One lesson worth drawing from our past is the ecological collapse on Easter Island. Arriving on the island via canoes in 400-700 CE, the Polynesians found themselves in a sub-tropical forest with tall palm trees, suitable for building homes, boats, and latticing necessary for the construction of statues. The vegetation of the island provided fuel-wood and materials to make rope. With their sea-worthy canoes, the inhabitants lived off a steady diet of porpoise.

It was the Easter Island society that built the famous statues and hauled them around the island using wooden platforms and rope with material from the forest. The construction of these statues peaked from 1200-1500 CE, probably when the civilisation was at its highest level. However, pollen analysis shows that at this time, the tree population of the island was rapidly declining, as deforestation took its toll. With the loss of their forest, the quality of life for the islanders plummeted. Streams and drinking water supplies dried up. Crop yield declined as wind, rain, and sunlight eroded topsoil. Fires became a luxury since no wood could be found on the island and grass had to be used for fuel. The islanders could not build boats for fishing and began to starve, lacking their former access to porpoise meat and having depleted the island's birds.

As life worsened, the orderly society disappeared and chaos prevailed. Survivors formed bands and bitter fighting erupted. Archaeological evidence suggests that violent civil wars and perhaps even cannibalism preceded the collapse of the civilisation. By the time Europeans discovered the island in 1722, it was a barren landscape with no trees over 10 feet in height. The small number of approximately 2,000 inhabitants lived in a state of civil disorder, were thin and emaciated.

Drawing our wisdom from micro and macrocosmic relationships, the lesson from Easter Island is a clear example of what could happen to planet Earth if we continue to build our lives on unsustainable strategies and ideologies. The Easter Islanders worshiped their ancestors who had become gods. Anthropological studies have shown that the social and economic power of a clan chief was measured by the size and number of statues he had, so there was fierce competition between clan chiefs to build the biggest and the best statues. As more and more people built ever larger statues, increasing numbers of trees were destroyed and the people were not spending enough time growing food or fishing. When the trees were gone, the land lost its fertility, people starved, and there was no wood to build boats to escape.

The Easter Islanders worshiped their ancestors and their egos. In our modern world we have our equivalent: money, with the luxury and fame it brings. The way to accumulate money is by working, which often comes with long hours and stress. These are the old views and lifestyles on which we have built our world. These ideologies and ways of life will not support our children's future. It is imperative that we build our nest on different values, envision a life-supportive future, and live it today, so that our children can grow and build a sustainable, fertile, and harmonious one when their time comes.

Thomas Kuhn & Paradigm Shifts

Quantum leaps in civilisation come with a colossal price: violent social, religious, and political upheavals. Take the example of the shape of Earth. All the way back to Mesopotamian times (6000 BCE) we believed that the world was flat. Sometime in 330 BCE, Aristotle provided observational evidence that the world was round, and though more and more people began to accept his hypothesis, it was highly contested by scholars for another 18 centuries, until bold navigators from Spain sailed around the earth in 1522.

In another instance in history, Claudius Ptolemy in 150 CE postulated that Earth was the centre of the Universe (the geocentric view). Despite its theoretical inconsistencies, his system influenced astronomy for 1,400 years, until in 1543 Nicholas Copernicus published his theories on the heliocentric Universe. Copernicus was denounced as "the fool who will turn the whole science of astronomy upside down." Galileo Galilei reintroduced the theory of Copernicus that Earth revolves around the Sun in 1633. He was silenced by the Roman Catholic Church because his scientific theory contradicted their interpretation of several scriptural passages. He was placed under house arrest from the time of his trial to the end of his life in 1642. The important point is that Copernicus and Galileo did more than just overturn a belief of medieval philosophy with their heliocentric theory: they expanded society's focus from God and the afterlife to the awareness of the Universe, its vast possibilities, and people's role in it. Such epochal moments in history are known as *paradigm shifts*.

The term was first used in 1962 by Thomas Kuhn in his book *The Structure of Scientific Revolutions*. Kuhn was talking about how scientific developments change the perspectives through which we see the world, and give rise to a new worldview and way of life. In addition to the examples above, the discoveries of Isaac Newton in 1687 changed our perspective of the physical world. They gave rise to the industrial revolution and, subsequently, to our modern age of science and technology. Similarly, another paradigm shift happened in 1905, when Albert Einstein published his paper on special relativity, which challenged the set of rules laid down by Newton on the laws of force and motion over 200 years before.

Again, from a micro and macrocosmic wisdom, there are personal paradigm shifts. Within a particular time in history, new confounding and disturbing information impact the perspective of individuals at a personal level. As a critical mass of personal paradigm shifts (with similar information) is achieved, a cultural shift happens. The steps below help us understand the process of paradigm shifting.

i. New data is gathered in our on-going quest to answer every question. Some of this data does not fit in our existing paradigm.
ii. Eventually, so much ill-fitting data is accumulated that it challenges the older framework; it gets the boldest minds to think and offer alternative hypotheses.
iii. A crisis develops as older answers are unable to adequately address the recurring problems. However, some of the alternative hypotheses demonstrate promise, offering a new understanding—a new paradigm.
iv. The new paradigm generally gets a lukewarm reception; some laugh, while others are hostile to it. From the old paradigm, the new data seems bizarre and unfathomable.
v. Those who hold the old view go to their graves unconvinced of the value of the new view, while others are able to embrace it. Finally, a critical mass of those who accept the new paradigm is achieved and the new view takes dominance, ushering in a new historical age.
vi. In time, the process repeats itself.

Many would say that the world today needs to change, and when we look at what is happening around us in the framework of this process, we know we are between steps 4 and 5. In the light of the overwhelming challenges that we are facing, perhaps the question we need to ask is, "What core beliefs, values, and worldviews must we re-examine to support the new ecological world we want to manifest?"

The Major Assumptions of Our Time

The Merriam-Webster dictionary defines *assumption* as the act of taking for granted or supposing that a thing is true. Over time, the totality of our experiences becomes our worldview, and within it, there is a labyrinth of interlocking assumptions that affect our opinions, decisions, and actions.

Based on the studies of the Institute of Noetic Sciences, drawing from multiple disciplines such as economics, politics, sociology, biology, new physics, and religion, researchers have come up with a list of assumptions that are affecting global cultures and ecological health:

- growth is good; more is better;
- economic wealth is the truest sign of progress;
- "the market" is the most reliable measure of value;
- individual selfishness serves the common good;
- we live in a world of scarcity;
- humans are superior to other creatures;
- the earth is ours to exploit;
- the world consists of *us* and *them*;
- people are intrinsically bad;
- technology or God will save us.

Many of our planetary and human problems can be traced to the assumptions outlined above. Throughout this book, there is much evidence to support these assumptions and we will further present additional data to reinforce and consolidate the discoveries presented here.

Climate and Environmental Deterioration

Global warming, rising sea levels, exhaustion of marine life, deforestation, the extinction of animal and plant life, the collapse of coral and wetlands biodiversity, the depletion of fresh water, precipitated weather disasters such as acid rain, droughts, and floods caused by pollution, crop failure, rising water, and airborne diseases are all telling us that the way we are living is stripping the earth of its resources and inhibiting its capability to rejuvenate. They are also telling us that living on ecological credit and filling our land, air, and seas with pollution are burning the candle on both ends. In part, this is linked to our mindset of "more is better" and living a lifestyle that is based on *wants* as opposed to *needs*.

Self-Interest, Capitalism versus Holism

The rise of capitalism and modern technology has accelerated globalisation and consumerism. Coupled with the reasoning that "the pursuit of individual self-interest would inevitably produce the greatest good for the whole," it has largely driven nations and individuals to serve themselves first and be better than others. With a rapidly depleting planet, the assumption about scarcity strengthens, which promotes competition as opposed to cooperation. This essentially compels people to think more from an "I" versus "we" perspective, "us" versus "them," and embrace separatism rather than holism.

Holistic living is about looking at the planet and all its creatures as one living organism. With our current view that the world is made up of non-living matter, the planet becomes something to exploit and therefore our economic, politic, and social systems are symbiotically created by applying that worldview.

GDP as Measurement of Success

Most countries use the GDP metrics to measure their progress and development in a year. Just as a reminder from a previous chapter, GDP is the sum of private consumption + gross investment + government spending + (exports − imports). Using such perimeters does not truly measure the well-being of a nation, for it does not include factors such as natural resources and environmental protection, poverty, education, or health management. Using the GDP as a proxy of a nation's well-being has been criticised by many Nobel Laureates: it is misleading, un-ecological, and unsustainable.

Started by an international conference in 2007 titled Beyond GDP, a new economic and social development metric has been set up, known as the *Genuine Progress Indicator* (GPI), to measure true progress, the wealth, and well-being of nations. It includes income distribution to the poor, value of work contributed to childcare, education, and social welfare, cost of crime management, cost of natural resource depletion and degradation, increment of leisure time, and many other *true* measurements. So far, the response from countries around the world is slow, resistant, and uncooperative.

Conflict and Arms Sales

Over the past 10 years, terrorism, political upheavals, ethnic conflicts, resource protection and the US invasion of Iraq have increased worldwide military spending by 34%, to more than a trillion dollars. The United

States is responsible for nearly 50% of that total. The countries with the largest percentage increases in military spending are China (165%), Saudi Arabia (94%), and India (82%). Developing countries continue to be the primary focus of foreign arms sales activities. Arms kill more than half a million men, women, and children on average every year. Many thousands more are maimed, tortured, or forced to flee their homes. The uncontrolled proliferation of arms trade fuels human-rights violations, escalates conflicts, and intensifies poverty.

The need to protect ourselves with more firearms and weapons of mass destruction has been fed by the human difficulty to embrace the idea that we are but one family. Moving our collective mindset from "us and them" to "we" is the major challenge of our time. A global shift to the mindset of "we" would allow us to free up hundreds of billions of dollars used on military defence, which take up a huge chunk of most countries' resources.

More neighbourly collaborations would emerge and true multilateral sustainable solutions could be instituted across several regions of the world. Indeed, the global shift to a "we-consciousness" is possibly the toughest quest of our time, and one that would give credence to popular phrases like "We are all in this together," "We can't make it all alone, by ourselves," or "It's all of us or none."

Age of Adolescence and Accountability

Notable authors on spirituality and science like Carl Sagan, Thom Hartmann, Gregg Braden, and many others talk about the evolution of man currently being similar to the age of adolescence. In worldwide lectures, evolutionary theorist Duane Elgin, author of *Voluntary Simplicity* (1993), asks audiences, "If you were to imagine the human species as a single human being in the process of development from infancy to old age, where would you place our species on the development spectrum?" Without exception, the developmental stage that receives the greatest number of votes is adolescence. Metaphorically, it signifies that we (the human species) are struggling through the teenage period, slowly groping toward adulthood.

In Christianity, this is also known as the age of accountability. It is the time when a person enters the age of reasoning, experimentation, and exploration, beyond the grounds of familiarity and security. Collectively speaking, the challenge is to apply mature wisdom in the context of an immature global society. In this age, do we individually have the maturity and courage to take personal responsibility for the change necessary to ensure the future well-being of our planet? Or are we going to naively take a passive position of hoping that something other than ourselves, like the government, technology, or God will save us? And while we take that stance, are we continuing to lead a life that depletes, degrades, and destroys humanity and the planet?

Redefining Success

How Do We Measure Our Success?

Everyone is striving to be happy, whether they are conscious of it or not. Focusing on the 20% of the people in the world who are consuming 80% of the world's resources, and also the people (60%) who are hungrily striving to be at the top of this 20%, their definition of success would be: a successful career that provides lots of money. And with that, people believe they would be happy.

Many definitive studies have been done over the past two decades on the relationship between money and happiness. Most of them concluded that once we have sufficient material resources to provide for our needs, more money fails to make us happier. One of the noteworthy research studies was carried out by two Princeton professors, Alan Krueger and Nobel Laureate Daniel Kahneman, in 2004. According to them, people with above-average incomes are relatively satisfied with their lives, but are rarely happier than others in moment-to-moment experience; in fact they tend to be more tensed and stressed, and do not spend time in particularly enjoyable activities. All in all, their research confirmed that the link between income and happiness is mainly an illusion.

So, What Truly Makes Us Happy?

While humanity is ever searching for the answer to this question, perhaps we can start by examining what *doesn't* give us happiness. In 2005, John de Graaf, David Wann, and Thomas Naylor co-authored the second edition of a book named after a disease of extreme consumerism, *Affluenza*. Living a time-poor, stress-rich lifestyle, a lack of inner security often propels people to try to fill the resulting void by surrounding themselves with materials that do not fulfil them, nor sustain the planet. Annie Leonard, creator of the popular YouTube video on consumerism (watched by over 7 million) called "The Story of Stuff," talks about a vicious lifestyle where countless people (perhaps hundreds of millions worldwide):

- spend long hours at work;
- go back home tired and watch TV; media makes people feel inadequate and imperfect;
- go shopping to feel good (which includes eating unhealthy food);

- have to work even harder (the cycle spirals into deeper levels of unhappiness).

In short, work, money, watching glamorous TV programmes, and buying stuff we don't need will not give us true happiness. In addition to consumerism, there is also the problem of how products are being made and sold. No one would consciously vote for human exploitation, health endangerment, unethical practises, and environmental annihilation. But, the reality is, we do use trivial, decadent, and toxic things made by grossly underpaid and exploited workers in third-world countries, sold by profit-hungry mega-corporations. So how we choose to spend our money is our vote for what exists in the world.

A growing number of international sociological and psychological studies have uncovered a number of things that support our well-being and happiness:

- developing meaningful relationships;
- pursuing meaningful goals and enjoying the process;
- taking delight in and enjoying the simple things in life;
- keeping the mind and body healthy;
- maintaining work-leisure balance;
- volunteering and assisting others;
- enjoying sense of humour.

Can Simplifying Our Lives Bring Us More Happiness?

Our never-ending quest for a richer understanding and appreciation of happiness has brought us to the realisation that our outer material possessions and the adoration we get from others do not equate to inner happiness. *More*, *bigger*, *newer* are not necessarily *better*. In fact, the levels of mental illness, depression, and discontentment in developed, richer countries are among the highest in the world. The true wisdom rests in how we define *enough*, in the realisation that when we cut down on our *wants* (when we buy/have LESS) by being clear of what we really need, we get to experience true happiness (MORE).

A mass exodus has already started to happen and is currently still happening: millions of people around the world are living their lives in accordance to the wisdom of "live simply so that others can simply live." This is a motto propagated by Mahatma Gandhi, a simple man voted by the United Kingdom as the Man of the Millennium. People who have congruently walked their talk have come up with a body of knowledge and practical tools to live simpler and more sustainable lives, not only to benefit themselves, but their local communities and the world. Also known as the "voluntary simplicity movement," this earth-healing exodus isn't based on fear, resignation, or a perception of scarcity, but a redefinition of abundance.

According to Duane Elgin, author of *Voluntary Living and the Living Universe*, simple living is about "living in a way that is outwardly simple and inwardly rich." So this movement is all about celebrating life, developing self-reliance, living without unnecessary wants, in harmony with the environment, and developing one's inner-richness (spirituality). It is about rediscovering a way of life that is in harmony with the whole world and awakening to the joy of simple living. When we simplify our lives, we have more time for things that truly give us happiness. This is the wisdom of "less is more."

Can We Truly Be Happy in the Midst of Suffering?

> All men have a mind which cannot
> bear to see the suffering of others.
> —*Mencius, Confucian philosopher*

Recent advances in science have enabled us to rediscover our humanity. According to neuroscientists, caring instincts such as empathy and predisposition to trust and to be trustworthy are built not only into our psychology, but our biology as well. Using fMRI machines, Christian Keysers, a neuroscientist at the University of Groningen, in the Netherlands, confirmed that when subjects observed faces that expressed disgust, the disgust cortex of their brain was activated. Daniel Goleman, in his book *Social Intelligence* (2006), states that the neural networks for feeling and action share a common code in the language of the brain. This shared code allows whatever we perceive to almost instantly lead to appropriate action.

Many psychologists and neuroscientists now believe that mirror neurons in our brains give us the ability to place ourselves in the shoes of someone else. A study carried out by Decety and Lamm in the field of Social Neuroscience (2006) monitored the brain patterns of participants while they were imagining themselves receiving a pinprick. They were then directed to watch someone else receiving a pinprick. Both scientists discovered that the same neurons were activated in both instances. After several other experiments, they concluded that not only are humans able to mirror the pain in others, but that they also strive to soothe their pain.

In another study by social psychologist Daniel Batson, individual participants observed other participants receiving electric shocks when responding incorrectly during a memory test. Observers were told that the person they were observing had experienced shock trauma as a child. They were then given the opportunity either to leave the experiment or to volunteer to receive shocks on the other person's behalf. Those who reported feeling compassion for the other stayed on and received a number of shocks for them.

To counter those who would assert that such apparently altruistic behaviour was an effort to gain the approval or admiration of others, Batson clearly stated that participants offered to help others in distress *even* when their actions remained anonymous. The research suggests that the impulse to respond to others in need is wired into our most primal structures of care and connection. Daniel Goleman further explained in 2006 that "although people can also ignore someone in need, that cold-heartedness seems to suppress a more primal, automatic impulse to help others."

In short, our quest for quintessential happiness can only be truly fulfilled when we see improvement in the lives of others, when our good fortunes are shared, when our existence makes a difference to others' lives, when our endeavours build a world where people experience peace, love, joy, and beauty.

Part III
THE SECOND PATH—TRANSFORMATION AND GROWTH

The Hopi pictograph, in unison with other ancient prophecies mentioned in Part II, shows us a second path, which is a shift out of business-as-usual. To understand how this shift can happen, I have looked into other sources of wisdom. Although our research spans many scientific disciplines, it is the unification of ancient wisdom, Gnostic, shamanic, mystic, and esoteric spiritual knowledge that made the tonnes of scientific data discernible, meaningful, and useful.

Before we embark on a transformative personal, planetary, and possibly divine discovery, let us share with you a powerful learning in our journey:

> Everything is a metaphor for everything else.
>
> —*Gregory Bateson*

Bateson, an anthropologist, cyberneticist, and social scientist, said that we are intimately connected to everything around us. By looking at the story of our own lives, we can find the metaphor of what is happening around the world. Similarly, when we look at what is happening around the world, we can explain what is happening in our lives. A similar wisdom can be found in the book *A New Earth*. The author, Eckhart Tolle, shared with us an important learning he gained from just observing two ducks that fought and later floated on peacefully. He said,

> We are a species that has lost its way.
> Everything natural, every flower or tree,
> and every animal have important lessons to teach us
> —if we would only stop, look, and listen.
>
> —*Eckhart Tolle*

Bateson's and Tolle's quotes just above complement very well the proverbial saying, "When the student is ready, the master will appear." The master is not necessarily a teacher, guru, or scientist. It can be your cat or dog, your children, a bus advertisement, a book, a tranquil moment, a plant you are watering, or shapes of clouds in the sky. In short, everything around us can be our teachers. In fact, the focus is not on who or where the teacher is, but on our readiness. So, in order for us to understand the choices we have before us, I propose that we tune in to our immediate environment and daily experiences to find the answers.

What does it mean to *be ready*? It means to be open, be at ease, and be clear of our intentions, of what we positively want to manifest in our world.

In countless scientific studies, scientists have concluded that what we intend affects what we experience. According to Von Neumann, a renowned quantum physicist, to have a complete understanding of events in nature, we are required to take into account two simultaneous processes, one pointing to the dynamics of the physical reality, and another referring to the dynamics of human consciousness or intention. The interaction of these two processes manifests what we experience in our world. He said:

> What we discover has already been influenced and
> preselected by the nature of the question itself
> and its underlying intention.
>
> —*Von Neumann*

So get yourself ready, adopt a state of openness, ease, and clarity, and let us begin the process of creating a new collective intention that propels us to design a new life-supportive world.

CHAPTER 12
WHAT DOES IT ALL MEAN?

MAN-MADE OR NATURAL?

We have reached an unparalleled, dire situation where multiple emergencies converge, as we have discussed so far; and we have never faced such a planet-wide multi-system crisis with a population of seven billion people.

One of the questions that have been raised by scientists and the public at large is, "Are all our geo-bio-socio-spherical problems man-made or natural planetary cycles?" Geologists and archaeologists have uncovered concrete evidence from ancient sites, soil and rock samples, as well as deep arctic ice cores that show how entire civilisations in the past have been displaced or wiped-out by natural catastrophes. There were floods, long droughts, ice ages, rising sea-levels, and violent volcanic periods. There was also clear evidence of a continuing rise and fall in Earth's magnetic field and the drastic reversal of the magnetic poles. Furthermore, there was archaeological evidence of ancient civilisations around the world which, due to lack of foresight and ecological wisdom, wiped-out their own existence. Through deforestation, changing their river systems, lack of proper sanitation and/or agricultural knowledge, they caused their own demise.

Earth, Human, & Celestial Events Are Connected & Cyclical

There is strong scientific evidence suggesting that such global catastrophes are connected to major human and celestial events. Human events refer to the rise and fall of super-powers and cycles of war and peace. Celestial events point primarily to the Sun's natural cycles, as in how much of its energy reaches Earth. What is important about these discoveries is that they have occurred more than once; they come in cycles. Scientists may argue about the frequency and the timing of the cycles, but one point they unanimously agree upon is that they do recur, and that is the way it has always been.

Science Confirms Human Contribution

On the other hand, reports from the Intergovernmental Panel on Climate Change (IPCC), as well as other similar scientific research bodies have confirmed the impact of human activities on global warming, land degradation, water shortages, depletion of food resources, and the decrease of biodiversity. Their studies involve diverse teams of scientists of up to 1,300 members coming from countries around the world. Conclusions were based upon years of research, of analysing complex data, and running intense computer climate simulations.

Their research and findings have been acknowledged and supported by the European Geosciences Union, the National Oceanic and Atmospheric Administration (United States), the Royal Meteorological Society (UK), the World Wildlife Fund, The Energy and Resource Institute (India), the Global Footprint Network, and numerous others. Many of the details presented in this book about what is happening with our weather, land, water, and food are just tips of the iceberg; indisputably, we *have* contributed to the dire straits we are in.

Controlling and Reducing Greenhouse Gas Emissions

What can the world do about this? One important step is to reduce greenhouse gas emissions. The United Nations and its panel of climate change experts have been having intensive talks with the governments of almost every country around the world since 1992. Under the UN Framework Convention on Climate Change (UNFCCC), they set out a framework for action aimed at stabilising atmospheric concentrations of greenhouse gases. As of November 2009, 187 states/countries have signed and ratified the Kyoto Protocol which was formed in 1997.

Up until now, even though some progress has been made, the world's major industrialised nations are still severely off target. From the outset, the requirements of the protocol were criticised by environmental groups, as the targets were too low. The latest scientific studies are advocating emission reductions of up to 25-40% by industrialised countries before 2020. The current protocol demands a reduction of just 5%.

Future of Humanity versus Gross Domestic Product (GDP)

This is a complex problem concerning many nations and the key concern is how the reduction of carbon emissions is going to impact the growth of each economy. In other words, each country's GDP is more important than the impending climate catastrophe. With the global recession that started in 2007, the major discussion in Copenhagen in December 2009 was a challenging one. The thought was that with the major learning from the Kyoto Protocol, the UN Climate Chief, Yvo de Boer, and his team could better formulate more workable strategies. Also, with increasing pressure from the citizens of many countries, it was thought to be important that governments commit to taking action on arresting global warming. The eventual result was disappointing, with no firm target set for limiting the global temperature rise, and no commitment to a legal treaty and no target year for peaking emissions. Countries most vulnerable to climate impacts didn't get the deal they desperately needed.

The search for a meaningful climate change commitment continues, while the world is speeding toward total ecological collapse.

Adapting to the Destructive Climate Changes

The destructive effects of climate change are already being felt in many parts of the world and the United Nations Foundation is asking countries to adapt to on-going "unavoidable" changes. Suggestions include improving response strategies and the management of natural resources to cope with future food shortages, building climate-resilient cities, strengthening international and regional institutions to deal with weather-related disasters and the increasing number of climate-change refugees.

In short, more devastating geo-bio-socio-spherical changes are coming. And the intervals between the changes are getting shorter. Be it natural or man-made, we must stay vigilant, updated, and be prepared. Most of all, we must change the way we lead our lives. The basic approach is two-pronged: to work on restoring nature and to reduce our carbon footprint.

Beyond Scientists, Governments, & Corporations

Other questions often asked by the general public are: "Aren't these huge global problems in the hands of scientists, governments, and corporations?" "What has anything got to do with me?" "How could I possibly make any difference?"

The problems that we are facing today are simultaneous, as several systems are rapidly deteriorating; this is clearly beyond any government, corporation, or group. Governments around the world are coming out with more drastic strategies to reduce greenhouse gases and revive the planet. Corporations are now facing increasing pressures from environmentalists and most of all, the public at large: citizens around the planet are demanding that businesses be carried out in an environmentally healthy and socially ethical way. Making changes in huge organisations requires time because many systems and factors are involved and they need to be considered carefully. Scientists of various disciplines are working with governments and corporations to come out with ecological and sustainable solutions.

Individuals Make the Biggest Impact

Still, each individual has a very important role to play. Many seem to believe that the individual can't really do much about the dangers facing the planet anyway, so why bother? The awareness, intentions, and consciousness of the individual have a major impact on reality. Even in terms of day-to-day activities, the choices of the individual have a tremendous impact on the planet.

The heart of the problem is in the way each of us lives his/her life. Our daily existence affects the world at geospherical, biospherical, and sociospherical levels, and these living systems are influencing each other. A change in one affects the rest in unseen, powerful ways. Tim Flannery, author of *The Weather Makers: How Man Is Changing the Climate and What It Means for Life on Earth*, has challenged readers to make deep cuts (80%) in carbon emissions to keep the world from reaching critical threshold points. An Australian mammalogist and chairman of the Copenhagen Climate Council, Dr. Flannery is not just referring to governments and corporations, but equally important, to individual citizens of the planet.

Impact of Buildings

A recent study based on the US Energy Information Administration's 2008 report made an important discovery. Though it is based on the United States, it can be mapped across to other developed or rapidly developing nations. The data revealed that up to 38% of all energy consumption and greenhouse gas emissions comes from residential, commercial, and industrial buildings. If individuals in the United States were to collectively convert half of their light bulbs to compact fluorescent ones, they would reduce 42.4 million metric tonnes of carbon emissions per year. Turning off home computers alone, when not in use, would cut carbon emissions by 8.3 million metric tonnes per year.

Impact of Commuting

The next largest source of carbon emission is transportation, which accounts for 34% of US carbon emissions. Today's internal combustion engines are inefficient at converting fuel into motion: cars waste up to 85% of the energy from the fuel in their tanks, losing a huge portion as heat. If individuals collectively drove 20 fewer miles every week, they would reduce 107 million metric tonnes of carbon emissions per year. If individuals collectively improved their cars' gas mileage by 5 miles per gallon, they would cut carbon emissions by 239 million metric tonnes per year!

Impact of Our Daily Living

Many developed countries in the world have a similar situation. The largest amount of carbon emissions comes from buildings, followed by the transportation and industrial sectors. Statistically, these two sectors contribute to 70% of each developed nation's carbon emissions. In other words, we (collective individuals) are adversely impacting the global climate by the way we live in our homes, work in our offices, drive to work, and recreate in shopping malls.

Most Important People on the Planet

It is we ourselves who are the most important people on the planet, regardless of our language, race, or religion. The generations living today—especially those who choose not to ignore, blame others, or wait for others to fix problems—are going to determine the fate of all the generations in the future. Conversely, these same generations living today are the ones who are the most dangerous to the planet, because they hold the power to tip the scale toward destruction or transformation.

Without detonating weapons of mass destruction we are still on the path of devastation and degeneration because, as individuals, if we do nothing and continue to live the way we've been living, we will continue to experience what we have been experiencing so far: a world of increasing scarcity and intensifying strife.

Therefore, this is an issue about responsibility on a planetary level. Are we (collective individuals) taking responsibility for the state of the planet today? Also, are we mature enough to stop focusing on who has done more or less damage to the planet, and start working toward supporting each other in reducing our carbon emissions and consumption to heal our planet?

The Point of No Return

For good reasons, many researchers on the subject of world shift emphasise the importance of evolving our consciousness quickly. One of the critical concerns that scientists have today is, "Have we reached the point-of-no-return and how do we know?" The point-of-no-return is a theory on how things might go terribly wrong for the planet if a relatively small warming of Earth upsets the normal checks and balances that keep the climate in equilibrium. As the atmosphere heats up, more greenhouse gases are released from the soil and seas. Plants and trees that absorb carbon gases may die due to the extreme climate conditions, and as the vicious cycle continues, a runaway effect may be created. Global climate will get to a point where the atmosphere is so different, that most of life on earth would be threatened. The tipping point in many scientists' view is a mere 2° Celsius rise in global temperature; it is the maximum limit that humanity can risk.

Though no one knows for sure, most scientists would agree that we are very close to the tipping point and we are running out of time because of the multi-faceted problems we are confronted with. To put it differently, every minute we delay taking the necessary action to avert catastrophe is bringing us closer to the point of no return.

The Critical Focal Point

In short, in order to overcome this global crisis, we need more governments to cut down carbon emissions and energy consumption in their respective countries, many more business corporations need to de-emphasise profits and focus on social and environmental responsibilities, and more individuals, hundreds of millions more, need to question and push governments to shift priorities on ecological farming and developing renewable energy sources that would replace the old ones. On a personal level, billions need to start cutting down their consumption habits and lead a simpler, more organic life. It all boils down to shifting our values, life priorities, perceptions, and behaviours. How can we make all these planetary-wide, difficult, and complex changes within a short period of time if we don't collectively evolve our consciousness to a global level?

> Without a global revolution in the sphere of human
> consciousness, nothing will change for the better
> and the catastrophe toward which the world is headed[...]
> will be unavoidable.
>
> —*Vaclav Havel, the first president of the Czech Republic; in 2005 voted fourth in the World's Top 100 Intellectuals*

What Is the Collective Global Shift in Consciousness?

A Living and Conscious Planet

As mentioned in the first chapter, consciousness is defined in the dictionary as "a sense of one's personal or collective identity, including the attitudes, beliefs, and sensitivities held by an individual or group." Evolutionary and systems theorists define it in a much broader and richer way. Gregory Bateson, for example, proposed to define the mind as "a systems phenomenon characteristic of living things." He also emphasised that mental characteristics are manifested not only in individual organisms but also in social ones and ecosystems. It would then be appropriate to consider that the earth is not a plethora of organisms, but one single entity that is able to reflect back on itself as one interactive and evolving consciousness. Dr. James Lovelock formulated this supposition in 1982 to create what is known as the Gaia Hypothesis, in which he asserts that the earth functions as one superorganism. A world-renowned environmental scientist and futurist, Dr. Lovelock, in his latest book, *The Vanishing Face of Gaia: a Final Warning* (2009), sends a dire warning to the world.

Waking Up to Oneness

Parallel to Dr. Lovelock's ecological hypothesis, Ken Wilber, Ervin László, Peter Russell, and many other leading evolutionary scientists postulate that the earth and its inhabitants are waking up to a new global consciousness that is transpersonal, collective, and integral. From the spiritual domain, Sri Aurobindo, a Indian sage, foretold the coming of a new collective consciousness which he termed "supraconsciousness." At this point, according to Sri Aurobindo, sustainable solutions to humanity's problems become available in the context of a radical transformation of human life into a form of divine existence.

PAUSE: HISTORICAL SHIFTS IN CONSCIOUSNESS

Returning to the notion that our survival depends on evolving our consciousness, history has given us a great deal of evidence that our consciousness has shifted in the past. With each civilisation, the people who lived in them had a different awareness, different worldviews, states, values, priorities, and behaviours. Jean Gebser's (1905-1973) theory on the evolution of consciousness provides a useful reference. Gebser's ground-breaking work, *The Ever-Present Origin*, ploughed through historical evidence of many fields such as poetry, the visual arts, architecture, philosophy, spirituality, physics, biology, and other natural sciences, and he saw traces of the emergence and collapse of various structures of consciousness throughout history.

According to Gebser, for our species, we can trace the evolution of its consciousness through five distinct periods. These evolutional mutations involve structural changes in both mind and body and, therefore, they are not just simple paradigm shifts, but rather fundamentally different ways of experiencing reality. We awaken to the fifth structure by remembering the previous four co-present structures contained within each of us. Gebser identifies these fives periods as the Archaic, Magical, Mythical, Mental, and Integral structures of consciousness. Below are brief descriptions of the key characteristics and time periods of each structure.

- Archaic Human (5 million-200,000 BCE): the early man, *Homo habilis*, who made stone tools; language was not developed yet; like animals, he had a simple awareness and was egoless; he didn't distinguish between self and others/the world or between soul and nature; neither space nor time existed (zero dimension); lifestyle/culture were nomadic;

- Magical Human (200,000-10,000 BCE): *Homo sapiens sapiens*; more advanced tools; primitive art and rituals; a rudimentary self-sense emerged and man awakened to his own finiteness and vulnerability; he communicated primarily mimetically, not linguistically; he still lacked the concepts of space and time and believed he lived in a one dimensional, unitary world; lifestyle/culture involved hunting, gathering, music, and dance; he still did not distinguish himself apart from nature;
- Mythical Human (10,000-2000 BCE): the Neolithic Age; man settled in permanent villages and started growing crops; clear separation from nature; major life institutions were established, such as family, education, religion, government, and military; language development was the hallmark of the mythical era; story-telling (myths) governed the collective mind; two-dimensional awareness and time were cyclical, driven by seasons; lifestyle/culture involved agriculture and mythic communication (gods and symbols); man was still egoless but on the way to selfhood;
- Mental Human (Earlier Phase 2000 BCE-1500 CE): the philosophic age (with Socrates, Plato, Aristotle, etc.); logical abstraction became the process of structuring mental activities; man used his mind to master the world around him; the consciousness of "I" was concretised; religion primarily moved from polytheism to monotheism;
- Mental Human (Later Phase 1500-2010/2011): the Renaissance Age put man in charge of his own destiny; egocentrism was born; man stepped out of the mythical two-dimensional circle into three-dimensional space; he discovered the power of different perspectives; time became linear (past—present—future); the world is ruled by scientific experiments and mathematical proof; lifestyle/culture involves industry, technology, and science; people are driven by materialism and individualism;
- Integral Human (2012 onward): a structure of evolution that we are gradually unfolding; there is a new relationship to space and time and we are moving into a fourth dimension; we see the past, present, and future not as separate but as a whole; all previous structures of consciousness are available to us, therefore, we are not stuck in any of them; multi-levels and multi-faceted knowledge interweave into one conscious system; finding lasting solutions to three-dimensional problems becomes

easier; there is a transparency between self, others, and nature; self is part of a larger conscious-spiritual whole that also nourishes the self; lifestyle/culture is communal and based on collective consciousness; the planet and its inhabitants are embraced as one ecosystem.

The Shift to Integral Consciousness Has Already Begun

According to many social/evolutionary scientists and mystics, such as Don Beck and Chris Cowan (Spiral Dynamics), Ken Wilber (Integral Theory of Consciousness), Ervin László (Systems Theorist), Arjuna Ardagh (Translucent Revolution), and Paul Ray (Cultural Creatives), the shift to an integral humanity is already happening. Based on current estimations, 1% of the people in the world are operating and living in this emerging consciousness. The important question is: will this trend continue to grow widely and quickly enough, before the world reaches the point-of-no-return? A more important question is: are you (the reader) willing and ready to be part of this global-shift movement today?

Getting a Clearer Picture

To gain more clarity, we need to take a bird's eye view of all the scenarios we have discussed up to this point. Some of the key messages are:

- the way we live our lives has brought us to where we are today;
- if we continue to live our lives in the same way, the path to a world of strife, decay, and destruction is certain;
- not doing anything different, ignoring the warnings, continuing our lives in a business-as-usual manner means choosing the path to destruction;
- the way forward, no matter how much research we do, will contain a high level of uncertainty;
- rather than staying with a familiar way of living that leads to destruction, moving forward mandates that we experiment with uncertainties;
- in every crisis lies the opportunity to live life differently: in the dismantling of the old, we get a chance to create the new;
- since in every crisis lies the opportunity to create the new, how we collectively choose to see the future is crucial;
- our vision of the future must include (a) unity and holism, (b) moving away from materialism and toward humanism

and ecological activism, (c) living simply and organically, (d) developing collective and spiritual consciousness, (e) redefining and realigning our life purpose and priorities.

Two Paths

It is apparent that there are two principal contrasting paths before us: one that leads to a breakdown of our geo-bio-sociosphere and another that leads us to a breakthrough—a new world of harmony, cooperation, and sustainability. The decisions that we collectively make in the narrow window of time now presented to (most especially, 2012 – 2017) and our subsequent actions to 2030 are going to determine the course of our evolution. The livelihood of future generations lies in the hands of those of us who are alive today. Every minute we delay being part of the solution, we are bringing ourselves closer to a world of strife, suffering, and decay. To further elucidate the severity of the crisis that we are in, let us contemplate on the words of these luminaries:

> The continued momentum from our past actions will unfold over decades or centuries; toxic chemicals that permeate our water and soil, and the build-up of greenhouse gases will take their toll for years to come. [...] And if we examine more carefully [...] we can find points of leverage where simple, gradual changes might halt or even reverse our contribution to this cataclysm.
>
> *—Daniel Goleman, internationally acclaimed psychologist and author of* Emotional Intelligence *(1995)*

> We need money for education of children and healthcare, but not for a life of luxury... In my own case I never use a bathtub, only shower. Whenever I leave my room I always turn off the light. Taking care of the environment should be part of our daily life.
>
> *—14th Dalai Lama, Tenzin Gyatso, revered Buddhist leader and 1989 Nobel Peace Prize winner*

> Without a doubt the outmoded world of materialism is leading to greater unhappiness, through pollution, overpopulation, lack of nourishing food and water, and the loss of natural habitats. [...] Timely change through a shift in consciousness can bring about a new model of happiness [...]. Such a shift is already occurring—now it needs a critical mass.
>
> *—Dr. Deepak Chopra, physician and Ayurvedic philosopher, credited by* Time *magazine as the prophet of alternative medicine*

Besides the climate and the economy, there are hosts of other "unsustainabilities": poverty, inhuman living conditions, hunger, crime, terrorism, war, intolerance, and the revolt and frustrations they entail. [...] We either change in time, or suffer the consequences.
—Dr. Ervin László, scientist and humanist, author of over 85 books, and recipient of four honorary PhDs, and the Goi and Mandir peace awards

Decision-makers have not faced up to just how close the world may be to the climate "tipping point." [...] Are you ready to say, "Okay, kids, I inherited this house, but I neglected to maintain it, so you will have to worry that the roof might collapse at any time"? That is not the type of legacy we want to leave to our children.
—Mikhail Gorbachev, former President of the Soviet Union; recipient of several honorary PhDs and the 1990 Nobel Peace Prize

From within the dying population [of a caterpillar], a new breed of cells begins to emerge. [...] Clustering in community, they devise a plan to create something entirely new. [...] Out of the decay arises a great flying entity—a butterfly—that enables the survivor cells to escape from the ashes and experience a beautiful world. [...] That is where we are today.
—Dr. Bruce Lipton, internationally recognised authority in epigenetics, author of Spontaneous Evolution

We Must Exercise Our Choice Now

Unequivocally confirmed by scientists, politicians, and spiritual teachers, we are in the midst of a life-threatening global crisis. This crisis is growing and intensifying every day while we go about talking, commuting, working, eating, shopping, and sleeping. Within this crisis, there are two paths for *all* of us, and therefore we must make a deliberate conscious choice, one that is not made just with our heads, but more importantly our hearts; and we must exercise it in our daily lives. The greatest power on earth is in each and every one of us, when everyone acts with the collective in mind and the collective acts cohesively as one mind. We can take whatever we want to do individually and multiply it by 7,000,000,000! Stay tuned to the global consciousness; be actively involved in the worthwhile WE movement. The sooner we take action, the more feedback we will obtain to evaluate our actions. In other words, we have more time to get intelligent.

Keep It Simple and Start Experimenting

There are tonnes of information on this subject and one can approach this with thoroughness or start experimenting with simple actions right now. Too often we look for complex explanations for our questions, but quintessential wisdom is always simple. It was Einstein who said, "When the solution is simple, God is answering."

Here are 9 mindsets to help us crystallise what we have discussed so far and increase our momentum in supporting the world-shift movement.

- Be clear about what we *want* and what we really *need* in our lives.
- Live simply, so that others can simply live.
- De-emphasise money and work; centre our lives around important connections, loved ones, personal growth, volunteerism, and spiritual development.
- Vote for what we want to see in our world with what we buy daily.
- Be deliberate in cutting down carbon emissions and energy consumption.
- Make time to educate ourselves about what is happening; actively support world-shift causes.
- Communicate, connect, and offer support; spread the word, gather with like-minded people, and uplift each other with life-supportive activities.
- Build a sacred relationship with Mother Nature.
- Evolve our consciousness; make time to meditate, contemplate, or pray; connect to the universal consciousness.

CHAPTER 13
WHAT IS STOPPING US FROM EVOLVING?

It is imperative that we ask ourselves this question: "What is stopping us from evolving into a healthier and more harmonious world?" Certainly, there is no shortage of relevant technologies or ideas that could help us improve our dire situation. But why are we not seeing larger life-supportive movements around the world, in our own communities and countries? By asking the question "What is stopping us?" we hope to uncover the out-of-awareness forces that are holding us back. We cannot change those things that we are not aware of. As long as we are unaware of the invisible forces, they continue to drive us to a future of strife, suffering, and very possibly extinction.

Questions provide answers. In our search for deeper wisdom, we must examine the thoughts and feelings that structure the questions we ask. This leads us to the realm of paradigms; it is our most fundamental way of valuing, perceiving, thinking, and doing associated with our worldview. The term *civilisational analysis* points to the comparative study of the patterns and long-term dynamics of macro cultural-social-historical events on modern transformations. In the Axial Age (800-200 BCE), for instance the quest for human meaning and the rise of a new class of religious and philosophical beliefs in China (Confucianism and Taoism), India (Hinduism and Buddhism), the Middle-East (Zoroastrianism and Judaism), and the Western world (Hellenism and Platonism), have had a profound influence on human thought, even up to the present.

A civilisation's worldview shapes the way we see the nature of reality, our feelings of social connection and what we value in life. Therefore, when a civilisation shifts from one worldview to another, it is not just a change of ideas, but it is a deep and pervasive reorganisation of its core of existence.

> It is difficult for the matter-of-fact scientist to accept the view
> that the substratum of everything is of mental character.
> But no one can deny that mind is the first and most direct thing
> in our experience, and all else is remote inference.
>
> —*Sir Arthur Eddington, British astrophysicist*

> The external world is only a manifestation of the activities of the
> mind itself [...], the mind grasps it as an external world simply
> because of its habit of discrimination and false-reasoning. The
> disciple must go into the habit of looking at things truthfully.
>
> —*the Buddha*

MAJOR BLOCKS TO OUR ADVANCEMENT

The intention of this segment is to provide an understanding of how our current paradigm is holding us back from evolving into a healthier and more harmonious global civilisation. Civilisational paradigms persist until they generate problems that cannot be solved. These problems then become the ground of perturbation that produces the next paradigm. The dominant mindset in our current paradigm is *scienceism*. This is a worldview that natural science has authority over all other interpretations of life, such as philosophical, religious, mythical, spiritual, or humanistic knowledge. This idea is analogous to Jean Gebser's Mental Human theory that was discussed earlier. Science can no longer solve our multi-system converging problems unless it evolves into a more spacious, inclusive, and integral consciousness. According to David Bohm, the eminent physicist, "The human endeavour to live according to the mindset that everything is treated as separate, such as religions, sciences, nations, families, etc., has, in essence, led to the extreme global crisis that we are confronting today."

Science has not only provided practical tools and means to our daily living, it has also shaped our beliefs. People immersed in science often believe that science is capable of describing all reality, or it is the only true way to acquire knowledge. Sri Aurobindo says that the premise of science is that the physical senses are our only means of obtaining knowledge. In other words, science would not accept the existence of anything that is supernatural, supraphysical, or extrasensory. This severely limits the extent of our learning and, therefore, our growth.

Presented below are the current evolution inhibitors; they are the mindset and filters that block us from evolving to a healthier world. They are drawn from several sources of ground-breaking work by Duane Elgin, Ervin László, Bruce Lipton, Gregg Braden, the Institute of Noetic Sciences, and many others. These are not isolated, separate proposals; they interpenetrate and reinforce each other.

a) Scienceism: we will look at the effects of scientific materialism, reductionism, and determinism. This addresses the question "How do we know if something is real when we can't see, hear, feel, and measure it?"
b) Survivalism: here we get into the theory of evolution. Is the axiom "survival of the fittest" true? This addresses the question "Do we compete or complete to exist?"
c) Humanism: we delve into the heart of human nature; are we capable of caring, giving, and supporting others? This addresses the question "Is human nature selfish and cold-hearted?"

The three evolution inhibitors have produced a collective worldview of: "Unless I can see, hear, or feel something I won't believe it." "Where is the science behind what you are saying?" "What has God, Spirit, or my intention got to do with this issue?" "In order to survive today, we must be faster, smarter, and more hard-hearted; there's no room for softness!" "People are fundamentally selfish; they will save/benefit themselves first and watch others suffer." "I don't think people are basically kind, caring, and good; the only person I can trust is myself."

If we took such worldviews and multiplied them by seven billion people, we could easily imagine what the world would be like. In fact, it is right before our eyes today. Of course, it is a matter of intensity; not everyone believes or thinks like that. The question is, how many of us truly believe we are much more than what we have become? How many of us believe that we can rise above our programmed values, beliefs, reactions, and negative experiences that are not ecological? One thing is for sure: our desire for a better world must be supported by a different worldview and consciousness. Let us begin to free ourselves by taking a closer look at our current evolution inhibitors.

SCIENCEISM

The scientific era began in 1543 when, at the end of his life, Copernicus published his theory on the *Revolutions of the Heavenly Spheres*, which proposed that the Universe was heliocentric.

René Descartes further strengthened the scientific paradigm in 1649 by dividing the cosmos into two realms: (a) the realm of things that could be physically measured—the world of time, space, matter, and (b) the realm of thought—the world of consciousness and spirit. With his theory on natural philosophy, Descartes focused on the world of matter.

In 1687, Isaac Newton formulated the mathematics that successfully explained the phenomenon of our world. His theory on gravity, the three laws of motion, and other theories confirmed the movement of planets. These theories can be used to calculate the orbits of our satellites and put a man on the moon. Since then, three main worldviews of Newtonian philosophy have shaped how humanity has studied the Universe.

- Materialism: the only thing that exists is matter, and all things are composed of material substances or are the result of material interactions. The things that count in life are those that can be measured. If you can't see, hear, or feel something, it is not real, reliable, or trustable.
- Reductionism: this means that any complex system is the sum of its parts and, therefore, to understand it is to strip it down and

study its components, without looking outside the system; and that is the only way to make something work.

- Determinism: this is the view according to which every event in nature, including human thoughts and actions, is causally determined by an unbroken chain of prior events; we can determine and manipulate the result of nature.

These scientific discoveries, components of what we are here calling scienceism, ruled our consciousness up until the late 1890s and, to a large extent, still rule our thinking today. Science seemed to have found all the answers to our questions about our physical world. The Universe was like a machine that behaved linearly and was measurable, predictable, and controllable. However, the incompleteness of these worldviews was established by Sir Joseph J. Thomson, a British physicist, who demonstrated in 1897 that the atom was not the Universe's smallest particle, as Newton had claimed. The atom was made up of even smaller things, which were later known as electrons, protons, neutrons, etc.

Max Planck, the father of quantum physics, revealed through his experiments that electrons could jump from one energy state to another, spontaneously, without intermediate energy values. This established that things in nature did not move in a linear manner. Rather than focusing on the parts, Planck's discoveries, along with those of other scientists, such as Ludwig von Bertalanffy, Edward Lorenz, and David Bohm, point to the need of approaching nature from a concept of holism rather than reductionism. All matter exists in a vast quantum web of connections, and a living thing is an energy system involved in a constant transfer of information with its environment. Rather than a cluster of individual, self-contained atoms and molecules, objects and living things are now properly understood as dynamic processes in which parts of one thing and parts of another continuously trade places.

Einstein contributed to the rise of new physics in 1905 with his studies on the photoelectric effect, which showed that nonmaterial light waves displayed physical qualities that were previously associated with matter. This strange behaviour of matter behaving as light and light behaving as matter questioned the foundation of Newton's physics. In 1927, Arthur Compton developed a method for simultaneously observing individual scattered x-ray photons and recoil electrons. Together with the work of other scientists like Louis de Broglie and Niels Bohr, Compton's research confirms that all particles (material) have wave-like (nonmaterial) properties and vice versa.

While the findings of the above Nobel Laureates destabilised the worldviews of materialism and reductionism, Werner Heisenberg, Max

Born, and Ilya Prigogine's discoveries dethroned the worldview of determinism. Heisenberg's famous uncertainty principle states that it is not possible to measure simultaneously both the velocity and the position of an electron. The more precisely the position is measured, the more unclear the value of its velocity, and the other way around. Heisenberg's theory asserts that an electron, or any other subatomic particle—the ingredient that makes up all matter—is unpredictable. Thomas Young's famous double-slit experiment, first conducted in 1803, and subsequently fine-tuned by other scientists up to the recent 21st century, proves the indeterminability of photons.

Furthermore, the establishment of virtual particles in Quantum Mechanics states that no matter what we try to measure in a given situation, there are always some unknown factors (virtual particles) that will influence that situation. Even though virtual particles appear very briefly in our world, they interact with other particles and influence their behaviour. The spontaneous appearance of virtual particles is random, so their numbers and properties are unpredictable. And without the possibility of providing the position, course, and speed of the virtual particles, Newton's Law of determinism becomes more problematic.

The above discoveries clearly show that we must change our scientifically entrained worldviews. The belief that the Universe is completely governed by scientific materialism narrows our world of possibilities, especially the realm of finer energies, higher dimensional realities, consciousness, and spirituality. Instead of just breaking things down to components and details, we want to open ourselves up to more integral approaches, to see other knowledge as jigsaw pieces in a bigger picture.

We can celebrate the indeterminism of things, knowing that there will always be more than one way to interpret and achieve outcomes. Rather than isolating and disconnecting from nature, we can acknowledge and embrace the truth that we are connected to all things and that maintaining harmony and the well-being of nature and people of all nations, ultimately benefits the individual.

> All matter originates and exists only by virtue of a force [...]. We must assume behind this force the existence of a conscious and intelligent mind.
>
> —*Max Planck, Nobel Laureate physicist*

> No matter what a deluded man may think he is perceiving,
> he is really seeing Brahman (the Creator) and nothing else [...].
> He sees mother-of-pearl and imagines that it is silver.
> He sees Brahman and imagines that it is the Universe.
>
> —*Sri Adi Shankara, Indian spiritual teacher*

SURVIVALISM

Few scientific discoveries have altered our worldview as much as the theory of evolution. In 1858 Darwin's *Origin of Species* astounded a world that was primarily ruled by monotheism and creationism. The identification with the idea that we are evolving beings living in an evolving world has greatly changed our understanding of ourselves and our Universe. However, most of us are not aware of the full story behind this monumental event in history. In December 2008, *National Geographic Magazine* featured an article entitled "The Other Darwin," which talked about an important figure in history who could have steered the course of human consciousness very differently, yet due to some shady arrangement, he fell into obscurity after his death in 1913. His name was Alfred Russell Wallace.

In the 1800s, Wallace and Darwin were working on their theory of natural selection through separate avenues. After his voyage on the Beagle, Darwin spent more than 20 years solidifying his theory in his home in England. During that time, Wallace travelled to different parts of the world, collecting and studying animal species in their habitats. Wallace, a self-taught naturalist, studied and collected no less than 126,000 specimens in his lifetime.

Both of these men were greatly influenced by the work of Charles Lyell, a scientist who laid down the principles of geology, and by Robert Malthus, an economic philosopher whose major academic contribution was entitled the *Principle of Population*. Malthus suggests that life would be an on-going struggle because animal and human life reproduce geometrically (2 → 4 → 8 → 16, etc.) while vegetation reproduces arithmetically (1 → 2 → 3 → 4, etc.). In other words, animal and human population would ultimately outstrip the earth's capability to produce food. When the world comes to this stage, it would be filled with poverty, vice, and misery; it would be a world where only the strongest and most ruthless could survive.

Darwin incorporated the Malthusian idea into his theory of evolution and expressed it as "the survival of the fittest," meaning that in the face of struggle and scarcity, it is the competition between individuals of the same species to survive and reproduce that preserves the favourable variants. In other words, it is in the elimination of the weakest and selective breeding in society that the fittest versions are preserved. Wallace's interpretation of the Malthusian concept was different. In Ternate (Indonesia) he asked himself, "Why do some die and some live in the face of scarcity?" His answer was that those variants that best fitted their circumstances, survived. In other words, Wallace emphasised environmental pressures on varieties and species forcing them to become adapted to their local environment. The difference between these two interpretations has tremendous impact on our worldviews today.

As history unfolded, in 1858 both Darwin's and Wallace's theories were presented to the Linnean Society of London at the same time. However, through a series of complicated manoeuvres among Darwin, Lyell, and their upper-class friends, Darwin was positioned as the major contributor, while Wallace was listed as a junior. Many people would say "So what?" According to Dr. Bruce Lipton, author of *Spontaneous Evolution* (2009), "In a Wallacean world, we would improve in order not to be the weakest, but in a Darwinian world, we struggle to acquire the status of being the best." Over the past 150 years our worldview has been reinforced with messages like "beat the others," "eradicate your competitors," and "every man for himself." Collectively, this has promoted a civilisational paradigm of competition rather than cooperation.

Stated another way, in Wallacean terms, the law of the jungle is non-survival of the non-fittest. So, to survive we don't need to be the fittest, we need to be fit enough to adapt to the environment. Therefore, staying tuned and responsive to the environment is crucial to our survival.

When we include the insights of Professor Timothy Lenton on this subject, we receive a powerful message. Lenton proposes that the entire earth is an intelligent, self-regulating organism and, therefore, evolutionary traits that support it tend to be reinforced and those that undermine the environment tend to be restrained. This might suggest that any organism that does not support the whole (the earth) would eventually be unselected by the earth's self-regulating processes. A humanity that destroys the balance of the planet would not be selected to continue and would go the way of the dinosaurs.

Over billions of years, Earth has gone through many cataclysmic changes and it is still here. Clearly, the planet itself does not need rescuing; it is the surface inhabitants who do. The certainty of our existence is no longer a question of how strong we are in competing; it is a question of how cooperative we are in supporting each other, including other species and the earth. It should be clear now that we exist not to *compete* but to *complete*.

Our task must be to free ourselves from the prison by
widening our circle of understanding and compassion to embrace
all living creatures and the whole nature in its beauty.

—Albert Einstein, Nobel Laureate physicist

> True happiness comes not from a limited concern for one's own well-being, or that of those one feels close to, but from developing love and compassion for all sentient beings.
>
> —*the Dalai Lama*

HUMANISM

What is human nature? Do we have the capacity to care, give, and selflessly help others? These questions challenge the core of our beliefs as they relate to our worldview of a friendly, safe, and kind world or one that is cold, dangerous, and cruel. According to Richard Dawkins, author of *The Selfish Gene* (1976) and *The Blind Watchmaker* (1986), evolution is best viewed as the result of genes in action and selection at the level of organisms or populations almost never overrides selection based on genes. He further explains that genes that get passed on are the ones that serve their own implicit interests. In other words, Dawkins asserts that we are born selfish. He also believes that natural selection favours those who cheat, lie, and exploit. Altruism, he claims, is fundamentally inefficient because it impedes natural selection.

Throughout the 20th century, scientists painted the picture that human beings are basically self-centred with a primary focus in life to achieve maximum wealth and pleasure. From the ideas of Thomas Hobbes (materialism), Adam Smith (*Wealth of Nations*), John Stuart Mill (utilitarianism), and others, the largely accepted view is that we do not have inborn traits of compassion or altruism; people only act for the well-being of others when it serves their own interests. However, there is an increasing amount of research that demonstrates our traits of empathy, kindness, and the strong tendency to trust and be trustworthy.

A study carried out by neuroscientists James Rilling and Gregory Berns observed the brain activity of participants who were involved in helping others. The researchers discovered that the act of helping others activated the neurons in the caudate nucleus and anterior cingulated cortex, two areas of the brain that are associated with the experience of pleasure and reward. In other words, helping others brings the same pleasure we get from gratification of personal desire.

Dacher Keltner, Ph.D., a U. C. Berkeley psychologist and author of *Born to be Good: the Science of a Meaningful Life,* and his fellow social scientists have mounted up evidence that humans are successful as a species because of our nurturing, altruistic, and compassionate traits. He found that people with a particular variation of the oxytocin gene receptor are more adept at reading the emotional state of others, and are more likely to remain calm under tense circumstances. Oxytocin is secreted into the bloodstream and the brain, where it promotes social interaction, nurturing, romantic love, and other functions. According to their research, the tendency to be more empathetic may be influenced by a single gene. They discovered that friendly touches stimulate activation of the vagus nerve, a bundle of nerves in the chest that calm fight-or-flight cardiovascular response, and trigger the release of oxytocin, which enables feelings of trust.

Research by Darlene Francis and Michael Meaney reveals that sympathetic environments—those filled with warm touch—create individuals better suited for survival and reproduction. Rat pups who receive high levels of tactile contact from their mothers in the form of licking, grooming, and close bodily contact, show reduced levels of stress hormones in response to being restrained, explore new environments with greater gusto, show fewer stress-related neurons in the brain, and have more robust immune systems as mature rats. According to Keltner, "sympathy is indeed wired into our brains and bodies, and it spreads from one person to another through touch."

Robb Willer, another researcher at U. C. Berkeley, states that the more generous we are, the more respect and influence we wield. In a recent study, Willer and his team gave each participant a modest amount of cash and directed them to play games of varying complexity that would benefit the public good. The results, published in the journal *American Sociological Review*, showed that participants who acted more generously received more gifts, respect, and cooperation from their peers and wielded more influence over them. His findings suggest that anyone who acts only in his or her narrow self-interest will be shunned, disrespected, and even hated.

Are people predisposed to trust and cooperate with others? Economics Professor Paul Zak created a trust game in which Participant A is given a chance to transfer a segment of money to Participant B, who will automatically receive three times the amount that Participant A sent. While doing this, A knows that B will have a chance to reciprocate the gesture by sharing the winnings with A. Traditional economic models would predict that A would not risk any money because B has no incentive to share the bounty. Surprisingly, over repeated experiments, Zak's research showed that in 75% of the cases, A will transfer the money to B. In 90% of the cases, B will return some of the winnings to A. Zak drew blood samples from participants B during the experiment. In cases where A extended trust to B, B experienced a significant increase in oxytocin levels. The conclusion was that when someone is extended trust, they feel close to that person, which compels people to want to return the goodwill by being trustworthy.

In another game devised by Ernst Fehr and Klaus Schmidt, an allocator is given a sum of money and is told to keep a certain amount and apportion the rest to a second participant, a stranger who they called "the responder." The responder can either accept or reject the offer. If rejection is the choice, neither player receives anything. Across 10 studies of people in 24 different cultures, the researchers found that 71% of the allocators offered the responder between 40–50% of the money. It showed that people around the world will sacrifice the enhancement of self-interest in the service of improving the welfare of others.

With more sophisticated and sensitive machines at our disposal, researchers around the world are showing us more and more that we have the predisposition to empathise, feel the joy of helping others, trust and cooperate with them. Overall, these findings support the hypothesis that humans have the biological and psychological capacity to be compassionate. The simple truth is that when we do good for others, we do good for us. We also know now that our nature is not cast in stone. Like everything else on earth, it is part of an unfolding evolutionary process. The complexity of the multi-systems problems that we face today requires more than ever that we amplify our cooperativeness and compassion and turn off our older evolutionary gene of competitiveness and aggression. It is time that we move up the evolutionary helix toward "Homo noeticus," as John White, the consciousness researcher, calls it. We become more collective conscious, we see the divinity in our world, we process more multi-dimensionally and integrally, and we create sustainable solutions by the way we live.

For all of humanity, it has been a long evolutionary journey
to where we stand today, on the brink of triumph or disaster [...].
Since we have the capacity for both moral and immoral
actions and the freedom to choose, our destiny lies within.

—Michael Shermer, science writer and historian

"I am torn," said the old man, "I have a wolf of compassion
and another of hatred fighting within me."
The boy anxiously asked, "Which one will win?"
"Whichever one I feed," said the old man.

—old American Indian saying

Transcending the Intellect

Over the last three centuries, spiritual knowledge and practice has often been received with scepticism. Being in the scientific paradigm, we value what we can count. Without qualitative proof that a system offers benefits, it is an uphill task to fight for social acceptance. We need scientific evidence of the results of spiritual practice, so that experts in fields like education, medicine, psychology, and psychiatry can seriously consider the inclusion of spiritual approaches in their work. One valuable example of scientific-spiritual work was carried out by Sir David Hawkins, M.D., Ph.D., co-author with Nobel Laureate Linus Pauling, back in 1973, of the influential book, *Orthomolecular Psychiatry*.

Hawkins's system, in which he presents what he called the Map of Consciousness, provides people with a framework and process to calibrate the levels of consciousness of anyone and anything, from the past to the present. Author of *Power vs. Force* (1995), Hawkins had 50 years of experience in medicine and psychiatry. In his research and observation, he came to see life in all its expressions as reflecting an inherent level of energy, from weak to strong. Using the muscle-testing method known as kinesiology, he developed an arbitrary scale ranging from 1-1,000 to denote levels of consciousness. The energy scale mirrors the consciousness level of all possibilities of animal and human life. On his Map of Consciousness, level 1 corresponds to the first detectable consciousness of life on this planet, like bacteria. It continues through the plant and animal kingdoms (humans included) to level 1,000—as the maximum possibility on earth.

Using the muscle-testing method of kinesiology, Hawkins spent over 20 years involving thousands of calibrations on thousands of subjects of all ages, personality types, and walks of life. He calibrated the consciousness of thousands of individuals, emotions, places, concepts, writings, intellectual levels, spiritual states, and teachers.

As one can imagine, Hawkins's work was a great accumulation of data whose interpretation has a far-reaching impact on humanity. The intent of this discussion of his work is to highlight its essential aspects that have implications for our current challenge in transcending our civilisational paradigm. One of the crucial features of his research is that the scale he devised increases logarithmically to the base of 10. This means that a number on the scale actually represents ten to that power: 1 on the scale is 10^1 (or 10), while 2 is 10^2 (or 100). So going from one number (integer) to the next is not adding 1, but multiplying by 10. Therefore, the rate of increase in power, as one moves up the scale, is astoundingly massive. Through his laborious work, Hawkins concluded that only a few people in all humanity have reached 1,000, and these are the great avatars like Jesus, the Buddha, Krishna, and Zoroaster.

Of great importance was his understanding that each thought, feeling, and action was permanently recorded in an ever-present field of consciousness that extends beyond time and space. It is similar to the notion of Carl Jung's collective unconscious, the Akashic Field in the words of Ervin László, and Nature's Mind in Edgar Mitchell's terminology.

Another important proposition that Hawkins came to on the basis of his consciousness calibration technique was that a consciousness level of 200 (the level of courage) separated truth from untruth (falsehood). When muscle-tested, all subjects at energy levels below 200 (e.g., 175: pride, 150: anger, 125: desire, 100: fear, etc.) went weak. Everyone became strong in response to the life-supportive field of a consciousness he measured at 200 and above (e.g., 250: neutrality, 310: willingness, 350: acceptance, 400: reason, 500: love, etc.). Level 200 (courage), Hawkins proposed, is the level of empowerment, where life is seen as exciting, challenging, and stimulating. Problems which bring people's energy measurement to below 200 are an impetus to those who have evolved into true *power*, as opposed to *force*.

The model has great implications, even if we would not be able to verify the exact numbers that Hawkins assigned to various states. According to his system, *power* is aligned to divine truth and *force* is pegged to untruth (falsehood). Force is like movement: it is exhaustive and therefore it needs to be fed with energy. Power is like gravity. Gravity does not move against anything; its power moves all objects within its field, but it does not move. In other words, force consumes; power, in contrast, energises, supplies, and supports. Therefore, power is associated with that which supports life, whereas force consumes life.

Especially relevant to our discussion here is the understanding that Hawkins had about consciousness linked to levels in the 400s and 500s. According to his research, 80% of the world's population is below 200 and only 4% cross over to the 500s (love). The percentage of people in the world who reach level 540 (unconditional love) is smaller still, 0.4%. The level of the 400s is the realm of reason, intellect, logic, science, medicine, and, essentially, the Newtonian worldview. Therefore, in a society or country calibrated at 400, knowledge and education is highly valued. The level of the 400s is often the major stumbling block to higher levels of consciousness development. The 400s are dominated by linear thinking, objectivity, and focus on content, while the 500s are the realm of non-linear thinking, subjectivity, and context-awareness. Thus, transcending the intellect/reason/science dimension is the movement from what the world considers objective to the experiential subjectivity.

Below is an excerpt from Hawkins's book, *Transcending Levels of Consciousness* (2006):

> Reason does not of itself provide a guide to truth. It produces massive amounts of information and documentation, but lacks the capability to resolve discrepancies in data and conclusions [...] Reason is limited in that it does not afford the capacity for the discernment of essence or of the critical point of a complex issue [...] Transcending this level is relatively uncommon in our society (only 4%), as it requires a shift of paradigm from the descriptive to the subjective and experiential [...] The limitation of the mind is evidenced by its structure in that the functional ego is linear, dualistic, and dominated by the Newtonian Paradigm of causality that reinforces the illusion of a separate, personal "I" as a self-actualizing causal agent.

With multi-system problems converging, scientists around the world are not able to resolve the discrepancies in the massive complex data. It is quite apparent that science alone is incapable of explaining and resolving the problems that we are facing today. To transcend this level, we must open up to the realm of non-linear thinking and subjective experiences, and adopt an all-inclusive consciousness. This means we need to open up to non-scientific disciplines such as meditation, intentional healing methods, traditional healthcare techniques, energy psychology, shamanic practices, spiritual principles and exercises, and ancient and new cosmological systems. Also, it is important that we move our focus from "I" to "we," to expand our collective awareness and operate more out of this awareness in our lives.

Though our intellectual and scientific mind is able to gather, sort, classify, recode, and recombine information—in order for it to transform then into identifiable knowledge—we must turn it into our inner experience. It is through one's daily practice and realisation that one begins to confirm the value of the knowledge. According to Hawkins, the transformation of intellectual knowledge into divine wisdom can only come about through some form of spiritual practice, such as meditation, contemplation, or prayer. The evolution from the 400s to the 500s, in his terms, is like the timeless path from doing to being; this is made possible by surrendering to the divine the need to control things. By placing our faith in spiritual intention and surrendering to the divine, we move from thinking about something to becoming—as in becoming the change we want to see in our world.

The level of the 500s involves the subjective realm of love and it is facilitated by the intention to break away from the constraints of self-interest. The nature of love is to give and assist, while the human ego wants to be served in its quest for benefits. Increasingly, as one moves beyond the 500s, the appeal of simplicity, peace, and inner silence becomes dominant. More and more, people in the 500s value all of life and they are concerned for the welfare and happiness of others. They also experience more moments of intuition, seeing the hidden subtleties that are influencing situations. They can understand instantaneously, without relying on sequential symbol processing.

Loving energy does not come from the mind, it radiates from the heart. At the Institute of HeartMath, scientists have investigated the heart in communication with the brain and the body, by way of both an extensive neural network and an electromagnetic field. They report that the electrical component of the heart's field is about 60 times greater in amplitude than the brain's. The heart's magnetic field is 5,000 times stronger than the brain's, and it appears to transmit not only physiological but also psychological and a form of social information. As it progresses to higher levels, this loving energy becomes more and more unconditional, forgiving, unchanging, and permanent.

As we mentioned earlier, we are all part of an all-encompassing, ever-present field of consciousness that extends beyond time and space. In *Power vs. Force*, Hawkins states that in this field of consciousness, every improvement we make in our private world improves the world at large for everyone. We are all floating on the collective consciousness of mankind, so any enhancement we make comes back to us. In other words, we all add to our common buoyancy (energy field) by our efforts to improve life.

In short, the more people move toward the levels of willingness, acceptance, understanding, and especially love, joy, and peace, the sooner we will see our world heal.

According to Hawkins's research, the world currently calibrates at the consciousness level of 206. It is clear that the global problems that we are facing now can't be solved at that level. We must raise our consciousness individually, and collectively move toward the 500s.

This is where we collaborate with greater acceptance of differences, resonating the emotional state of forgiveness, transcending national and religious boundaries, and cooperating with a worldview of harmony.

We need to remember the logarithm effect: the Universe recognises those efforts and intentions that truly support life; the higher the consciousness, the more energy is available to counterbalance the levels below. Problems that cannot be resolved at the level of 206 would be easily worked out at higher levels.

PAUSE: A PARABLE OF TWO FARMERS

Parables are the preferred communication tool of avatars in most spiritual traditions. A parable is a short story used to illustrate a lesson or moral; instead of using scientific, literal figures, these teachers chose to convey crucial ideas and messages through a parable. Below is a story adapted from the works of Lee Carroll, the renowned author of 11 Kryon books. One of Carroll's major accomplishments was when he was asked to present his powerful parables in New York before a United Nations chartered group known as S.E.A.T., the Society for Enlightenment and Transformation. It was so well accepted that the group invited him back six more times.

> Two famers spend all their time working hard on their fields, trying to be successful, and every year their crops yield plentifully, which allows them and their families to have a good life. One day, a wise man comes to their fields saying he has an important divine message to deliver. Both farmers are curious and listen intently; they find out that God loves them and because of their diligence, He is giving them the power to increase their crops tenfold!
>
> To enable this power, they have to discard their old crops, thoroughly remove parasite roots and impurities, and reseed immediately. God will change the seasons to support them. As their fields are in this moment filled with crops that are ready to be harvested and sold at the market—which would secure their provisions for this and next year—, the farmers are reluctant to the new idea. The first one discusses it with his family and they conclude that God would not bestow disasters upon them, so they agree to the idea and do what they are told. The second farmer does not believe the wise man and continues to harvest as usual.
>
> Soon, unexpected rains and powerful winds come that help the first farmer's crops grow luscious, but destroy the ones of the second farmer, who loses everything and goes on to live in uncertainty and fear.
>
> Being grateful to God, the first farmer shares his abundance with the second one. Together, they work their

lands until the second farmer is able to plant his seeds according to the new changes in seasons.

Just like the two farmers, many of us work very hard to make a profitable living year after year, and this takes up most of our time. The parable is a good metaphor for our lives in many respects:

- "getting rid of all old crops" signifies eliminating all old ways of doing things;
- "removing all impurities" means ungluing from our attachments, including those things that we know intuitively are wrong or do not support the new life we want;
- "reseeding immediately" is equal to starting to act upon the new life we want, beginning to grow with the newly required actions;
- like the farmers, we know that changes are going to happen soon and, even though we have never experienced this before, we need to have faith;
- "the promise of future rewards (tenfold harvest)" signifies our future ability to co-create the good life we want on our planet; but for this to happen we need to believe it and make changes;
- "the destruction of the second farmer's fields by the rains and winds" connotes that the new earth will not support even the most successful old strategies;
- "the generous sharing of the first farmer's abundance and the humble acceptance of the second farmer" tells us that judgement and pride need not stand in the way of progress.

The essence of the parable revolves around the challenge of processing new world-changing information, going against our programmed beliefs and behaviours, and acting on yet-to-be established new knowledge. This is where the world is at right now. Our old ways of valuing, thinking, and acting have helped us to evolve to where we are today, but they are also the major evolution inhibitors stopping us from bringing forth a better world.

Embracing an integral worldview, harnessing the energy of acceptance and compassion, moving from an "I" to a "we" consciousness, letting go of old paradigm values and lifestyles, and the willingness to act on new, life-supporting knowledge are going to determine our future. We are increasingly realising that these new paradigm requirements are not only scientific but also spiritual. In fact, spiritual knowledge pre-dated the scientific era by at least 2,000 years.

This chapter is laced with many relevant scientific and spiritual quotes. The two distinct fields of knowledge strive to understand reality by using two different approaches to examine it. The intention is to make obvious that even though science and spirituality have been viewed as antagonistic for the past 250 years, more and more today, luminaries from both sides are finding their knowledge and wisdom complementary. The emerging synthesis of science and spirituality is offering us a new lens through which to perceive our changing world. It is with these new studies and applications of matter in spirit and spirit in matter, that we may find the means to transcend the paradigm we are in.

> Even though the realms of religion and science in themselves
> are clearly marked off from each other, nevertheless there exist
> between the two strong reciprocal relationships and dependencies [...]
> The situation may be expressed by an image: science without religion
> is lame, religion without science is blind.
>
> —*Albert Einstein, Nobel Laureate physicist*

> True reconciliation proceeds always by a mutual comprehension
> leading to some sort of intimate oneness. It is therefore through
> the utmost possible unification of spirit and matter that we shall
> best arrive at their reconciling truth and so strengthen
> the foundation for a reconciling practice in the inner life
> of the individual and his outer existence.
>
> —*Sri Aurobindo, renowned Indian spiritual teacher*

CHAPTER 14
WHAT NEW KNOWLEDGE CAN HELP US?

To the mind that is still, the whole Universe surrenders.
—Lao Tzu, Chinese philosopher

Modern science may have brought us closer to a more satisfying conception of the relationship between spirit and matter with the concept of complementarity. It would be most satisfactory of all if physics and psyche could be seen as complementary aspects of the same reality.
—Wolfgang Pauli, Nobel Laureate physicist

These two insightful descriptions of our world come from two different masters—one a spiritual philosopher and the other a scientist—even though separated by 2,500 years, both arrived at the same wisdom through different methods of understanding the cosmos. Science has shaped our lives and our world. It has also become our way of thinking, learning, valuing, and trusting what exists in our world and how it works. Yet, despite its vast oceans of knowledge, science has great challenges in telling us what connects the physical Universe to the living world, the living world to human society, and human society to the realm of the mind. However, over the last few decades, with the development of super sensitive equipment, our understanding of how the world works has changed dramatically. The more experts from different fields work together, the more they appreciate and are excited about the new integrative discoveries they are making and the possibilities these bring.

Integral studies and experiments in the fields of physics, biology, medicine, ecology, evolution, systems theories, psychology, philosophy, and spirituality are giving rise to an important field of knowledge called "spiritual science," where science and spirituality are seen as complementary. This chapter is dedicated to exploring this new knowledge and the presuppositions coming with it that could help us bring forth a healthier and more harmonious world. It is hoped that this new knowledge can help us get out of the box (paradigm) we are in, so that we are *not* looking at new problems with old eyes.

FIELDS OF STUDY WITH SPIRITUAL DISCOVERIES

Volumes of books and research articles are exploding around the world on the subject of spiritual science, uncovering the similarities between ancient knowledge and scientific principles and methods. Criss-crossing among many disciplines, this subject includes the study of:

- the effects of meditation on the human brain and on hormonal changes in the body;
- the effects of spiritual beliefs and prayers on the psychological, emotional, and physiological health for oneself and others;
- the benefits of spiritual, hands-on energy healing of critical illnesses (e.g., AIDS, cancer, cardiac problems, etc.);
- the impact of human intentions on nature (plants, water, bacteria, etc.);
- the influence of human beliefs on gene behaviour;
- the impact of human observation on the smallest building blocks of the Universe: atoms, electrons, wave-particles;
- the correlation between electromagnetic fields around Earth and human events;
- an intelligent and responsive field of consciousness that permeates the Universe—foretold by ancient spiritual traditions;
- the relationship between human consciousness and the geophysical, social, and biological conditions on earth;
- planetary evolution and its connection to ancient spiritual knowledge, indicating an intentional, rather than a random, meaningless Universe.

Many of the studies referred to above have already been shared in this book. There are numerous fields of study that are contributing to the knowledge of spiritual science, such as quantum physics and consciousness, evolutionary biology and epigenetics, general systems theory and research on macroshifts. These fields of study are increasingly confirming the presence of an intelligent order in the Universe that is out of our day-to-day awareness. At a surface level, many of these new discoveries appear to overturn and to contradict old assumptions, but, in truth, they subsume them.

Think about this: air, vapour, water, and ice are different realities that exist at the same time; on the surface they look and feel different, but they are actually part of each other and intimately connected. The existence of one does not nullify the other. The higher order subsumes the lower one: air is the lightest, most intangible, and most versatile, so it subsumes the lower and more dense realities. The hidden messages of these studies tell us that there are better and simpler ways for us to live more healthily, joyfully, and harmoniously. At the same time we are currently steeped in our civilisational beliefs, and it is like trying to break ourselves free from a tar pit. We would require new images of reality and the invocation of a different form of energy to move. To activate this latent inner-technology in all of us, we need new information; and spiritual science offers us promising possibilities.

Quantum Physics and Consciousness

Quantum physics is the study of the behaviour of matter and energy at the atomic, sub-atomic, and nuclear levels. Prior to the discovery of quantum physics in the early 1900s, the mechanistic science of Newtonian physics (dating from 1687) appeared to explain quite adequately almost every phenomenon in our world. With the splitting of the atom in the late 1800s, the foundation of the Newtonian world was shaken. The sub-atomic particles that arose from splitting the atom did not behave like conventional solids. In other words, what was happening in the microscopic realm mismatched the macroscopic reality—a world governed by scientific materialism, reductionism, and determinism. Since then, the ongoing discoveries in quantum physics have provided humanity with a new, deeper, and wider understanding of our world. The ramifications of quantum physics are vast. Highlighted below are some of the crucial discoveries relevant to this chapter.

Matter and Observer Are Inseparable

This discovery simply states that on the sub-atomic level, the very act of observing something changes it. Many scientists used to believe that the world outside of us was unaffected by our observation. In other words, whether we were observing something or not, that something was going to remain what it was. This paradigm was shattered with the advent of quantum physics. Science has always been on a quest for the most basic substance of reality. Up until the early 1900s numerous scientists believed that the Universe was made up of physical particles. Yet these apparent physical particles were also proven to be immaterial waves by other scientists.

So, is the Universe made up of particles or waves? The surprising final answer—coming from Nobel Laureate scientists—was that if the scientist designed an experiment that registered particles, he noticed particles; if the scientist created an experiment to detect waves, he detected waves. In other words, the observer determines the reality perceived!

Another experiment that supports this point is the famous double-slit experiment, first conducted by Thomas Young in 1803. He found that when coherent light passes through a barrier with one opening, it displays one band of light on the wall. However, when two slits are used, instead of displaying two bands of light, as particles would, it shows an interference pattern, as waves would. This experiment was repeated many times by numerous other scientists and the results were always the same. Was the coherent light in Young's experiment waves? So it seemed. Only waves could form interference patterns on the wall. This assumption did not make sense until experiments were performed in the 1900s, when only one electron (a subatomic matter) was emitted at a time. Logical reasoning told the scientists that a single electron could not be a wave and it could only pass through one slit and not both slits at the same time. Yet, when single electrons were emitted, a wave interference pattern built up on the screen, as if each electron passed through both slits at the same time!

In order to unravel this mystery, scientists at Israel's Weizmann Institute conducted the same experiment in 1998 with a slight twist; they placed a sensitive instrument in front of the double slit to observe the movement of the electron. With the supersensitive instrument turned on for both paths of the electron to the slits, the interference patterns disappeared. The act of observing an electron changed its behaviour! In other words, our act of observing something, in this sub-atomic domain at least, changes the very thing that we observe. A video clip cleverly presents this phenomenon at http://www.youtube.com/watch?v=3_BzTMeV4HI, as a cartoon.

> Everything material is also mental and
> everything mental is also material.
>
> —*David Bohm, renowned quantum physicist*

> The objective world arises from the mind itself.
>
> —*the Buddha*

By now it should be clear that it is impossible for us to simply be bystanders in our world. As physicist John Wheeler said, we live in a "participatory Universe," not one where we manipulate or force our will on it, but a Universe that we co-create. Both science and spirituality describe a relationship that gives us the power to influence how matter behaves in our world. Therefore, we have the power to influence our reality by simply changing the way we perceive it. Now, most classical scientists believe that the way electrons and photons behave has little to do with how we live our everyday lives. The ancient traditions, however, suggest that it is because of the way things really are at the unseen subatomic level that we can change our world. At a collective level, the questions that we ask, the intentions that we uphold, the attention that we give to something, the emotions that we invoke every day ultimately create the kind of world in which we live.

The Oneness of Everything

In 1964, John S. Bell presented a theorem that profoundly influenced the advances of science. It culminated in the acceptance by the scientific community of the term *quantum entanglement*. This scientific discovery—first made in the 1980s by Alain Aspect—proved that particles can influence one another instantaneously despite being separated by vast distances.

To demonstrate this, scientists split a single photon into two separate particles, creating twins with almost identical properties. These twins had reciprocal spin states—when one span upwards the other span downwards. The scientists then fired the twins, at the speed of light, in opposite directions. At one of the endpoints, one of the twins was forced to spin in the opposite direction. Astoundingly, even though the other twin was 10 kilometres away, it instantaneously reversed its spin! The experiment was carried out several times by other scientists, over greater and greater distances, and the result was the same every single time.

This discovery poses a conundrum, because in Einstein's theory of relativity, nothing travels faster than the speed of light. But these experiments clearly showed that in order for one of the twins to know the change in spin of the other, something must have travelled faster than the speed of light to inform it. Or could it be that in reality they were somehow still connected, i.e., were entangled? Is the space between the twin particles is an illusion? Were they never separated in the first place?

A good example to explain the mechanics of quantum entanglement is the idea of holograms, which are three-dimensional representations of objects, recorded by a special process. A holographic recording (of an object) consists of the pattern of interference created by two beams of light encoded onto a photographic plate. When a beam of light is shined onto the photographic plate, the object encoded reappears in a three-dimensional form. Interestingly and importantly, if the holographic plate is cut into many pieces, each piece—no matter how small—will still have the full view of the entire original object. Thus, each portion of the plate contains information of the entire object.

Many renowned scientists in the last 40 years have believed that we are living in a holographic universe. Going back to the quantum entanglement experiment, the signal from one particle never travelled to reach the other; they were instantaneously connected to each other like a hologram. A change in property anywhere in a hologram means a change everywhere.

One of the most important underlying messages of the hologram is that in life, a little change here or there can suddenly change everything. A seemingly small change in one place can dramatically shift an entire pattern. Another important interpretation of this principle is that any change we wish to see in our world—be it to heal a loved one or create peace in war-torn countries—doesn't have to be sent from our hearts to the places where it's needed. Once our prayers are felt inside of us, they are already everywhere!

Science is showing us that we live in an inseparable world, a reality that is fully interactive, interconnected, and interdependent. In light of this fact we cannot not influence each other, therefore, we must choose to be life supportive. When we collectively utilise our potential and resources to truly listen and support each other, the whole is greater than the sum of its parts.

The Miracles of Positive Intention and Emotion

Across history and cultures, people have claimed to have been healed by another person's caring intention, even at a distance. Healing people at a distance is now known as *distant healing intention* (DHI). This includes prayers, spiritual healing, intentionality, energy healing, shamanic healing, and nonlocal healing. Very often, healers involved in distant healing share the conviction that their process includes contact with an indescribable spiritual dimension. Over the past four to five decades, researchers have developed techniques for measuring possible distant healing effects. The best experiments have employed rigorously controlled designs that rule out all known conventional sources of influence, such as environmental factors, physical manipulations, suggestion, and expectancy.

Russell Targ and Harold Puthoff conducted experiments in the 1970s on the transference of thoughts and images. They placed a "receiver" in a sealed, opaque, and electrically shielded chamber and a "sender" in another room, where he or she was bombarded with bright flashes of light at regular intervals. Both sender's and receiver's brainwaves were registered on electro-encephalograph (EEG) machines. As expected, the sender's brainwave reflected the bombardment of flash-lights. However, after a brief interval, the receiver also began to produce the same patterns! This indicates that thoughts can be sent to or received from another person without any form of communication perceivable by the senses.

A remarkable DHI study was reported in the *Western Journal of Medicine* by Fred Sicher in 1998. In this study Sicher was working with advanced AIDS patients and he incorporated a wide range of healing practitioners from different healing, spiritual, and religious traditions.

Healers received a photograph of their patient with that patient's last name and T-cell count. Each patient in the treatment group received healing efforts from one healer at a time, one hour per day, six days per week, for ten weeks. After six months, the prayed-for patients had acquired significantly fewer new AIDS-defining illnesses and had lower illness severity, fewer doctor visits, fewer hospitalisations, and improved mood scores as compared to the control patients. These were highly significant outcomes, given that AIDS at the time of this study had a gloomy prognosis and no effective treatment.

Would healing intentions have any impact on human DNA? This question spurred Leonard Laskow, M.D., Glen Rein, Ph.D., and Rollin McCraty, Ph.D., to conduct a series of experiments between 1993 and 2000. The first set of experiments involved measuring the growth of tumour cells in a culture. Dr. Laskow began with a "loving healing presence" through a transpersonal alignment and a conscious heart focus. This process enabled Dr. Laskow to vibrate at the same frequency as the cells, creating a state of entrainment. His overall approach was to "focus lovingly" on whatever he wanted to change. Combining intentional focus imagery with a coherent heart-based resonance, Dr. Laskow was able to achieve a 21% to 39% inhibition of growth rate in tumour cells. The phenomenal results were achieved with the intentions of "Return to the natural order" and "Let God's will flow through these hands." In an adjacent room, a non-healer was simultaneously duplicating the same experiment, but was reading a book instead of focusing on the culture. The results showed that the growth rate of the tumour cells continued to increase.

Next, the researchers used human placental DNA dissolved in water, which had distinctive characteristic patterns when measured with ultraviolet spectroscopy. Using specially designed mental and emotional state management techniques, trained practitioners focused their emotions and intentions on beakers containing human DNA. Using sensitive scientific equipment, the researchers were able to examine the chemical composition and visual structures of the DNA. The results were undeniable: the practitioners' emotions and intentions changed the shape of the DNA, even though the practitioners didn't even touch it! The experiments revealed that the coherent emotions of the practitioners (such as love, gratitude, and compassion) caused the double-helix of the DNA to unwind (relax and elongate). The unwinding and winding of the DNA are important in DNA replication, repair, and other basic cell functions. The experiment proved that a loving state can produce physiological effects at the DNA level. It also provided vital information that only those individuals who were able to maintain coherent electrocardiograms (ECGs) could alter the DNA conformation.

Even more astoundingly, the researchers speculated that the effects might be due to the proximity of the DNA to the participants' hearts, since the heart generates a strong electro-magnetic field. So they carried out a similar experiment at a greater distance of half a mile from the DNA samples. The effects were the same! Five non-local trials showed the same effect, all with statistically significant results.

Combining what we have covered so far in the quantum physics and consciousness segment, we now have evidence that our thoughts and emotions have the power to create and heal our bodies and our world. Presuppositions like "an observer has the power to influence reality," and "a change in anything changes everything instantaneously," plus "life-supportive thoughts and emotions heal our world" can literally help us to break out of our current paradigm of isolation, fragmentation, and self-absorption. The union of science and spirituality is bringing us to the threshold of a new world—a collective consciousness where coherent intentions and emotions become our way of life in an increasingly rejuvenating world.

The Awesomeness of No-Thing

According to NASA, the Universe is made up of 70% "dark energy," 25% "dark matter," and 5% normal matter. In other words, 95% of the Universe is empty. The commonly held view was that matter occupied space and moved about in it, space serving as a backdrop or container. This classical concept has been radically revised in light of the discoveries made in the last several decades. Scientists are beginning to learn about the nature of the *quantum vacuum*—the subtle energy-sea that underlies all of space and time.

Space is now considered to be the primary reality, instead of matter, because space is not empty; it is filled with unlimited potential. Without this quantum vacuum, no electrical, electromagnetic, gravitational, and nuclear energies can propagate. It would be like throwing a stone into a dried-up pond and hoping to see ripples. In this analogy, the quantum vacuum is the medium that carries the ripples. This no-thing space is also known as the "zero-point field," so named because it contains energy at ground state or vacuum state at the temperature of absolute zero on the Kelvin scale. The energies of this field persist even when all other energies vanish. All matter in the Universe is generated from the zero-point field. The energy potential of this field is tremendous. According to physicist John Wheeler of Princeton University, the energy density per gram/cm^3 is

10^{94}, which is more than all of the matter in the entire Universe! So space is not empty, it is filled with incredible matter-equivalent energy potential.

> The atoms or the elementary particles [...] form a world of potentialities or possibilities rather than one of things or facts.
> —*Werner Heisenberg, Nobel Laureate physicist*

> In Buddhist emptiness there is no time, no space, no becoming, no-thingness; it is what makes all things possible; it is a zero full of infinite possibilities, it is a void of inexhaustible contents.
> —*Daisetsu Teitaro Suzuki, Japanese Buddhist scholar*

The matter that occupies the Universe we live in emerged when the quantum vacuum became unstable; this led to the cosmic explosion known as the Big Bang. In other words, the quantum vacuum existed prior to the Big Bang! The Vedic spiritual scriptures of India (thousands of years old) talk about the existence of a force that underlies creation. This force was there before the beginning and it is called *Brahman*. It is identified as "the unborn [...] in whom all existing things abide." The same text states that all things exist because "the One manifests as the many, the formless putting on forms."

The existence of the quantum vacuum was established by the experiments of Vladimir Proponin at the Institute of Biochemical Physics of the Russian Academy of Sciences. He first removed all the air from a specially designed temperature-controlled tube, creating what is thought of as a vacuum. Despite the emptying, Proponin knew that something would remain inside, and that was photons—the stuff that our Universe is made up of. Using sensitive equipment, the researchers identified the locations of the photons, which were all over the tube. They then inserted human DNA into the closed tube. Proponin found that the photons arranged themselves differently in the presence of the living material. Through an invisible force, the DNA shifted the locations of the photons. What Proponin discovered next startled him; the arrangement of photons remained unchanged for a long period of time, even after the DNA was removed! He concluded that a new field structure was triggered in the physical vacuum. This "phantom effect" seems to be a manifestation of an overlooked substructure in the energy-sea that permeates the Universe.

Scientists are now contending with the possibility of an unknown intelligent force in the quantum vacuum. It has the capacity to store, connect, and propagate information coherently, instantaneously, and everlastingly. The renowned physicist David Bohm called it in-formation, meaning a process that actually forms all living and non-living things in the Universe; it is the in-formation in nature.

The Talking Universe

With the most cutting-edge studies in particle and field theories, scientists are uncovering a new foundation that generates all things without themselves being generated by other things. This primordial foundation, this virtual energy sea (the quantum vacuum, as we have mentioned) contains a new factor that is neither particle nor energy: it is information. Analogically speaking, it is like the DNA of the Universe. This information governed the Universe at its birth, and, therefore, directed the evolution of its basic elements into complex systems.

According to Edgar Mitchell, the founder of IONS, information is ever-present everywhere and it existed before the Big Bang. The quantum vacuum is the holographic information processor that records the historical experience of matter. It is obvious that the information here is not static but active; it is "in-formation." Ervin László, one of the world's foremost systems theorists, states that this in-formation links particles, organisms, ecologies, solar systems, entire galaxies, as well as consciousness non-locally and trans-personally. It is a subtle, instantaneous, ever-present, and non-energetic connector of all things. Through the process of wave interference in the quantum vacuum, vortices are created. These vortices store the information that the waves carry and pass it on to other vortices when they interact. In short, the Universe is a memory-filled cosmos where everything in-forms and interacts with everything else instantaneously and constantly.

Ervin László calls the quantum vacuum the Akashic field, having taken the idea from Indic philosophy and cosmology. *Akasha* (ākāśa) is the Sanskrit word for *aether*, in both its elemental and mythological meaning. Akasha cannot be perceived through ordinary senses, but only through spiritual practices. In Indic philosophy, the Universe is made up of two things: one is akasha and the other one is called *prana* (prāṇa). Everything that is formed evolves from akasha. It is the eternal source of all forms, the realm of promise, potential, paths to be walked, and the primal source that creates and nourishes the four elements (fire, earth, air, and water). Prana is the primal energy, and when it acts upon akasha, all known forms arise. The quantum vacuum encompasses the processes of akasha and prana; it is the womb of all matter and energy in the Universe.

Like all things in the Universe, human thoughts, perceptions and emotions create waves and, therefore, information-carrying vortices. These waves and vortices interfere with the waves created by other people, giving rise to complex holograms in the Akashic field. Generations after generations of humans have left their holographic traces in the Akashic field. The collective holograms of families, communities, cultures, and nations create super-holograms of all people; this is the collective in-formation of humankind. This collective in-formation can be tapped. We can tune our consciousness to resonate with the holograms in the Akashic field (very similar to the theory of resonance). Tuning forks and strings on musical instruments resonate with other forks and strings that are tuned to the same frequency. It is important to note that the resonance effect does not occur when the instruments are not tuned to the same frequency level, or harmonic range.

The long-held question of whether our minds can affect the physical world has motivated scientists all round the world to find answers. As early as the 1960s, scientists started off with simple mechanical processes like flipping a coin over and over again to see if it would match the experimenter's intention. The research then evolved into computerised processes, where bits of information were tossed between ones and zeros. In both methods, the experimenters were able to obtain results matching their intention a statistically significant number of times.

In the 1990s, two scientists, Roger Nelson, Ph.D., and Dean Radin, Ph.D., started to use Random Number Generators (RNGs) to track special events where groups of people shared common intense states, such as motivational workshops, religious prayers, theatrical presentations, etc. These RNGs are devices that rely on quantum-level processes to generate random numbers. These devices have passed an extensive array of stringent tests for randomness. Unless influenced by external factors, the RNGs would continue to generate random numbers. If the numbers begin to show patterns of non-random orderliness, or coherence, then an anomaly has been detected: in most cases, the studies showed that multiple minds holding the same intense thoughts created some kind of deviation from the norm of the equipment.

In 1997, Nelson placed RNGs all over the world, had them run continuously, and compared the results with highly emotional global events. This was later known as the Global Consciousness Project. Nelson directed all the RNGs in the world (36 of them) to download their continuous stream of random data into one central computer via the Internet. Together with Radin, the researchers studied the vast data and contrasted them with the biggest news, to see if they could find any statistical connection. By 2006, the RNGs had grown to 60 in number, and the researchers had correlated the data with 205 top events. These memorable events included the death of Princess Diana, the NATO bombing of Kosovo, the millennium celebration, Pope John-Paul II's pilgrimage to the Middle-East, and the global peace demonstration against the United States invasion of Iraq.

The most indicative data came from 11 September 2001. Nelson, Radin, and two other scientists studied the data that then came in from 37 RNGs. According to the 4 individualised analyses, the effect on the RNGs during the plane crashes was unprecedented. The global mind simultaneously reacted to the horror. Nelson and Radin eventually published a summary of their analysis in the prestigious physics journal *Foundations of Physics*.

An additional confirmation of the effect of this terrible event came from another source. Two satellites orbiting the earth, known as the Geosynchronous Operational Environmental Satellites (GOES), detected a distinct rise in global magnetism. The increase in units of magnetic flux density was almost 50 nanoteslas above the normal baseline. This increase happened 15 minutes after the first plane hit the World Trade Center. Subsequent studies between Princeton University and the Institute of HeartMath led to the formation of the Global Coherence Initiative (GCI). Using a global coherence monitoring system, the GCI measured fluctuations in the magnetic field of the earth and its ionosphere. They discovered that the earth and its ionosphere generate a symphony of frequencies ranging from 0.01 to 300 hertz, and some of these match the frequency ranges of the human heart and brain.

The most startling discovery about the September 2001 tragedy was that the data from the days leading up to the event was normal, but the RNGs became abnormally coherent four hours *before* the first tower was hit! It is as if the world had felt a collective shiver several hours before the horrifying event happened. This suggests that the world had an unconscious knowing of the impending event. The work of Nelson and Radin also strongly suggests that certain random physical processes, such as the collapse of the quantum wave function, may not be random and indeterminate, as formerly assumed. In fact, they may be dependent on some factors (human intention), and, thus, may be influenced at a fundamental level under certain conditions. Researchers at GCI theorise that when large numbers of humans respond to a global event with a common emotional feeling, the collective response can influence the activity of Earth's energy field, which, in turn, affects all Earth's inhabitants.

This Akashic field is not perceptible in our ordinary day-to-day states; it can only be accessed through an altered state. Via meditative, contemplative, or prayer-like states, we can disconnect from our distracting five senses and tune into an untapped and powerful resource within each and every one of us. It is the equivalent of an Akashic modem to access a broad range of information that connects us to other people, nature, and the Universe. Instead of experiencing the world through our isolated tower of five windows (five senses), we open our roof to the Universe. This is the crucial and exciting phase in our current evolution: the awakening to our ability to connect to the universal consciousness. This is what the Chinese sage Lao Tzu meant when he said, "To the mind that is still, the whole Universe surrenders."

Evolutionary Biology and Epigenetics

We learned in school about genetic control, how genes primarily control the traits of life. However, the new science of epigenetics reveals that life is controlled by something above or outside the genes. In Greek, the prefix *epi* means *over* or *above*. Epigenetics is the study of the signals outside a cell that turn the genes on and off. Some of those signals are chemical, others are electromagnetic. Some come from the environment inside the body, while others come from outside the body. Epigenetics has shaken the foundations of biology and medicine to the core, because it reveals that we are not victims but masters of our genes.

The Biological Instruction Centre

A biological breakthrough was made in 1953 when James Watson and Francis Crick discovered the double helix structure of DNA and its molecular building blocks, called nucleotide bases: adenine, thymine, guanine, and cytosine. A gene is represented by a length of DNA code that contains the nucleotide base sequences needed to make a specific protein. Protein molecules are the building materials of the cell, and they are responsible for an organism's physical and behavioural traits.

Based on this discovery, Crick developed the concept called "the central dogma" which has been accepted by scientists and taught in universities around the world ever since. This concept dictates the flow of information in cells, as in information → copier → proteins.

The DNA and its nucleotide base sequences represent the information. The RNA (ribonucleic acid) represents the copier; its function is to transcribe the information from the DNA and translate it into protein molecules. The essence of Crick's concept—that resulted in a civilisational misperception—is that information in cells only flows in one direction, i.e., DNA to RNA to protein. This means that protein cannot send information about life's experience back to the DNA. Therefore, environmental information cannot change genetic destiny.

The interpretation by some of Crick's central dogma was that our genes control our physical and behavioural traits. In other words, that our genes control our health, physical capabilities, intelligence, emotional quotient, etc. Since we cannot change our genes, we are victims of our own heredity. Strangely, this concept was accepted by many scientists, even though they did not know how Crick arrived at this conclusion. This concept reigned supreme for 40 years.

Reversing the Central Dogma

Howard Temin made an important discovery in the 1970s, which led to his receipt of a Nobel Prize. Temin showed that certain viruses carried the enzymatic ability to reverse the flow of information from the RNA back to the DNA. At that time, Temin was declared a heretic, because he went against the central dogma. It was later discovered that HIV and hepatitis viruses use the same process to thrive in the human body. Temin's contribution to humanity was huge. Through reverse-processing, hereditary changes can be made by design or environmental influence, and not only by accidental mutation, which was a commonly accepted Darwinian theory.

Another landmark finding entitled "The origin of mutants" was brought to the world's attention in 1988. It was published in the prestigious British journal *Nature* by an internationally renowned geneticist, John Cairns. He chose bacteria with a crippled gene that couldn't digest lactose. He then put this bacteria into cultures in which the only nutrient was lactose and expected no bacteria to grow, because they could not metabolise the lactose. He was proved wrong! A large number of cultures showed growth of bacteria colonies. In his study, a stressful environment somehow spurred the bacteria to mutate. *The experiment clearly showed that environmental stimuli could be fed back into an organism to rewrite genetic information.* Furthermore, his experiment showed that out of five possible different mutations, all the living bacteria expressed the same type of mutation, and only the genes responsible for metabolising lactose mutated. This clearly demonstrates that *the mutation is not random, but purposeful.* Just like Temin's, Cairns's finding was initially hailed a heresy in evolutionary biology by both British and American scientists.

A recent discovery in 2007 by Professor Sydney Pearce of the University of South Florida established that a sea slug (Elysia Chlorotica) extracted genes from algae and slipped them into its own DNA. This is the first proven instance of gene transfer between multi-cellular organisms. If the slug has one meal of algae early in its life, it can use the borrowed genes to produce its own chlorophyll and live out its life (if it chooses) relying solely on sunlight. These discoveries by Temin, Cairns, and Pearce contradict the Darwinian theory which states that evolution is driven by random mutation (lucky events).

According to Bruce Lipton, Ph.D., who is currently one of the world's foremost experts in epigenetics:

> Genes are not emergent entities, meaning they do not control their own activity. Genes are simply molecular blueprints. And blueprints are like design drawings; they are not the contractors that actually construct the building. Epigenetics functionally represents the mechanisms by which the contractor selects appropriate gene blueprints and controls the construction and maintenance of the body. Genes do not control biology; they are used by biology.

In other words, external environmental factors such as geomagnetics, electromagnetics, all forms of vibration—including our thoughts and emotions—can turn our genes "on" or "off." It has been clearly shown here that the influences on human consciousness are not limited to humans, but are also interwoven with other species, the planet's frequencies, as well as the all-encompassing Akashic field.

In short, we are co-evolving with nature.

Epigenetics has also revealed that offspring could inherit traits that are a result of their parents' experiences. This was demonstrated by Randy Jirtle, Ph.D., and Robert Waterland, Ph.D., of Duke University in their experiments on a group of yellow, obese, pregnant agouti mice, who were prone to diabetes, hypertension, and cancer. The mice were given methyl-group-rich supplements (folic acid, vitamin B12, and choline). Studies have shown that these supplements can attach themselves onto a gene's DNA and change the binding characteristics of regulatory chromosomal proteins. The result showed that mothers who received the methyl group supplements produced lean brown mice, even though their offspring had the same agouti genes as their mothers; in other words, the agouti gene was "turned off" through the methylation effects of the supplements. The agouti mothers who had not received the supplements produced yellow mice pups, which ate much more than the brown mice pups. They eventually ended up weighing twice as much as the lean brown pups.

Another related discovery was made by Michael Skinner, Ph.D.; female gestating rats exposed to environmental compounds such as endocrine disruptors would reduce their male offsprings' fertility. Dr. Skinner, a molecular biologist at Washington State University, made it clear that the life experiences of grandparents and even great-grandparents alter their eggs and sperm so indelibly that the change is passed on to their children, grand-children, and beyond. It is called "trans-generational epigenetic inheritance."

In his study of exposing rats to vinclozolin (fungicide) and clusters of atoms called methyl groups, he was able to identify and analyse the "on-off" settings of sperm DNA. Results showed that 16 DNA had been altered: it had turned on when the normal position was off, or off when the normal position was on. Those alterations appeared in the sons of mothers exposed to the fungicide when they were pregnant, in the sons of the sons, and in the sons of the sons' sons. The point is not that this fungicide causes problems in animals and humans, but that an environmental exposure can leave its mark on DNA for at least four subsequent generations.

In the old paradigm, genes predetermined outcomes. Now, everything we do, eat, or smoke can affect our gene expression and, therefore, future generations. This can suggest that we have the power to influence our own evolution. The future well-being of our children and our species resides in the way we lead our lives today.

Paradigm Shift in Healing

In epigenetic studies, both mechanical- and human-generated electro-magnetic fields can produce a shift in gene expression. Bernard Grad, Ph.D., of McGill University, demonstrated that waters blessed by energy healers would significantly promote faster and larger growth in germinating seeds. Conversely, he also proved that water held by psychotically depressed patients would retard or drastically slow down growth of germinated seeds. On the mechanical side, Ciba-Geigy, a large pharmaceutical company, was able to grow trout with distinctive hooked jaws that had been extinct for 150 years, using only electrical effects on the fish eggs.

Electromagnetism has increasingly been used for healing in the last 40 to 50 years. The Pulsed Magnetic Stimulation is a machine that delivers timed magnetic pulses at controlled strengths to parts of the human body. It has been used on people with Alzheimer's, epilepsy, and Parkinson's disease. It also helps diseased or damaged cells to utilise oxygen, thus accelerating their recovery. A normal cell has an electrical potential of about 90 millivolts. An inflamed (wounded) cell radiates 120 millivolts, and a cell that is in a state of degeneration may drop to 30 millivolts. By entraining the targeted cells to the PMS machine, they can be brought back into a healthy range. Electrical changes are not only seen in targeted cells but also in groups of cells, in organs, and in whole systems.

Entrainment is similar to the concept of resonance: it happens when a system that has a certain periodicity is pushed by another type of periodic waveform in a synergistic way. It is like a child on a swing, where the pumping action of the legs and distribution of the weight to gravity is in harmony with the swinging action; the child will gradually increase the swing to a higher and higher degree; when the amount of energy displaced by the pumping action has reached the swing's highest zenith, resonance occurs. In biological terms, many resonances can happen in various ways. Viruses, bacteria, cells, and DNA all have resonance frequencies, and they all affect the organism and environment within which they live. In neuropsychology, a certain set of neurons could be sensitive to a resonant condition caused by certain ideations and thought patterns, which get amplified through reiteration. These resonances alter the brainwaves and, therefore, influence the ocean of quantum-wave potential that we are swimming in all the time. This explains the law of attraction, as in what we focus on, tune in to, and feel in our hearts, we manifest through vibration.

Harold Burr, in his 1936 work on *The Electro-Dynamic Theory of Life*, clearly suggested that diseases show up in a person's energy system before manifesting as symptoms. A professor of Yale University, Burr believed that physical diseases could be treated by restoring balance to a person's energy system. To support this theory, Dr. Jacques Benveniste (1992) demonstrated the effects of electromagnetic fields by exposing a human heart to the electrical frequency of histamine molecules, instead of injecting the person with histamine. When this was done, the heart rate speeded up. Benveniste then exposed the heart to the electrical frequency of atropine, and it reduced the flow of blood in the coronary arteries, just as the organic compound would have done.

While the human genome project was well publicised, another, perhaps more important one was virtually unknown: the mapping of all the proteins expressed in all of our cells at all points during our development and adult life. It is called the Human Proteonome Project (HPP). Currently, scientists estimate that we have about 100,000 proteins, and they can be assembled in many different combinations, such as insulin, serotonin, dopamine, etc. The number of known combinations is four million, and it is still rising.

Numerous studies on the power of belief and its impact on health clearly show that the potential for self-healing is almost unlimited. It has been clinically proven in placebo studies that a fake pill or bogus medical procedure has the power to activate healing proteins in our bodies. The phenomenon is related to the perception and expectation of the patient; if the substance is viewed as helpful, it can heal, and if it is viewed otherwise, it can cause negative effects.

In 1998, Irving Kirsch and Guy Sapirstein used a meta-analysis to analyse 19 clinical trials, which involved 1,460 patients receiving anti-depressant medication and 858 receiving placebo sugar pills. The experiment revealed that 75% of the effect attributed to the perceived use of antidepressants was due to the placebo effect. In other words, the clinical benefits from antidepressants were largely due to the beliefs of the patients and not to the actual chemical ingredients of the pills.

In another study done by Dr. Thomas Luparello, 40 asthmatics were exposed to the same placebo twice, and each time they were told a different substance was being administered. First, the participants were told they were inhaling a bronco-constrictor as part of an industrial air pollutant study, while they were actually inhaling saline, an inert substance. Of the 40 asthmatics, 19 experienced difficulty in breathing. In other words, 47.5% reacted to the placebo. The same patients were then told that the inhaler contained a bronco-dilator (still using saline solution), and their condition was reversed: they breathed much better! The belief of the patients opened and closed their bronchial airways.

There are many scientific studies on the power of placebos, and they are all, more or less, giving the world the same message: when we change our beliefs and attitudes, we change our biology. Our bodies are capable of synthesising many different substances (more than four million protein combinations), with drastically different molecular structures within our organs. This activity can be activated or released through our beliefs, intentions, and emotions.

The benefits of meditation and prayer have increasingly been studied by scientists over the last four decades. Instead of popping a pill or undergoing surgery, more people are turning inwards and are developing rapport with their subconscious—the key to unlocking spiritual power within. Scientists have used sensitive equipment to study the effects of meditation. The electro-encephalography (EEG) measures the electrical activity in the brain. Positron Emission Tomography (PET) provides a three-dimensional image of functional processes in the body, and the Functional Magnetic Resonance Imaging (fMRI) measures changes in blood flow in the brain and spinal cord. Scientists have also done detailed analyses of the substances released into the body by the brain and the vital organs.

In 2005, studies done at Yale and Harvard Universities showed that meditation increases grey matter in the brain and slows down brain deterioration. Richard J. Davidson of the University of Wisconsin discovered that mindfulness training over eight weeks significantly increases activity in the "happy-thoughts" part of the brain and greatly enhances the immune function. Dr. Andrew Newberg studied the brains of Tibetan Buddhists in meditation and Franciscan nuns in deep prayer. The activities in their brain suggested that they were experiencing a merging with all that is around them—they were one with everything.

Another study showed a release of dopamine, which is a neurotransmitter that produces positive emotions and is also associated with the reward system of the brain. This explains the positive feelings expressed by meditators. Many studies on meditation have confirmed that during meditation EEG activity between four regions of the brain synchronises. Meditation makes the brain more coherent, which means more can be done with less effort, and this promotes optimal physiological functions.

All in all, the studies strongly suggest that meditation helps people to attain a state of restful alertness, higher comprehension and creativity, decreases anxiety, irritability, and moodiness, improves the ability to learn and remember, and promotes self-actualisation, emotional stability, and overall feelings of vitality and rejuvenation. The neurochemicals released by the brain do not come with a long list of side effects, as do many of the remedies of allopathic medicine. According to the *Journal of the American Medical Association*, a meta-analysis done by Lazarou, Pomeranz, and Corey on 39 studies from US hospitals (from 1966 to 1996) revealed there were approximately 2,216,000 people hospitalised due to adverse drug reactions in 1994, of which 106,000 died. It is important to acknowledge with deep awareness that the best medicines are those that we prescribe to ourselves *via the quality of our consciousness.*

The picture of healthcare is changing; more and more people are moving away from surgery and drugs, and toward healing by energy and intention. While classical cell biology concentrates on the physical molecules that control human biology, epigenetics focuses on the electromagnetic pathways through which signals, in the form of intentions, emotions, and beliefs, can affect our biology.

With an evolving global consciousness in the domain of healthcare, it is possible that the "doctor" of the future will prescribe the following:

- set intention three to five minutes every morning and visualise how you are going to walk through your day with life-supportive values, such as gratitude, integrity, compassion, etc.;
- eat less, walk more; eat raw vegetables and fruits, cut out meat; avoid preservatives and highly processed food;
- meditate once a day for at least 15 minutes; connect with the world; feel peace, love, joy, and hope;

- make room in your daily life to help those less privileged and give your blessing to everyone you meet;
- be aware of your body throughout the day; consciously choose to stretch and relax your tension spots;
- pause more often throughout your day; tune into the moment; be open to messages from surrounding and inner promptings, and use them to guide your actions;
- make time to be with nature: forests, parks, ponds, seas, etc.; empty your thoughts and open your senses;
- set aside a minimum of two hours to connect and bond with loved ones;
- take your supplements (vitamins and herbs) regularly;
- give your best in whatever you undertake; surrender in trying to control those things you cannot change;
- acknowledge yourself before you sleep; give thanks for all the blessings and learning you've experienced.

General Systems Theory & MacroShifts

When scientists study any phenomenon in the world, they try to understand the interactions of its components and how these change over time. In scientific terms this is known as Systems Theory. In 1928, biologist Ludwig von Bertalanffy proposed an over-arching process of studying all systems theories (the system of systems), which is now known as General Systems Theory (GST). A system can be closed, where its interactions occur only among its system components and not with the environment. Or a system can be open—when it receives input from the environment and releases output to the environment. Open systems tend to move toward higher levels of organisation, while closed systems can only maintain or decrease in organisation.

In GST, there are no systems within the Universe whose boundaries are impermeable to everything. This theory, therefore, looks at the world in terms of relationships and integration. Systems are integrated wholes whose properties cannot be reduced to those of smaller units. Instead of concentrating on basic building blocks or substances, the systems approach emphasises the principle of organisation. Every organism, from the smallest bacterium through the range of plants, animals, and human beings—our family, society, and the planet—is an integrated whole and, thus, a living system.

The main goal of GST is to study the general principles of system functioning so that it can be applied to all types of systems in all fields of research. As von Bertalanffy points out, "there are many instances where identical principles were discovered several times because the workers in one field were unaware that the theoretical structure required was already well developed in some other field." GST is therefore trans-disciplinary and quite integral in its approach; it brings together theoretical principles and concepts from ontology, philosophy of science, physics, biology, and engineering. Applications are found in numerous fields, including cosmology, sociology, political science, economics, organisational theory, and psychotherapy. GST forces us to broaden our perspective, to consider how each decision and action affects the multi-nested open systems we live in.

A theory is the more impressive the greater the simplicity
of its premises, the more different kinds of things it relates,
and the more extended is its area of applicability.

—Albert Einstein, Nobel Laureate physicist

As in science, so in metaphysical thought, that a general and
ultimate solution is likely to be the best which includes
and accounts for all, so that each truth of experience
takes its place in the whole.

—Sri Aurobindo, renowned Indian spiritual teacher

Bertalanffy believed that the need for a general systems consciousness is a matter of life and death, not just for us, but also for all future generations on our planet. Our civilisation is experiencing enormous difficulties due to a lack of ethical, ethological, and ecological criteria in the manifestation of human desires, which is currently primarily concerned with making larger profits for a small minority of privileged humans. He advocated a new global morality, an ethos which does not centre on individual values alone, but on the adaptation of mankind as a global system to its rapidly changing environment.

Due to the vastness of this knowledge, we have selected to discuss at this time these key principles or topics: Chaos Theory, Strange Attractors, Fractals, Bifurcation and Dissipative Structures, and Macroshifts.

Chaos Theory

This is a study that explains and models the seemingly random components of a system. It connects our everyday experiences to the laws of nature by revealing the subtle relationships between simplicity and complexity, and between orderliness and randomness. Examples of chaos in action include random changes in weather, the changing populations of animals, the rise and fall of civilisations, the propagation of impulses along nerves and much more. A unique feature of the chaos theory is its sensitivity to initial conditions; infinitesimally small changes at the start could lead to dramatically large changes in the system. This theory is used to anticipate order out of chaos and to uncover the underlying elements of chaos that produce order. One of the most important discoveries from the chaos theory is that a relatively small, well-timed and well-placed jolt to a system can throw an entire system into a state of transformation.

Strange Attractors

An "attractor" is a state in which a system eventually settles. For example, a marble that swirls around a bowl eventually settles at the bottom of the bowl; the point at which the marble settles attracts the marble. On the other hand, "strange attractors" represent trajectories upon which a system runs from situation to situation, without ever settling down. In order to model these random patterns that occur in nature, these trajectories were represented in computer-generated space. Some of these trajectories are attractors, which means that they attract nearby trajectories to converge on them.

The most famous strange attractor is called the "butterfly effect," discovered by meteorologist, Edward Lorenz. He was also known as the father of Chaos Theory. While inputting starting conditions in a computer (to simulate weather patterns), he rounded one of the numbers to three, instead of six decimal places. The small difference (.506 instead of .506127) produced a dramatically different simulation of weather patterns. He came to propose that a chaotic system has a sensitive structure, and that the smallest fluctuations at certain times can hurl its trajectory into a radically different path. The development of this concept led to iconic questions such as, "Can the flapping of a butterfly's wings in Australia set off a hurricane later in the Atlantic?"

The cultural equivalents of attractors would be chiefs, tribes, nations, and what gives us identity, like religion, class, and worldviews. The information we have reviewed so far shows that a shift in human worldviews, priorities, and consciousness, can create a butterfly effect in our current sensitive and chaotic world.

Fractals

Fractals are non-regular geometric shapes that have the same degree of non-regularity on all scales. Just as a stone at the base of a foothill can resemble in miniature the mountain from which it originally tumbled, so are fractals self-similar whether you view them from close up or very far away. This phenomenon was discovered by mathematician Benoit Mandelbrot in 1974, while he was working at IBM. Within the overall shape, there lies a repetitive pattern, whose exquisite substructure may not look exactly the same, yet it repeats itself in all scales. Fractals beautifully characterise the essence of chaos.

Fractals are found throughout nature. Trees, mountains, clouds, lightning, coastlines, cauliflower, and blood vessels are examples of fractals. At any scale of organisation, a tree is made out of repeating self-similar patterns in a range of different sizes: the shape of the trunk is similar to the shape of a branch, which is similar to that of a twig. Or we may consider the embryonic development of all vertebrates (including humans). The embryo passes through a series of similar structural stages. A human embryo, for example, shape-shifts from one that looks like a fish embryo to that of an amphibian embryo. It continues morphing until it takes the shape of a reptilian embryo, followed by that of a mammal, and finally takes on the form of a human.

Fractals offer us a new powerful way to look at the relationships between our environment and ourselves. The wisdom of fractals parallels the ancient Hermetic message of "as above, so below," which has been around for at least 2,000 years. This suggests that if we discover patterns at one level, we can apply them at all levels, that by studying the fractals in one system we can apply them in many, if not all, systems. Examples of conceptual fractals in our world are bifurcation and dissipative structures, which exist in many levels of evolution, such as the biological, social, and technological ones.

Bifurcation and Dissipative Structures

In the 1970s, Robert May, a mathematical biologist studying animal populations, discovered that when population growth parameters are at high levels, the system would break apart, and the population would oscillate between two alternating values. His work confirmed the notion that biological systems are governed by the dynamics of chaos. This splitting of one single path into two was dubbed "the process of bifurcation." It means that an evolutionary path of a system forks off; thereafter, the system evolves in a different way. Complex systems, such as biological and social ones, evolve through bifurcation. The bifurcation period offers an unprecedented opportunity to decide the system's future. The selection between the two paths is ultimately decided by forces happening within and outside the social systems that are wavering back and forth. This period is very similar to the deeper meanings uncovered in Chapter 2 on the word *crisis*. As we said:

"In Greek, the word *crisis* means a turning point, where a choice has to be made relying on one's ability to discern. From the medical dictionary a *crisis* is the turning point in an acute disease—it is the point of change rather than status quo. In the Jewish Kabbalah, a crisis is a sign of the end and also the threshold for a new beginning [...]. From these descriptions, it is obvious that in a crisis, two different scenarios will be simultaneously presented and one has to make a discerning choice."

Ilya Prigogine, a Nobel Laureate chemist was the first to show that conditions that give birth to structures are "far from equilibrium." Using population studies as an example, if the birth rate of a population suddenly rises, strange things can happen—the population is far from equilibrium. In such a system, new dynamic states of matter and energy exchange emerge; Prigogine called these states "dissipative structures." A dissipative system is characterised by the spontaneous appearance of chaotic attractors, where interacting elements are given the environment to self-organise differently. In social evolution, for example, such dissipative events create new, higher order systems (suprasystems), of which previous systems become functional subsystems. These suprasystems are largely irreversible. According to Prigogine, irreversibility has a constructive role in system evolution: it brings forth new forms.

In summary of these various concepts and their implications: at the brink of a critical instability in evolution, fluctuations that were previously rectified by self-stabilising feedback within the system run amok; they break the system's structure. The system goes into a period of chaos. The bifurcation is either the path of disintegration of the system (breakdown), or the path of reorganisation, where the system becomes resistant to the fluctuations that destabilised the prior system (breakthrough). In the breakthrough scenario, the system becomes more robust in information processing, has greater efficiency in the use of free energy, as well as greater flexibility and additional higher levels of organisation (suprasystems).

Macroshifts

This is a term coined by Ervin László, one of the world's leading general systems theorists and futurists. He states that as nature and human society evolve, they alternate between periods of relative stability and periods of increasing, and eventually critical, instability. When the instability reaches the critical point, the system either collapses or evolves to a new condition of dynamic stability. These critical junctures are called "macroshifts," and they happen in all segments and levels of society: the rich and poor, the economic and political systems, the private as well as the public sectors. Laszlo says that macroshifts follow a recognisable pattern; they unfold in four major phases.

i. *The Trigger Phase:* A series of hard technological innovations trigger the macroshift. These are means to attain greater efficiency and effectiveness, however, ecology is not taken into consideration.

ii. *The Transformation Phase:* In the second phase, new technologies rapidly proliferate, there is an increase in the production of resources, the population grows more quickly, and a more complex society develops. These changes bring about the need for new skills and organisations, which place a greater deal of stress on society and the natural environment.

iii. *The Decision Phase:* In this third stage, society's massive changes reach a critical point. Adverse consequences start arising as the complexity and development of society meets the environmental degradation. Social and cultural chaos takes centre stage, with the majority of the population wishing to preserve old-established values; more and more people are looking for alternatives in a predominantly unstable society and global system.

iv. *The Chaos Point Phase:* Here the entire system is dangerously unstable and on the verge of collapse. The old worldviews, structures, and processes are no longer capable of keeping up either with the high number of current demands, or with the multi-system dysfunctions. Change is unavoidable because the system is not sustainable anymore. Two different possible paths are created for the future:

    - *breakdown*: this is the path of strife, scarcity, and destruction. Change is difficult here as people, governments and corporations are too rigid in their worldviews to make a change. Social upheavals and violence spread because people and organisations are unable to cope with social complications and environmental deterioration;
    - *breakthrough*: this path evolves into a more cooperative, robust, and creative consciousness. The worldview of a critical mass of people matures in time. People start living and embodying a new set of life-supportive values and responding positively to environmental and social stress. A new civilisation is born.

Laszlo's macroshift phases paint a clear picture. Rapid technological innovations put pressure on existing established structures and institutions and eventually set off a macroshift. More robust, creative, and energy-efficient suprasystems can only manifest when a critical mass of life-supportive new beliefs takes hold in the global consciousness. It is apparent that technology is the driver of macroshifts in society, while the values and consciousness of groups of people decide the outcome. At such times, the highly unstable global systems become hypersensitive to influences, very much like Edward Lorenz's butterfly effect. The collective consciousness of the planet could tip the macroshift forces toward breakthrough or breakdown; the outcome is still unknown.

Ancient Fractals

Systems theories are tools for understanding the world. But these new sciences contain understanding that has been indigenous to non-Western societies for thousands of years. Natives in India, for example, have been using the Kolam pattern (fractals) on carpets, rugs, and floors to bring prosperity to their homes. They use coarse rice flour to draw the pattern with the intention of inviting ants, birds, and other small critters to eat it; it is a daily tribute to harmonious co-existence. The Chinese Eight Symbols (Bā Guà) is a system of trigrams constructed into 64 four pairs (the results being hexagrams), found in the *I Ching* and other texts. These patterns represent movement and transformation of the elements in nature, and how the energy of life flows through the cycles of Yin and Yang. The system is associated with or used in astronomy, geography, geomancy, martial arts, anatomy, and medicine.

Similarly, Islamic art and design use simple fractal patterns to generate complexity, and to serve as mental tools to focus the mind on the contemplation of the infinite.

More than just an aesthetic decoration and petty ritual, the insights of chaos and complexity are found in Hindu, Taoist, Buddhist, and Islamic teachings. Humility before nature, diversity and connectivity of life, generation of complexity from simplicity, the need to understand the whole to understand a part, are concepts not only believed in, but acted upon daily by Indians, Chinese, and Middle-Eastern people. In these cultures, the spiritual templates (fractals) of life are lived; they are a way of being that exchanges and harmonises with life, and are not just used to obtain material gain.

> At any stage of the theoretical construction there exists a hierarchy of laws [...]. Thus in the practice of scientific research the clear-cut division into a priori and a posteriori in the Kantian sense is absent, and in its place, we have a rich scale of gradations of stability.
>
> —*Hermann Weyl, 20th-century mathematician*

One must reckon that each successive cause in the hierarchy of causes

among the elements dissolves back into its own cause [...] until the
whole mass of effects dissolves finally into the Absolute,
the highest and most subtle cause of all.

—*Shankara, Indian philosopher and sage*

PAUSE: THE FIVE EYES

Studies in fields like quantum physics and human consciousness, epigenetics, and general systems theory are still in their infancies. Like everything else, when new knowledge is being introduced, many people cannot resist trying to refute it. At the same time, it is this tendency to challenge contradictory ideas that brings about a new awareness. Many of the human structures and processes that enable us to learn, innovate, and thrive are based on our past experiences; and here lies our main evolutionary challenge.

- How can we see beyond the limitations of our physical eye, so that we can assess a situation more wisely?
- What can we do to create more space in our hearts and minds for new possibilities to grow?
- How do we take steps into our desired future based on ideas that we can't see, hear, or feel?
- What we value and detest drives our actions; what would happen if we released our attachments?
- What do we need to understand so that it is possible for us to help others who do not support our values?

Many of these questions are beautifully and powerfully explained by the well-known Buddhist teachings of the Five Eyes. This Buddhist text, the *Diamond Sutra*, was translated into Chinese in the 5th century and has helped millions in their quest for spiritual wisdom. This is one of the most frequently chanted sutras in Buddhist monasteries, and it is said that if a person embodies just four lines of the sutra within his spiritual practice, he will be abundantly blessed.

In Buddhism the Five Eyes refer to the Physical, Heavenly, Wisdom, Dharma, and Buddha Eyes. This wisdom goes beyond the anatomical function of the eye. In the English language the word *see* denotes discover, ascertain, witness, understand, recognise, examine, judge, and discern. One can easily notice that the eye goes beyond capturing and processing visual data; it shapes our sense of reality and who we are. What is written below is an adaptation and expansion of Dr. Chia Theng Shen's talk given at the Temple of Enlightenment in the Bronx, New York in 1969.

Physical Eye

Despite its tremendous influence on human thinking and actions, the physical eye has many limitations. We are only able to perceive a limited range on an electromagnetic spectrum chart; we cannot see infrared or ultraviolet wave lengths. Prior to our invention of sensitive instruments, the world that we thought we knew was at best incomplete, and at worst untrue. The Buddha pointed out these limitations more than 2,500 years ago, without the microscope!

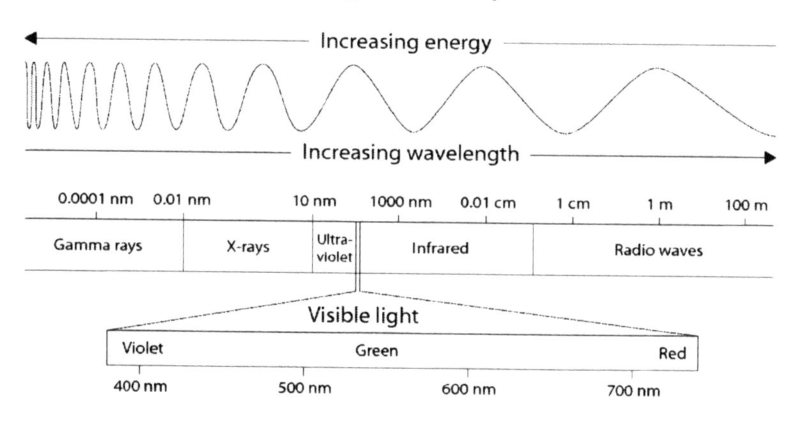

Determining our world through our physical eyes is like being in a small room and looking out at the world through a tiny window. With this limited view all our lives, we insist that our world is complete, real, and true, or that it is more perfect, valid, and genuine than the world of others.

Heavenly Eye

According to the Buddha, one can see everything with the Heavenly Eye and no obstruction can block its view. In modern translation, there are two ways to attain it. One way is through *dhyāna*; loosely translated this means *meditation*. The other way is through the use of instruments, such as microscopes, telescopes, infrared sensors, etc. These instruments can be considered extensions of the eye. It is easier for people today to relate to the concept of the Heavenly Eye with the use of modern instruments to see things that otherwise could not be seen. Again, without the use of such instruments, the Buddha was able to see subtle energies, tiny particles, and higher-dimensional realities through years of meditation.

The Buddha realised that although people's ability to see is limitless, the ability is limited by the physical eye. In other words, through meditation people can transcend the limitations of their physical eye. Here are some quotes from history's luminaries who share the Buddha's insight:

I shut my eyes in order to see.
—Paul Gauguin, French artist, painter, and sculptor

A great thought begins by seeing something differently,
and it is with the shift of the mind's eye.
—Albert Einstein, Nobel Laureate physicist

Things don't change. You change the way
you look at things. That is all.
—Carlos Castañeda, American anthropologist

When the doors of perception are cleansed, man will
see things as they truly are: infinite.
—William Blake, British poet and painter

Wisdom Eye

To appreciate the Wisdom Eye, we need to first familiarise ourselves with a fundamental principle in Buddhism, *shunyata (śūnyatā)*, which means *emptiness*. Emptiness is a state of bliss that transcends the challenge of change and impermanence. The Buddha concluded that human suffering is related to, or originates from, the physical body and the concept we call *self*. In other words, when we achieve detachment from our thinking, emotions, body, things, and concept of self, we are free from suffering.

There are several ways to look at the insights of the Wisdom Eye, and the first one is through the process of disintegration. Using an electric fan as an illustration, if we take out the motor, can we call the motor an electric fan? The answer would be "no, it is just a motor." Now, if we take out the blade, can we call it an electric fan? Again: "no, it is just a blade." When all the parts are separated, where is the fan? We can thus see that an *electric fan* is a name assigned to a group of things put together. In actuality, the true essence of the fan is emptiness. This process of disintegration can be applied to anything in the world, and it will lead us to the same conclusion. The Buddha applies this process to himself. In his imagination he removes his arm from his body and asks if the arm could be called the human body or self? The answer is "no." The Buddha takes every piece of his body apart, and none of the pieces can be called the human body or self. When all the pieces have been removed, where is the self? In some texts, the Buddha is said to propose that not only is the physical body empty in this sense, but even the concept of *self* is empty.

Another way to look at this insight is through the process of integration. With the discoveries in new physics, scientists have confirmed that all known things in the Universe are made up of particles, such as atoms, protons, quarks, etc. Up until today, there seems to be no end to how small a particle can be. In fact, it is said that all known forms in the Universe are made up of the immeasurable formlessness. All matter in the Universe is generated from the zero-point field, also known as the quantum vacuum. In the language of the Buddha, this is called emptiness. Amazingly again, 2,500 years ago the Buddha concluded that everything in the Universe can be understood in terms of emptiness.

One more way of understanding this concept is through the process of penetration. To explain this process, we need to use modern metaphors. One can look at a man with an infrared device, x-ray machine, microscope, and electron microscope. Firstly, we would see the man consisting of red, yellow, and green colours (infrared). Secondly, we would see the man without his skin and flesh, just bones (x-ray). Thirdly, we would see the man's molecular structure (microscope). Lastly, we might see disappearing and reappearing waves or strings of energy (electron microscope). This subtlest and most elemental nature of the man is something which we cannot see nor grasp without special instruments.

The important point to note here is that the man did not change; we changed the way we looked at him. We turned the man from flesh to a bunch of flashing colours, then bones, followed by molecules and lastly, emptiness. When a man reaches this state, everything in the Universe, including himself, is seen as emptiness, therefore, it is proposed, all suffering thus disappears. All of those who reach the level of the Wisdom Eye are said to experience profound bliss.

Many followers of the Buddha reached this stage, and the Buddha gave all of them a stern warning that even though they have realised that the Universe is impermanent, unreal, and empty, they must not cling to the state of emptiness, because it too can be lead to attachment, and thus bring suffering. They must continue their journey to the next level.

Dharma Eye

One who has only experienced the physical eye will assert that physical things are real. One who reaches the Wisdom Eye will profess that all manifestations are phantoms, impermanent, and unreal, and that only emptiness is permanent, real, and of value. Now, a man is said to attain the Dharma Eye when he understand that, although all forms of manifestations are transient, they are not separate from emptiness, they are part of emptiness, and they are real within the consciousness of that realm. The full realisation of this wisdom naturally is associated with the generation of unconditional love and compassion.

A short story helps explain the Dharma Eye. A giant castle is on fire, and only one door leads to safety. Many men, women, and children are frolicking in the castle, and only a few are aware of the dire situation. The people who are aware try frantically to get out, and the way out is long and difficult. Through the suffocating smoke and scorching fires, they eventually get out to safety. Lying on the ground, gasping, and enjoying the fresh air, they are grateful and overjoyed with their narrow escape from death. However, one of them remembers that many people are still trapped inside or probably unaware of the terrible fire. He knows that even if they are aware, they do not know the way out. Without paying attention to his own tiredness and safety, he goes back into the fire and smoke, again and again, to lead the others out of the burning castle. This person is said to have attained the Dharma Eye. This kind of consciousness is not driven by a need to demonstrate anything to others, but it is an inward deep appreciation of compassion, which emerges from the realm of emptiness and the openness to and recognition of the suffering of all beings. This compassion is non-discriminating and unconditional, and these people are detached in their acts of helping others.

The Buddha Eye

It is difficult to use words to convey the far-reaching meaning and power of this level. The Buddha Eye is the most perfect of the Five Eyes. It surpasses the other four. The Buddha Eye follows these descriptions:

- it is non-dual; it has no subject or object;
- it is infinite infinity, so it has no space;
- it is instantaneous and spontaneous, so it has no time;
- it is all-inclusive and complete, so it has no nothingness.

At the level of non-duality (the Buddha Eye), there is no subject and no object. We can refer to the earlier descriptions to assist our understanding. At the level of the Wisdom Eye, the beholder of the Wisdom Eye is the subject and the emptiness is the object. At the level of the Dharma Eye, the beholder of the Dharma Eye is the subject and the various levels of the Universe are the objects. At the level of the Buddha Eye, it would be erroneous to say that the Universe is the object and the Buddha is the subject because, the Buddha is the Universe and the Universe is the Buddha. It would also be a mistake to say that the Buddha possesses the Buddha Eye, because the Buddha Eye is the Buddha and vice versa. On the level of non-duality there is observing but no observer, as subject and object are one.

The concept of infinity is difficult to fathom for most, if not all, human beings. To simplify this explanation, we can rely on mathematics. The first dimension is represented by a line, the second by a plane, and the third by a three-dimensional space. All shapes and objects in the lower dimensions are accessible by higher dimensional entities, but not the other way around. In other words, all entities in the higher dimensions are imperceptible to lower dimensional entities. Within the first, second, and third dimensions, there are already infinite shapes and sizes. Now, what would happen if we were to imagine the fourth, fifth, and sixth dimensions? The M-Theory, a branch of theoretical physics, has mathematically calculated 11 dimensions. A video clip on YouTube created by Rob Bryanton, author of *Imagining the 10th Dimension* (2006), can help to explain this complex science. In the Buddha's realm of consciousness, he sees infinite dimensions—the infinite infinity.

The concept of instantaneity and spontaneity simply states that time does not exist at the level of the Buddha Eye. A galaxy that is a billion light years away from Earth can be reached in just an instant. This concept has been explained earlier as quantum entanglement or the holographic Universe in the segment on quantum physics. Science has taken 2,500 years to catch up with the wisdom of the Buddha.

Last but not least, the concept of all-inclusiveness and totality is synonymous with omniscience. Omniscience is the capacity to know everything infinitely, including thoughts, feelings, life, and the Universe. According to the Buddhist Text Translation Society, "The Buddha Eye is like a thousand suns, shining everywhere, illuminating the one substance underlying all diversity. It shines on different things, but underneath, they are all one substance. With it one can see people, spirits, gods, and everything else, both physical and non-physical."

The notion of all-inclusiveness means that it is all-encompassing and, therefore, it includes nothingness. One is said to be omniscient when he transcends time and space, he dissolves without ego into all there is, he appreciates his unity with all, and in this way he is the mountain, he is the water, he is the cloud, he is the bird, he is one with the thousands of phenomena.

Crystallising the Buddha's Message

The connection between the Five Eyes and our current macroshift is apparent. At the very least, the Buddha's message helps to fully establish that our map of the world is not the only reality. In fact, he makes it very clear that our physical senses are limited. Although we are products of our experiences, we are capable of transcending our flesh-bound experience through awareness and inner transformation. Even though the levels and depths of the Buddha's knowledge could take us a lifetime to understand

and appreciate fully, here are some key pointers to help us begin to change the way we look at things.

- Though there is one unifying reality in the cosmos, it consists of many worlds, and in each world lies its own reality. Therefore, while we are striving toward collectiveness and wholeness, we need to pace and to experience and to appreciate the different levels of reality with compassion and love. At the same time, we must remain open to otherworldly possibilities.
- We need to make time to "expand our vision by closing our eyes" (to meditate) rather than relying on our physical eyes. We need to feel the stillness within us, to listen to our hearts, and to allow our eyes (and our understanding) to form new images—a vision that would paint for us a new way of being, connecting, and living.
- Wisdom abounds in every moment of our lives. We can use daily situations to retrain our eyes, and ourselves, to look at things differently, to look beyond what is obvious, especially when our habitual responses are triggered, to "pause" more often, to relax our eyes, and to let our inner clarity show us the way.
- We can understand that while others struggle with their own issues, we choose to remain in the space of emptiness (which brings with it a non-judgemental state of mind), and to continue to help them. We strive to sow the seeds of wisdom without the need to win others over. We pray for them, or wish them full happiness and health, and allow nature's plan to work its magic. This is the wisdom of co-creation.
- Situations don't change; we change the way we look at them and the way we respond to them. In doing so collectively, we change our world.

CHAPTER 15
WHAT NEW PRESUPPOSITIONS MUST WE UPHOLD TO EVOLVE?

THE POWER OF PRESUPPOSITIONS

Before any new technological or social change can take place, a new set of presuppositions must be embraced and put into use. All paradigm shifts in technology and society are built upon a series of assumptions, which at the time of the transition are often highly resisted, challenged, and misunderstood. In fact, many presuppositions cannot be proved when being introduced and formulated, because the technology and consciousness at that time is neither robust nor intelligent enough to understand and appreciate them, in part because they are not yet especially articulated and clear. Many technological and social paradigm shifts that have taken place have happened because more and more people accepted the new unproven presuppositions that came with the change because of the promise of understanding they hinted at, in the face of the earlier understanding that by proving itself inadequate was inviting replacement.

Presuppositions are defined as taking something for granted as true, or accepting something that has yet to be fully proven, as a foundation for a concept. In a branch of philosophy that deals with the nature and scope of knowledge, epistemology, such (epistemological) presuppositions are deep—and perhaps unstated—beliefs that form the foundation of a particular system of knowledge. As the foundation of an epistemology, they are assumed, and not proven in some sense or other.

Take the whole scientific enterprise, for example: according to Mariano Artigas, a well-known physicist, philosopher, and author of *The Mind of the Universe* (2000), there are at least four presuppositions that it is based on. The first is the existence of a natural world that is independent of human will. The second is the orderly character of this world. The third is the contingency of its natural order. The fourth is the human ability to know this natural contingent order. The existence of these presuppositions shows that science should be considered as human activity integrated within a broader context. In other words, empirical science is rooted in some ontological, epistemological, and anthropological presuppositions that are a necessary requirement for the very existence of science. For Artigas, these presuppositions are not proven within science itself.

Supporting this point are the assertions of Thomas Kuhn, the philosopher of science who first presented the term and concept of the *paradigm shift*. Kuhn said that scientists' preference for one paradigm over another is determined by a host of non-scientific and non-empirical factors. At the cusp of a paradigm shift, there will be competing accounts of reality which cannot be scientifically reconciled. Thus our understanding of science can never rely solely on full objectivity, but takes into account subjective perspectives as well, such as our evaluation of the relative neatness and range of explanation of competing theories.

Presuppositions in Scientific Shifts

A very important point to understand about the power of epistemological presuppositions is that by definition they cannot be proved within the field of knowledge in which they exist. Furthermore, these presuppositions are the fundamental assumptions upon which all of the other concepts and applications within the theory are proved. Taking a chapter from history, Euclid, the Greek mathematician of 300 BCE, built his entire geometry upon the concept of the point. He describes it as "an entity that has a position but no other properties; it has no size, no mass, no colour, nor shape." It is, of course, impossible to prove that a point really has no size, mass, colour, etc. However, when one accepts this presupposition, along with a few others, one can build a whole system of geometry. For example: "A line is the shortest distance between two points," "A rectangle is four lines connected together at equal angles," and so on. The conclusions of this system of knowledge are proved with respect to their adherence to the fundamental—but unproved—presuppositions.

Newton's mechanistic Universe ruled our worldview for more than 250 years, and probably still does so for many of us. The advent of the scientific age began with Newton's discoveries. His laws of motion and of universal gravitation literally changed the course of evolution on earth.

Among the many scientific principles that Newton developed, a key one stated that all the planets and other objects in the Universe moved according to a physical attraction between them, which he called gravity. This mutual attraction explained the orderly and mechanistic motions of the Universe. While he was able to formulate his law of gravity in his work, he never assigned the cause of gravity. In Newton's own words:

> [...] I have not been able to discover the cause of
> those properties of gravity [...] it is enough
> that gravity does really exist, and act according
> to the laws which we have explained, and
> abundantly serves to account for all the motions
> of the celestial bodies [...].

Newton's theories were built upon the existence of gravity, and they have been applied in almost every facet of our lives; yet his hypothesis on gravity was not proved; it was presupposed.

Another historical great is Einstein. His world famous elegant equation $E = mc^2$—which relates mass to energy—led to the development of nuclear energy. Einstein laid the foundation for original cosmological research and his work led to the discovery of black holes, wormholes, and gravitational lenses. His theory of general relativity paved the way for the development of the Big Bang theory, the event many scientists now believe was fundamental to the creation of the Universe.

Einstein achieved great success with his recognition and descriptions of the photoelectric effect and his articulation of the theory of relativity. Fundamental to his theories is the presupposition that nothing moves faster than the speed of light. In 1927-1930, Einstein was embroiled in a long-standing dispute with a group of physicists in Copenhagen over the predictions generated by a new branch of physics called "quantum mechanics." It predicted a bizarre phenomenon called "quantum entanglement" (mentioned in Chapter 14), where measurements performed on parts of a quantum system that are widely separated from each other can somehow instantaneously influence one another. From the presuppositions of Einstein's theory of relativity, particle interactions cannot propagate faster than the speed of light, and, thus, two spatially separated particles can never affect each other instantaneously.

In 1964, a brilliant Irish physicist, John S. Bell, published his work entitled "Bell's Inequality," which made it possible to construct experiments to directly test if quantum entanglement actually occurs. Other physicists began to perform experiments based on his hypothesis. After numerous experiments using many different approaches from then up to 1982, the results were clear: quantum entanglement existed. Nicolas Gisin's experiment in 1997 suggests that the information that travels between two inter-related particles is 20,000 times faster than the speed of light! Bell's Inequality was hailed by some physicists as the most profound discovery in science in recent times. Similarly to previous examples, the far-reaching impact of Einstein's theories and inventions were built on his thought experiments, calculations, and presuppositions.

Presuppositions in Social Shifts

With regard to paradigm shifts in society, the cognitive revolution is an important example. This is essentially the movement away from behavioural approaches to the psychological study of cognition as a central theme of understanding human behaviour. This movement started in the

1950s as a combination of psychology, anthropology, and linguistics developed within the field of artificial intelligence. Contrary to behavioural psychology, which is strictly based on verifying observable behaviour, cognitive psychology makes inferences that mental processes are testable through the science of reverse engineering. By the early 1980s, cognitive science became the dominant research line of inquiry in many fields of psychology.

According to Professor Alvin Goldman (1993) of Rutgers University, the expansion of cognitive psychology in the 1950s got us to examine the questions below.

- Is human behaviour shaped primarily by nature or nurture?
- Do human minds think with visual and other kinds of imagery, or only with language-like representations?
- Does a person's everyday understanding of other people consist of having a theory of mind, or of merely being able to imitate them?
- Is human action free or merely caused by brain events?

In other words, the theories of what drives human behaviour and how the human brain/mind works were re-examined with the rise of cognitive psychology. New presuppositions like "the mental world can be grounded in the physical world by the concepts of information, computation, and feedback," and "universal mental mechanisms can underlie superficial variation across cultures" are pillars of presuppositions upon which the tower of cognitive science can be erected.

Another major shift, this time in macroeconomics was the Keynesian revolution. This fundamental reworking of the economic theory of employment took place in the 1930s. It replaced the neoclassical theory of employment with the understanding of John Maynard Keynes that it is demand, not supply, that drives levels of employment. Keynes argued that governments should intervene to lessen severe unemployment.

The idea of using government spending to stimulate a depressed economy was not something people could easily accept. Many critics in the 1930s pointed to the fact that most governments were already either broke or in debt. To increase spending seemed imprudent, unless there was also an increase in taxation, which, of course, would have undermined the whole point of the exercise. As more and more countries adopted Keynes's presuppositions, which seemed illogical at that time, however, the global economic situation slowly improved.

In the 1960s, the movement for monetarism over Keynesianism marked a second macroeconomic shift. Monetarists held that fiscal policy was unimportant for economic stabilisation, in contrast to the Keynesian view that both fiscal and monetary policies were important. According to this theory, the best thing for the economy was to keep an eye on the money supply and to let the market take care of itself.

The theory of monetarism has several key tenets, such as "the control of the money supply is the key to setting business expectations and to fighting the effects of inflation," or "market expectations about inflation influence forward interest rates;" interestingly, the dictionary defines a *tenet* as "a principle, belief, or doctrine generally held to be true." It is a point of view put forth as a doctrine without having established a proven ground.

Over and over again, across many facets of human knowledge and experience, presuppositions are required to establish a new system of knowing and a new way of life.

Presuppositions in Ecology

Considering the gloomy situation our planet is in right now, the idea of understanding natural systems that support life on earth has become a survival imperative. The study of understanding the organisation of ecological communities and using that knowledge to create sustainable human communities is known as ecoliteracy.

Coined by David W. Orr and Fritjof Capra in the early 90s, ecoliteracy combines the sciences of systems, holism, and ecology, drawing together the necessary elements to build learning processes that deeply appreciate nature and the human role in it. In ecoliteracy, understanding the nature of the whole is more important than understanding the parts.

A distinguished author and professor of environmental studies at the University of Vermont, David W. Orr, declared that the goal of ecoliteracy is to point to the obvious fact that the disordered ecosystem that we are facing reflects a previously disordered mind, and, therefore, it is connected to the way we were educated about our relationship with the environment. In his words:

> The dominant form of education today alienates us
> from life in the name of human domination, fragments
> instead of unifies, over-emphasises success and careers,
> separates feeling from intellect and the practical from theoretical,
> and unleashes on the world minds of ignorant of their ignorance.

According to Fritjof Capra, a well-known systems theorist and physicist, "In the coming decades the survival of humanity will depend on our ecological literacy, which is our ability to understand the principles of ecology and to live accordingly." The young people of our time are born into a world of rapidly deteriorating ecologies, and if the knowledge of ecoliteracy is embedded in their psyches early, then it would certainly arm them with the necessary thoughts to guide their actions in building a sustainable planet.

The cofounder of the Center for Ecoliteracy in Berkeley, California, Professor Capra, said that in order to build and nurture sustainable communities that do not interfere with nature's inherent ability to sustain life, one must first understand the following principles of ecology:

- matter cycles continually through the web of life;
- most of the energy driving ecological cycles flows from the sun;
- diversity assures resilience;
- one species' waste is another species' food;
- life did not take over the planet by combat but by networking.

Here again, a *principle* is described as "a comprehensive and fundamental law, doctrine, or assumption, on which others are based." Considering what we have discussed this far, it is apparent that traversing the fields of mathematics, physics, cosmology, psychology, economics, and ecology, presuppositions are foundations for new discoveries, new knowledge and, therefore, new ways of living.

Our current maps of reality are holding us back from evolving. Our knowledge and assumptions about science, nature, survival, growth, and human behaviour often stop us from seeing deeper and higher levels of truth. Human beings have a strong tendency to preferentially treat information that confirms their preconceptions or hypotheses, independently of whether these ideas are true.

People often reinforce their existing maps of reality by selectively collecting new evidence, by interpreting evidence in a biased way, or by recalling information from memory selectively. However, even though we are the product of our experience, the Buddha assured us that we are capable of transcending it through deepening and widening our awareness (as mentioned in Chapter 14).

In summation, in order for a new paradigm to take hold, some yet to be fully understood phenomena have to be presupposed. These un-established, unfamiliar, and often counter-intuitive presuppositions are required for higher consciousness species to evolve. According to Thomas Kuhn, a paradigm shift is a distinctly different way of seeing the world; it involves not only re-evaluating the facts, but a redefinition of our values and presuppositions.

We need to focus on new presuppositions that support our desired future, to make them our way of life, and to turn them into a daily practice.

Attitudes of openness, active experimentation, raising one's awareness, and earnestly connecting to all life can make the new life-supportive presuppositions possible. When we see the fruits of our presuppositions, we naturally want to share our new way of living with others. In doing so, we expand the transformation from "I" to "we." As more and more people share their new experiences, a collective transformation begins to spread. With larger group transformation becoming more easily accessible, more individual transformations are stimulated. This "I-we" reinforcement will eventually bring forth a new paradigm, a new way of life that all of us truly want for ourselves, our children, and our planet.

Presuppositions as Strange Attractors of Meanings

To facilitate the assimilation and utilisation of new presuppositions for global shifts, one needs to consider the inner workings of how we make meaning out of our experiences. This is a vast field of knowledge to explain in this segment; however, a new field of knowledge known as Strange Attractors of Meaning (SAM) provides us a useful perspective of what needs to take place for new meanings to emerge in human minds and consciousness. According to Dr. Vladimir Dimitrov, a professor at the University of Western Sydney and author of *Social Fuzziology* (2002), the SAM is built upon the disciplines of Dynamic Semiotics (linguistics), Chaos Theory (system studies), and Fuzzy Logic (control systems).

In the mental space of humans where understandings are made, changed, and destroyed, there are spatial and temporal structures that make signs and symbols meaningful. The spatial characteristic of SAM is iconic; it is always attached to a core context. The temporal characteristic is symbolic, as symbols always suggest something else by reason of relationship, association, or convention. Once a certain sign makes sense to a person, this person can zoom deeper and deeper into the meaning of this sign. Although each level of meaning exploration may differ from other levels, there is similarity between the levels—just as with fractals. In other words, once a sign or symbol has been enriched, it can be generalised and/or applied in many levels of human experience.

According to Dr. Dimitrov, it is the strange attractors in the mental space of humans that make signs and symbols meaningful. A strange attractor is like a swirling vortex that propels and draws streams of thoughts and feelings. If there are no attractors of meaning behind one's action, the actions are meaningless; they are operating only at the physical level. New meanings emerge when new strange attractors are created in the mental space of individuals, groups, or societies.

Chaotic systems, such as the human mind and the collective consciousness, are ruled by strange attractors. Since chaos can occur in all size scales of human activity, people can use the instabilities (crises) in society to transform old attractors. This is possible because in a state of total crisis, the energies in a system become hypersensitive and the butterfly effect can occur: little changes can bring forth dramatic results. As we said, a strange attractor is like a swirling vortex: it influences and gathers all the external forces around it. However, the perpetuating power comes from within the vortex; it is holistic, self-organised, self-sustained, and it keeps the interacting masses of thoughts and feelings centred. In a global crisis, the butterfly effect gives incredible power to the minds and hearts of people who are congruently being the change they want to see in the world.

In chaos theory, the spontaneous creation of new attractors happens when an existing system's parameters are pushed beyond a certain critical value. The system loses its dynamic stability and enters into a state of crisis. The kinds of learning activities a person or group engages in can destabilise their own parameters through the intensification of certain critical values. The state of the learner or experiencer is also an important factor in promoting the formation of new meanings.

We can now see that new learning critically depends on the potential of SAMs to lose their stability and go into crisis. This crisis makes certain meanings of words or sentences "fuzzy." In complexity theory, when complexity increases in a social system, precise statements lose their meaning and meaningful statements lose their precision. In fuzzy logic, a mathematical concept developed by computer scientist Lotfi Zadeh, a statement can be true and false up to some degree. Complexity theory includes the principles of non-exclusion and non-isolation. Non-exclusion means that no options for future scenarios should be excluded, and non-isolation means not to isolate ideas down to one option. Conceptualised this way, fuzzy logic promotes more integral and less dualistic thinking.

Fuzzy logic creates malleable and easy to reshape frameworks in which an "either/or" mindset can be replaced by a "both/and" worldview. In other words, fuzziness can be seen as a way to make old meanings open for evolution or transformation. These SAMs at the human collective level are critical events to create societal or global transformation about the meaning of life.

It is the intent of the new presuppositions presented below to serve as SAMs for every global citizen. They have been cross-referenced and analysed through leading concepts of today's global-shift leaders. To name a few: Ervin László, Duane Elgin, Bruce Lipton, Gregg Braden, Barbara

M. Hubbard, Peter Russell, Edward J. Bourne, the Institute of Noetic Sciences, the Global Coherence Initiative, and many more. Also, these presuppositions (as you will see) are distillations based on the summations studied in this book.

A distinct shift in human consciousness must take place before a significant mass of people start changing their priorities in living their lives in a way that progressively supports life on this planet. The self-organising force of strange attractors—to create the butterfly effect—cannot come to life until participating masses and streams of ideas and emotions are: (i) constantly in motion—in chaos theory it means continually moving out of their equilibrium states (or comfort zones) and (ii) interacting intensively with each other, which suggests active networking, communication, and collaboration. Let us begin with an open state of mind to gain a clear understanding of the presuppositions for a global shift.

PRESUPPOSITIONS FOR A GLOBAL SHIFT

To facilitate the understanding of each presupposition, we have utilised the philosophical concept of a *datum*—a piece of evidence considered as fixed for the purpose in hand. What is taken as a datum may change as changes of theory and evidence arise.

(1) Datum: Wholeness

Worldview: all growing systems in the Universe are open. Thus, all systems are based upon the dynamics of relationship, exchange, integration, and organisation. Every organism, from the smallest bacterium through plants, animals, human society, the planet, and the Universe, is an integrated whole, one living system. Every thought, feeling, and action affects the multi-nested systems we live in.

Key words: connection, relationship, balance

Presupposition: We are part of the web of life that supports us; nature is constantly interacting and co-evolving with us. It is crucial that we connect, listen to, and cooperate with nature.

Holism is the idea that all the properties of a given system cannot be determined or explained by its component parts alone. Instead, the system as a whole determines the way the parts behave. The health of any system depends on how complete it is, as opposed to being divided and/or conflicted. Interestingly, the soundness (integrity) of a system is measured by its wholeness, its lack of division and of inner conflict or disharmonies.

More and more cosmologists and physicists agree that before the Big Bang happened 15 billion years ago, the Universe was made up of an immeasurable subtle field, known today as the quantum vacuum. This field of energy contains information (like DNA) which propagates all matter in the Universe. This finding in quantum physics supports the notion put forth by ancient traditions that invisible forms or spirits are responsible for moulding the physical Universe. With recent discoveries about the quantum field, we can now consider the non-linear nature of creationism as the supposition that an undefined intelligent energy existed before the creation of physical matter. The time coded process of turning matter into cells, followed by the assembling of cells into complex organisms, such as plants, animals, and humans, is the linear process of evolution. Thus, the existence of organisms on earth comes from both creation and evolution.

With this understanding, the former opposites and conflicts about our origin have been re-contextualised. The presupposition of creation versus evolution is no longer an equation of either/or, it is both/and. With boundaries and opposites melted into the whole, this holism is asking us to harmonise all these key polarities in our lives: family and work, personal and global, man and nature, masculine and feminine, intuition and rational, science and spirituality, etc.

The understanding of wholeness brings us to the appreciation that we did not come into existence through accidental mutation; we are a product of creation and adaptive evolution. We live on a mutually supportive and integrated living planet, the existence of which determines our well-being. As the most self-aware species on earth, we have tremendous power to tip the scale of evolution of the planet, and instead of exploiting it, we can co-create with Mother Nature. What we do to nature, we do to ourselves. Our vision of a desired vibrant future, besides seeing happy humans, must include harmonious co-creation with animals, plants, and the environment.

(2) Datum: Macro-Micro-Cosm

Worldview: the outer and inner world—the Universe and the human body—are one in resonance. When we experience crisis in our society and environment, it mirrors our inner world; we therefore have a part to play in it. The way we shape the world around us is in turn shaping our lives.

Key words: mirror, co-create, congruence

Presupposition: Our outer and inner worlds mirror and entrain each other. If we don't like what we see in our world, we can start changing it by examining our own worldviews and aligning our actions with our intentions.

In ancient Greece, macrocosm and microcosm were a general representation of seeing the same patterns reproduced in all levels of the cosmos, from the largest scale (macrocosm or Universe-level) all the way down to the smallest one (microcosm or sub-sub-atomic level). In this representation the mid-point is Man, who summarises the cosmos. In the Hermetic tradition, a philosophical belief from the 2nd century, they professed the wisdom of "as above, so below," which declares: "that which is below corresponds to that which is above, and that which is above, corresponds to that which is below, to accomplish the miracles of the one thing." It means that the macrocosm is the microcosm and through deep understanding of one, we can understand the other. Similarly, the ancient Chinese Taoist system believes that the changes in the climate, time of the day, and geographical locations affect humans biologically. Nature is, therefore, regarded as a macrocosm of the human body and vice-versa.

A modern equivalent, the study of chaotic systems, has given us fractals—irregular geometric shapes that can be split into parts, each of which is a reduced-size copy of the whole (self-similarity). Analogous to Russian nesting dolls, each smaller version of the doll is similar to, but not exactly the same as, the larger doll in which it is nested. This multi-level mirroring effect in fractals is truly the design principle of nature; the biosphere inherently unfolds nested self-similar patterns at every level of organisation.

"So what?" one may ask. With the understanding of the mechanics of fractals at one level, we can apply the knowledge at multiple levels. The patterns that we see in our relationships, weather, virus pandemics, stock market, war, social trends, or geophysical changes may be revealing to us the fractals that we need to understand in order to evolve to the next level. The principle of micro-macro-cosm means that when we adjust the microcosm—which is fully in our power to do—the macrocosm mirrors the changes. Therefore, when we raise our individual consciousness, we lift the whole world with us. Also, the path we have taken would become more obvious and easier to follow for those who are heading towards our destination.

(3) Datum: Collective Consciousness

Worldview: in a critical state of global crisis, only through a shift in collective consciousness can we avert the catastrophes that lie ahead. A more robust, creative, and energy-efficient suprasystem would manifest when a critical mass of life-supportive beings aligns and amplifies the new collective consciousness.

Key words: cooperativeness, unity, embracement

Presupposition: A world of compassion, collaboration, and resourcefulness would easily manifest through a united consciousness.

If a group of elite scientists wanted to build a more self-aware planet consciously, what would they do? They would probably develop more innervated technologies like sensors, cameras, microphones, transponders, GPS receivers, satellite surveillance, and seismic and electromagnetic monitors. They would, over time, pull all the data together and develop a program to analyse it and share it (on the Internet) so that more advanced self-aware technologies could be developed. Soon, the planet would be covered with these technologies. Critical information about the planet's health would be detected, analysed, and made instantaneously available to all parts of the planet. Information about the inhabitants' activities, such as wars, major achievements, disease outbreaks, economic upheavals, social governing policies, etc., would also be made available instantaneously everywhere.

All of these multiple and varied data would, at some point in time, be brought together to be cross-referenced, correlated and, therefore, completely new information would emerge. In the intense study of this complex information, recognisable patterns would emerge and show that these factors are intimately connected and sensitive to each other. The scientists would begin to see how global patterns would be reacting and responding to each other. These large patterns would begin to tell a story—the story of the planet—and the scientists would begin to anticipate trends that would affect all life on the planet. Some of the scientists would be optimistic about the future of the earth, but many would be very concerned.

This has already happened. This is where we are now. We have arrived at a point in our evolution where we are just beginning to be aware of the inter-influencing networks of the planet across multiple complex systems—geospherically, biospherically, and sociospherically. At one vital level, we are now aware of the interconnectedness and self-organising intelligence of the planet; we are indeed able to dissociate from it and to see how it works as a whole.

At another level, we are seeing more and more that our thoughts and feelings and the planet's life-supporting systems are interwoven. What's new is that we are coming to terms with the idea that our thoughts and emotions are influencing the planet's sociospherical and biospherical systems. Since all systems are intimately connected, the geosphere is definitely affected. We are beginning to recognise and acknowledge, with the help of several consciousness research institutes and independent scientists, the existence of a field of intelligent energy that connects all systems' information instantaneously, bypassing all space, time, and matter constraints.

As we have already exposed the evidence suggesting that a noosphere exists—this unifying field of consciousness—we must now bear in mind that it stores all past information of the Universe (even from before the Big Bang!) and it is accessible to any self-aware being. This means that it includes the thoughts, experience, and emotions of every living thing on the planet. More importantly, it includes the consciousness of—now—seven billion humans.

With this all-penetrating and all-encompassing field of intelligent energy, we are able to connect to each other. We are able to harness our desired thoughts and emotions through our intentions and moment-to-moment attention. We are already sharing our thoughts and experiences because the field naturally stores them, but are we accessing them? Because when we do, we are able to tune in to immensely rich information to solve our personal and global problems. Most of all, we are able to co-create with this field the kind of world that we've always wanted. The key is to resonate with and amplify this common desire together, as one global family.

In essence, collective consciousness pivots around the relationship between "I" and "we." The more individuals connect to "we", the more the collective can influence the individual. This is the power and potential of our collective consciousness: to be of one mind, purpose, and vision.

(4) Datum: Congruence

Worldview: bringing forth a desired world is ultimately an embodiment of our vision. Our world is a result of the congruence between our daily intentions and actions

Key words: embodiment, integrity, alignment

Presupposition: Be the change we want to see; lead by example.

Voted by a poll administered by the BBC not as man of the year, not as the man of the decade, nor of the century, but man of the millennium, Mohandas Gandhi was acknowledged by the British public as the man who has made the greatest impact on humanity in the last 1,000 years. Gandhi demonstrated how non-violence and the spiritual truth can be applied in international politics. He has inspired millions of people around the globe, including Nelson Mandela, Albert Einstein, Martin Luther King, Barack Obama, and many other luminaries.

Gandhi's contribution is not in the arena of science, military conquest, or literature; his legacy is a moral and spiritual one. His beliefs on the importance of spiritual truth and non-violence are not unique; they are preached by many other religions, some of which would bless their armies before they march into war. The difference is that Gandhi embodied his beliefs. The unity of his actions and beliefs was so complete that when a reporter asked what his religion was, he did not spout verses from scriptures, but simply replied, "You must watch my life; how I live, eat, sit, talk, behave in general. The sum total of all those in me is my religion." For Gandhi, to work was to pray. In other words, he lived his life as a prayer. He led by example.

The power of the congruence between his vision of a better world, his values, and his actions was calibrated at 760 on David Hawkins's Map of Consciousness. Based on Hawkins's map (with levels ranging between 0–1,000, where 1,000 is the highest level consciousness a human being can attain) 760 is the level of powerful inspiration where beings like Gandhi set in place attractor energy fields that influence all of mankind.

In the quantum vacuum field, where all consciousness is stored and connected, a single being at 760 is able to neutralise the negative energies of many people below the consciousness level of 200, which is life-depletive (negative energy), as opposed to life supportive (positive energy) at 200 and above.

According to Hawkins's non-linear system, as one moves up the levels of consciousness one achieves a logarithmic effect and it naturally counterbalances the negative energies below 200. To show the exponential counterbalancing effect: one person at 300 counterbalances 90,000 individuals below 200; one person at 500 counterbalances 750,000 below 200; and one person at 700 counterbalances 70,000,000 individuals below 200!

Such is the power of Gandhi's spiritual energy influence. On several riotous occasions in India, Gandhi would walk hundreds of miles through harsh terrains and weather to comfort troubled people and the rioting would stop. Lord Mountbatten, the British Viceroy at the time, said it would take 50,000 soldiers to achieve what Gandhi did in many parts of India. Most important of all, he achieved it without violence.

Some of Gandhi's everlasting principles and beliefs are elucidated in the following phrases:

> I do not believe that the spiritual law works on a field of its own.
> On the contrary, it expresses itself only through the ordinary
> activities of life. It thus affects the economic, the social,
> and the political fields.

What is faith worth if it is not translated into action?
Be the change you want to see in the world.

Gandhi's impact on the world can be best described in the words of Albert Einstein, "Generations to come will scarcely believe that such a one as this ever in flesh and blood walked upon this earth." Gandhi did not claim that he was a God. In fact, he openly declared that he was a common man, constantly working at his flaws. Among the many earth-changing messages he has given us, is the message about turning our ideals of a better world into our daily living, to turn our future ideals into a way of being—today. Our daily actions must be congruent with our vision of an ecological and harmonious world.

(5) Datum: values

Worldview: the things that we value in our success collectively shape our world. A new world comes into conflict with the old; thus, a redefinition of success is a must to facilitate this transition. We need to look at what truly makes us happy.

Key words: priorities, principles, what matters most

Presupposition: What we value in our lives shapes our world. Redefine our values to bring forth a life-supportive world.

The World Value Survey (WVS) is a worldwide non-profit network of social scientists studying changing values and their impact on social and political life. They carry out surveys in 97 societies representing almost 90% of the world's population. The ongoing research produces insights that look into the changes at the individual level that are transforming social, economic, and political life. The WVS provides information that helps decision makers to better understand and cope with changes around national and international security: religion, culture, and diversity, globalisation, gender issues, human development, and quality of life.

Before proceeding to talk about some of the findings of the WVS, it is desirable to bear in mind that values are defined as one's judgements about what is important in life. Along with worldview and personality, they generate behaviour. There are different types of values, such as ethical/moral, doctrinal/ideological, social, and aesthetic ones. Essentially, values tell people what is good, beneficial, important, useful, beautiful, desirable, appropriate, etc. They answer the question of why people do what they do. Groups, societies, or cultures have values that are largely shared by their members. The values identify those objects and conditions that members of a society consider important, desirable, and valuable.

In 2007, the WVS concluded that over the past generation, a shift in values has been occurring in a cluster of a dozen nations. They call this change "the postmodern shift." In these societies the emphasis is on shifting from economic achievement to post-materialist values that emphasise individual self expression, subjective well-being, and quality of life. In short, people are moving away from materialistic acquisition and protection and toward creative choice making and redefining the quality of life by opting for a life of simplicity.

Based on a deep inquiry process, Linda B. Pierce, author of *Choosing Simplicity* (2000), came to a number of conclusions about simple living from her research. Many of her interviewees talked about living their lives in an earth-friendly manner and each of them was doing this in their own diverse ways. One glaring pattern, however, did emerge: many of them commented on a strong relationship between simplicity and spirituality. Living simply has enhanced the spiritual aspects of their lives, and being in touch with their spirituality made their lives even simpler.

Tim Kasser, professor of psychology at Knox College, Illinois, gathered considerable research showing that the more materialistic values there are at the centre of someone's life, the more that person's quality of life is diminished. He found that people who placed a relatively high importance on consumer goals, such as financial success and material acquisition, reported lower levels of happiness and self actualisation. They were also prone to depression, anxiety, narcissism, antisocial behaviour, and lifestyle diseases. In his research, Dr. Kasser found that there is a weak connection between income and happiness, once a basic level of economic well-being is reached. The WVS 2007 Survey revealed that people in Vietnam, with a per capita income of less than $5,000, were just as happy as people in France, with their per capita income of $22,000. The overall message is that once a certain sufficient income is achieved, more income does not necessarily mean more happiness.

People who have evolved beyond materialistic existence (post-materialists) are focused on intrinsic values. Dr. Kasser defined these values as self-acceptance (pursuit of personal growth), affiliation (close relationships with family and friends), physical health (diet and exercise), and contribution (giving time, energy, and resources to the community). These are the values that lead to happiness.

In summation, here is a table and scale for us to gauge where we are, individually, in our evolution. Out of 100%, how much of our daily living is based on old/new world values?

Old World Values	%	%	New World Values
Material acquisition			Personal & spiritual growth
Consumerism—frequently buying new & better things			Clear understanding between needs & wants
Life ruled by clock (time poor)			Life navigated by compass (time spacious)
Life in the fast lane			Life of holistic self-care
Competitive			Cooperative
Self-interest, independent, disconnected			Care for others (connected to family, community, world)
Surrounded by things and offices			In regular touch with nature (animals, plants, etc.)
Relies on rational/analytical thinking			Balanced with intuition & open to new realities
Relies on will-power & medication			Proactively making time to rest without medication
Relying on quick & easy available food			Making time to eat healthily

Friedrich Nietzsche was a German philosopher whose critiques of contemporary culture, religion, and philosophy centred on a basic question regarding the foundation of values and morality. Nietzsche's work has caused us to evaluate both ourselves, and our interpretation of the meaning and value of religion and morality. When he said, "He who has a compelling why to live for, will stomach almost anything," he was referring to the fact that we live or die by what we stand for. Values determine the character of a man. Einstein's famous quote "Try not to be a man of success but rather become a man of value," supports Nietzsche's message. Embodying well-defined, life-supportive, and worthwhile values would change our lives; that is, when we congruently align our intentions and actions to them.

(6) Datum: Abundance

Worldview: the planet pays for the way we manufacture, market, transact, and transport our needs and wants. It is a severe price that all of us are paying now with our lives. We must take a deep look at and nourish the social, biological, and geological systems that support us.

Key words: reciprocity, equality, gratitude

Presupposition: How we use our natural resources determines the sustainability of the abundance we have. Abundance must include how our thoughts and actions nourish others and the environment.

The notion of abundance has been redefined and widely propagated in the last two to three decades. All teachers, gurus, and leaders of "abundance consciousness" talk about using a set of laws to change our mindsets and vibrations, so that we can change our destiny and quality of life. Some promote the notion that the Universe is like an infinite ocean and that all our needs and wants are like asking for a teaspoon of abundance from the Universe. Many of these abundance advocators—if not all—teach their followers the crucial skill of visualisation. They claim that by using this technique, anyone can interface with the Universe and make his or her life abundant. An advertising message from an abundance advocator could read, "You truly can experience physical, financial, relational, emotional, and spiritual abundance. It has all been provided. You only need to know how to begin attracting it!"

If one looks closely, this knowledge of tapping into the universal abundance is focused on individuals. The message is, "you can get whatever you want!" More often than not, the intentions behind what individuals are asking for and how they would impact their lives, communities, and environment have not been examined. Our unwise desire for as-soon-as-possible fulfilment of our individual needs and wants, more than anything else, has contributed to the dying planet today.

The intention of abundance consciousness is obviously not to destroy life on earth. Yet, unwittingly, it ends up being perceived as a means of fulfilling everyone's self centred needs and wants. More money, better looks, prestige and status, and more luxurious things can be obtained without concern and understanding of the price other people or environments are paying for their fulfilment. Without ecological intelligence, the joy and fulfilment of 15% of humankind steals the right to live of the other 85%. It also affects the environment and the biodiversity upon which human survival depends. A brutally honest and clear separation of one's needs and wants is the first crucial step to creating true abundance in one's life.

The fuller meaning of abundance includes richness and blessings in one's health, relationships, and time. This would mean that one's intention is directed at having more time to do meaningful things, to enrich desired relationship with loved ones and communities, and to enhance one's health through proper dieting, meditation, and exercise. These are experiences that make us happy as humans, yet they are often low on people's lists of priorities and are often taken for granted. If these priorities took precedence over one's life, then the meaning of abundance would be redefined.

Contrary to the constant outward display of the mind that is driven by grabbing and possessiveness, fuelled by a state of dissatisfaction and a worldview of lack, harnessing one's inner awareness of true abundance produces a vibration of appreciation, gratitude, and harmony. Amplifying such vibrations creates powerful attractors for universal resources to flow into one's life.

The widening gap between the overflowing rich and the starving poor violates the universal values of equality, caring, and sharing. This stands in the way of manifesting the true meaning of abundance: the resources of the whole are equally distributed, so that the vital functions of the whole are adequately nourished, allowing all to carry out carry out their crucial role and tasks in a healthy way. That, in turn, supports the whole. The keys to true abundance are equal distribution, central storage of surpluses, constant communication among members, and movement toward ecological efficiency. With the harmony of "I-we," the true power of abundance would effortlessly manifest.

(7) Datum: Metamorphosis

Worldview: the health of our future is determined by how fast we can redefine our current worldview about science, survival, human nature, and the purpose of our existence.

Key words: recontextualisation, surrendering, pushing boundaries

Presupposition: In order for a new beginning to materialise, we must end the old. Ending the old world must be accompanied by a clear and vivid vision of the new world.

From ancient wisdom to modern scientific inquiry, all large-scale changes on earth involve some form of radical transformation, regeneration, and/or transmutation. Alchemy is focused on a literal level on deconstructing and reconstructing matter; the transmutation of common metals into gold is analogous to the transmutation of the physical body into immortality. Far from being the work of superstitious fools, alchemy was the precursor of modern chemistry and was endorsed by two of history's greatest geniuses: Robert Boyle and Isaac Newton. Alchemists were trying to develop a universal solvent that would dissolve and coagulate any metal substances so that a new desired form could be created.

Also concerned with transformation, in mythological studies, the phoenix is a symbol for resurrection, and this is related to the transmutation process of alchemy. A phoenix is a colourful bird that dies in a self-created fire. In its pile of ashes, a phoenix chick is born, representing a cyclical process of life from death. As it is reborn from its own death, the phoenix signifies regeneration and immortality. Through the process of deconstruction, the hope of the future is reconstructed.

In a Christian text written by Saint John of the Cross in the 16th century, the term 'dark night of the soul' is a metaphor that describes a phase in one's spiritual life marked by an intense state of loneliness and desolation. Rather than resulting in devastation, the dark night is perceived by spiritual people to be a blessing in disguise. This desolation is intensified through the fact that all known alternatives have been tried and none works. Prayers and meditations lose their efficacy. No one understands the sufferer's emptiness and loneliness. In a state of total isolation and despair, an unknown peace emerges. It fills one's heart completely.

The dark night of the soul is essentially an intense purifying process that deconstructs one's ego. Serving only itself, the ego is convinced that it is the "doer," the creator of every thought and feeling, and the architect of every accomplishment. When the ego fully realises that there is nothing it can do to end the suffering, it dissolves, it gets transmuted, and a new being emerges. Very much like the phoenix rising out of the ashes, the disintegration of the ego brings forth a new being. At this point, a person knows clearly that he is a different being from what he once thought himself to be. One who passes through the dark night of the soul, sees others afresh, lives a new life with ease, and abides in one's own true nature. This is the unmistakable proof that one has entered a higher realm of consciousness.

In nature, the butterfly personifies the transformational processes of alchemy, the phoenix, and the dark night of the soul. From an egg to a caterpillar to a chrysalis, the butterfly, with each phase of transmutation and regeneration, ends its previous way of life and begins a completely new one. Richly discussed in Chapter 3, the butterfly reminds us that we must:

- be willing to let go of our life in the past and start completely anew (in our quest for a better life);
- embrace the uncertainty and pain as it serves to provide the strengths we need to begin the next phase of our evolution;
- break away from denser realms of material existence and move to a lighter realm of spiritual awareness;
- look beyond our current individual existence and have the courage to transform ourselves to serve a bigger and higher purpose.

All the above involve the willingness to embrace new worldviews and take bold steps to experiment with new ways of thinking, feeling, and living. We must recontextualise our priorities in life and align our mental focus, emotional compasses, and our actions daily. This includes exploring new life-supportive ways of eating, commuting, working, communicating, recreating, and home-living. It is important to allow time to reflect and meditate—to tune in to the collective consciousness. Envision a desired ecological world and translate that into writing so that it can be crystallised over time. Last but not least, be willing to make time for all the above.

(8) Datum: Resonance

Worldview: our worldview of how we evolve biologically has been dramatically redefined; electromagnetic, geomagnetic, and other vibrations can influence our genes and biology. Our thoughts and emotions are vibrations. We have far more power to heal ourselves, our environment, and therefore, our evolution, than we previously believed.

Key words: mirroring, entrainment, entanglement

Presupposition: The questions we ask, the intentions we uphold, the things we pay attention to, and the emotions we invoke everyday are ultimately vibrations that shape the kind of world we live in.

In science, resonance is the tendency of a system to oscillate with larger amplitude at some frequencies rather than at others. It is the state of vibrational adjustment analogous to a violin string in tune with a vibrating fork. Resonance occurs with all types of vibrations or waves. There are mechanical, acoustic, electromagnetic, nuclear magnetic, and quantum wave resonances, to name only a few.

Resonance occurs widely in nature. It is the process by which virtually all repetitive oscillating waves and vibrations are generated. Many sounds we hear, such as when hard objects are struck, cause a brief resonant vibration in the object. Light and other short wavelength electromagnetic radiations are produced by resonance on an atomic scale, like electrons in atoms. According to the father of neuropsychology, Donald Hebb, neurons that resonate together get wired together. Other examples of resonance processes include:

- timekeeping mechanisms in watches and clocks;
- acoustic resonances of musical instruments;
- electrical resonance of tuned circuits in radios and televisions that allow individual stations' broadcasts to be received; and
- material resonance at atomic levels that makes spectroscopic studies and applications (MRI, for instance) possible.

The notion of collective resonance was investigated by Rupert Sheldrake, Ph.D., in the late 1970s. A biochemistry graduate of Cambridge University (UK), Sheldrake proposed that there is a field within and around a living entity that organises its characteristic structure and pattern of activity. The hypothesis is that a particular form belonging to a certain group that has already established its collective morphic field, will tune in to that morphic field. This particular form will read the collective information through the process of morphic resonance. Sheldrake considers morphic fields to be a universal database for both living and mental forms.

Sheldrake's morphic resonance is a feedback mechanism between the field and the corresponding form of morphic units. The greater the degree of similarity, the stronger the resonance. This leads to the habituation of particular forms. Through this sympathetic vibration, an event or act can lead to similar events or acts in the future. Through the same process, an idea conceived in one mind can arise in another. Therefore, the existence of a morphic field makes the existence of a new similar form easier.

As discussed in earlier chapters, everything in the Universe—when broken down to their basic essence—consists of some form of energy and vibration. Our words, emotions, and thoughts are vibrations. Collectively (seven billion people), our consciousness is a rich and powerful morphic field. By being aware of this vibration and learning how to harness, amplify, and direct it through the process of resonance we can bring into existence the kind of world we want to experience. Across several schools of thought, the resonance process for manifesting desired outcomes involves primarily six steps:

- hold in mind a conscious intention of the desired outcome;
- make the representation of the intention vivid—as if it had already happened;
- be congruent; think, speak, and act in accordance with the intention;
- routinise the intention; through your daily living, create a fertile ground for the intention to manifest;
- release and allow the intention to manifest in its own timing—trust the process of the immaculate Universe; and
- be open to evidence that supports the desired outcome; acknowledge and appreciate the feedback daily.

(9) Datum: Choice

Worldview: we are both the destroyers and saviours of life on earth; the power of choice resides in the hearts of every one of us. This choice is made with the way we live our lives every day, the way we commute, eat, work, shop, recreate, and relate to others and the environment; this is going to determine the future of our planet and our species.

Key words: opportunity, self empowerment, life supportiveness

Presupposition: Making a deliberate choice to be life supportive is our most vital step to ensure our upward evolution.

The dictionary defines *choice* as "the act of choosing; the voluntary and purposive act of singling out from two or more options." This deliberate act often requires care and discernment. A moment of choice is also an opportunity for one to exercise one's right to take a certain course of action. More important than any other intentions of this book, is the recurring appeal of making a choice to be life supportive. This term is easy to define; it means to align one's thoughts, priorities, and actions to support ecological living in one's family, community, environment, and, ultimately, the globe. It is about being dedicated to the well-being of the planet via the reinvention of oneself. In a Universe where everything is connected, the betterment of one individual effort to be life supportive raises the quality of consciousness of the collective.

In order for a choice to be made, there must be first an awareness of a choice, i.e., the existence of better options. For many people, the awareness of other ways of living life more harmoniously and ecologically is the crucial first step. It is then followed by a period of studying, understanding, and evaluating the necessary steps to take. This is the juncture where most people get overwhelmed; there are so many issues to tackle and they all seem critical. However, when the dust settles, one would naturally choose those steps that are most convenient, and, in all likelihood, this would be an important and good start. This is also the phase where one's worldview and its inherent presuppositions come into conflict with the new information received. Primarily, this would be a conflict about beliefs and values. Fundamentally, it boils down to the issue of changing one's priorities, comfort, and way of life.

A famous Chinese saying encapsulates the heart of humanity's current problem: "Until one sees one's own coffin, one does not shed tears." Instead of constantly reacting only to dangers close at hand, the choice that every one of us needs to make is to look with farsightedness and the courage to empathise with the less fortunate.

As it was explained earlier, it is in human nature to relate to the suffering of others. However, as we evolve, some of us numb our hearts or look away. In truth, we must realise that "WE are all in this together." The problems we are facing in our world today are beyond the power of governments and corporations, as we said earlier; they are in the hearts and hands of each and every one of us.

Below are the phases involved in making the right choice.

i. Awareness: be constantly searching for better ecological alternatives.
ii. Study up: do the research, verify the details, strive to be clearheaded, and get to the heart of implementing worthwhile strategies.
iii. Metamorphosise: challenge your disbeliefs, be boldly open and willing to embrace new worldviews, and take concrete steps to experiment new ways of thinking, feeling, and living.
iv. Pace and lead: whatever internal and external strategies you implement, focus on sustainability, i.e., make sure you are able to make them part of your life, and incrementally increase your efforts to become more life supportive.
v. Evidence procedures: devise ways to measure your progress, e.g., cutting down shopping expenditures, reducing your carbon footprint, making dietary changes, increasing involvement with environmental and humanitarian causes, etc.
vi. Connect and share: share your ideas with others, connect with like-minded people, look for coaching partners, and be a mentor to others who are new to being life supportive.

(10) Datum: Nature's Cycle

Worldview: we cannot increase our consumption linearly in a cyclical system (nature) without producing toxic consequences. The land, air, sea, animals, and woodlands are intimately connected. They invisibly support all life, including us. A collapse in one element could trigger a rapid chain reaction; a different world can emerge any year, month, or day now.

Key words: interconnect, exchange, sacred reverence

Presupposition: A life-supportive and sustainable world is attained when we appreciate nature's cycles and align ourselves to them.

Nature's cycles have to do with how Earth renews itself. The living things within an ecosystem interact with each other and also with their non-living environment to form an ecological unit that is largely self-contained. An ecosystem contains within itself all the resources to regenerate itself. Sometimes, this renewal process is gradual and gentle. Sometimes, it is violent and destructive. Like everything else, we are part of nature's renewal cycle.

Though the point has been emphasised over and over again throughout this book, it is nevertheless important to crystallise this vital wisdom, to understand and embrace nature's cycles.

- The solar cycle affects the atmospheric temperature and electro-geo-magnetics of Earth. The lunar cycle affects ocean currents and tidal waves, as well as human and animal biorhythms.
- The balance of Earth's water cycle from ocean to cloud and to rain has been altered by man-made pollution; in turn, the currents of the ocean act up and alter the distribution of the Sun's rays and, thus, of the weather.
- Plant life turns carbon dioxide to oxygen. This cycle affects the atmosphere that regulates the Sun's rays, Earth's temperature, and rainfall.
- Animal biodiversity and its food chain cycle keeps plant, water, and land resources in harmony. To date, the damage to the planet's ecosystem is incalculable.
- In nature nothing is wasted; it moves from cradle to cradle. One species' waste is another species' food. Human expansion and consumption has turned natural resources into waste faster than Earth's ability to turn the waste back into resources.
- Human demand for affluent lifestyle goods and services impacts economic cycles and political trends. Besides the disastrous impact on nature's cycles, it also affects human priorities and social harmony. What humans value collectively, shapes the way they see and treat each other and the way they allocate resources to solve problems.

In short, celestial, Earth, and human events are intimately connected. The intricate patterns they weave tell us that we are in a constant cycle of exchange between receiving and giving, consuming and replenishing, contracting and expanding. To achieve harmony in living and non-living ecosystems, it is imperative that we consume less, recycle, and work with nature's cycles.

(11) Datum: Homo Chrysalis

Worldview: our world is currently shedding the tight skin of modernity and ungluing itself from the denser traditions for a more integral and lighter way of being in the Universe. Unless we proactively choose to be uncomfortable first, we will never experience a higher level of existence.

Key words: co-evolution, co-creation, collective transformation

Presupposition: The first step out of our old world is proactively to choose to live simply, and to be life supportive and aligned to the collective consciousness. Our world is a consequence of who we become.

With what we have discussed about nature's cycles, the concept of co-evolution becomes crucial to our current crisis. Instead of working against nature, we work with it. Up to this point, evolution on earth was unconscious (it happened without the conscious awareness of its inhabitants). However, the idea of co-evolution and co-creation has been growing among scientists and spiritual people in many fields. What it means is that the time has come for us to evolve with the planet consciously, and to transform and align ourselves deliberately to the changes that have happened.

Despite the numerous cataclysms the planet has gone through, life has prevailed; this means that those life forms that align with the new order of the planet survive. This time round, the focus is not just on the ability to adapt, but the wisdom to co-create with the planet; it is not about one forcing the other, but about the planet and its inhabitants working together to create a life-supportive world. Being the most self-aware species, we (humans) must consciously choose to transform ourselves for a sustainable future.

This brings forth a new evolutionary concept called *Homo chrysalis*. Briefly described, it is a short transitional phase that humans must go through between who they are now (*Homo sapiens sapiens*) and a new ecological and spiritual species in the near future. The release of scientific discoveries in the last 20 years is providing vital information as to where we came from and who we will be tomorrow. This emerging human species is called *Homo spiritus* by David Hawkins, *Homo luminous* by Alberto Villoldo, *Homo progressivus* by Teilhard de Chardin, and *Gnostic human* by the Indian sage Aurobindo.

If we continue to be who we have always been, then we will continue to experience the kind of world we are experiencing today. This new twist to an old saying is to highlight the perennial wisdom that our world is a direct consequence of our way of being. In other words, in order to bring forth a harmonious world we must redefine how we see our individual and collective purpose, the role we play in the world, and the way we relate to each other. All these redefinitions would then be translated into daily actions. They would guide our thoughts and feelings on what we give priority to in life.

The evolutionary concept of *Homo chrysalis* is connected to the metaphor of the butterfly. The emphasis on the chrysalis phase highlights the importance of cocoonship. As with the metamorphosis within the caterpillar's cocoon, we can expect the transformation to be intense and radical. In order for us to end our caterpillar-like existence, we must vividly envision an ecologically vibrant future (a butterfly existence). We must have faith in the dark and painful process of inner transformation. The inside-out metamorphosis of our world starts from the conscious ecological revolution of each country, family, and individual. In our increasingly prevalent worldview that everything is connected, the improvement of the individual raises the consciousness of the collective.

Stated another way, the *Homo chrysalis* phase is a conscious transformation at the individual level. Though it may feel as if one is in darkness, alone, unsure, and dealing with yet-to-be-comfortable choices, one needs to have faith that there are many others doing the same thing. With the Internet, the sharing of experiences and knowledge is becoming widely accessible. As one leads by example, one naturally shows the way to others and attracts other *Homo chrysalis*. Through this individual and group reinforcement, the increasing morphic resonance will make it easier for others to be awakened and shown the way.

With the knowledge gathered this far, the qualities associated with the *Homo chrysalis* include: the ability to solve problems more holistically and integrally, the capacity to access higher dimensional perspectives, and living a simple, environmental-friendly, and balanced daily life.

PAUSE: SPIRITUAL SCIENCE OF THE TIPPING POINT

This chapter is about embracing global-shift presuppositions and the progressive manifestation of congruent behaviours in all of us. It is important to acknowledge that the social sum total of everybody's little daily efforts doubtlessly releases far more energy into the world than the heroic feats of isolated individuals.

As mentioned earlier, we have arrived at a point in our evolution where we are increasingly acknowledging the inter-influencing networks of the planet, across multiple complex systems. At the same time, science is gathering evidence that consciousness is not confined to what is just taking place privately, within each individual.

Consciousness is an open system; it is shaped by language, society, and all our daily interactions. We are each an aspect of the collective consciousness of the world, and the contents of that consciousness are constantly being altered by the thoughts and emotions that each of us are expressing.

In social dynamic studies, a "critical mass" is the existence of sufficient momentum of a particular nature (e.g., political views, lifestyle demands, parental support groups, etc.) in a social system that becomes self-sustaining and eventually brings forth completely new developments. The levels at which a momentum for change becomes unstoppable is known as the tipping point. This is also the title of a book written by Malcolm Gladwell (2000). This is a dramatic moment in a social system when everything can change; all at once. Gladwell uses a medical model and he proposes his "three rules of epidemics" as factors that will determine if a social idea fades away or tips into a wide-scale proliferation:

- "the law of the few": people who are great connectors, information specialists, and/or persuaders;
- "the stickiness factor": the content of a message whose dramatic divergence from conventional wisdom makes its impact memorable;
- "the power of context": the precipitation of small changes in social groups and community environments that can cause a new idea to tip.

Time magazine in 2005 named Gladwell one of its 100 Most Influential People and *Newsweek* nominated him as one of the Top 10 New Thought Leaders of the Decade. His contribution helps us understand how trends start, spread, and are sustained. If we apply this concept to our vision of the new world, what might we see as a tipping point to carry us through this critical phase of our evolution?

The presuppositions highlighted in this chapter, when progressively embraced and practised by individuals, will help us evolve into the *Homo chrysalis* phase. As more people turn inwards to transform their outer world (millions are already at this stage), a tipping point could occur very soon (2012–2017).

A tipping point is similar to the butterfly effect. In a chaotic social system, the smallest push or fluctuation can hurl its evolutionary trajectory into a completely different path. In the societal equivalent of the butterfly effect, creators would be leaders of nations and religions, dominant worldviews and values, deeply rooted beliefs and presuppositions. A shift in human worldviews, priorities, and their underlying presuppositions can create a butterfly effect in our current sensitive chaotic world.

John Casti, Ph.D., a senior research scholar at the International Institute for Applied Systems Analysis in Luxemburg, said that no collective activities such as globalisation or trends in popular culture, like fashion, films, or books can be understood without recognising that this is how a group or population sees the future that shapes events. Feelings, not rational calculations, are what matters most. It is the non-physical realm that shapes the outcome of the physical world.

In his book, *Mood Matters* (2010), Casti tells the story of why human events happen the way they do and not some other way, showing how it is the collective mood of a population—its social mood—that biases the events that we witness. In other words, the mood of a collective, be it an institution, state, country, continent, or even the world, determines how that collective feels about the future.

In Casti's own words:

- "A happy public pushes happy songs up the charts, and a depressed public pushes depressing songs up the charts."
- "Optimistic people make a productive economy, and pessimistic people make an unproductive one."
- "It is the mood of the country that determines the outcome of elections."
- "Content and tolerant people make peace, while angry, fearful, and patriotic people make war."

The main message of Casti's book is that the future is predictable in exactly the same probabilistic way that the weather is predictable, and that thoughts cause actions and events, not vice versa. These individual moods are like bets people place about the future. The bets then self-organise into an overall collective social mood, which after an appropriate period of time give rise to wars, election results, styles in popular culture, etc. In light of his discovery, the moods that all of us uphold each moment influence the kind of future we unfold; our collective mood is creating the tipping point of the kind of world we wish to see.

In line with this quest for connecting and promoting a critical mass of human consciousness, the Global Coherence Initiative is a science-based enterprise uniting millions of people in heart-focused care and intention, to shift global consciousness from instability and discord to balance and cooperation. It is steered, advised, and monitored by well-known scientists, such as Rollin McCraty, Ph.D., Elizabeth Rauscher, Ph.D., Dr. Franz Halberg, Bruce Lipton, Ph.D., Roger Nelson, Ph.D., Dean Radin, Ph.D., and Marilyn M. Schlitz, Ph.D.

Led by renowned astrophysicist and nuclear scientist Elizabeth Rauscher, Ph.D., the GCI team designed and built a Global Coherence Monitoring System (GCMS) that measures the fluctuations in the magnetic fields generated by Earth and those in the ionosphere. The fields generated by living systems and the ionosphere interact with one another. There is a symphony of frequencies ranging from 0.01 hertz to 300 hertz, and some of the large resonances happening in Earth's fields are in the same frequency range as those of the human heart and brain. The GCI hypothesises that:

i. all living things are interconnected, and we communicate with each other via our biological and electromagnetic fields;
ii. not only are humans affected by such energetic fields, but, conversely, the earth's energetic systems are also influenced by collective human emotions and consciousness;
iii. large numbers of people intentionally creating heart-coherent states of care, love, and compassion will generate a coherent standing wave that can help offset the current planetary-wide wave of stress, discord, and incoherence.

An important focus of the GCI's efforts is to build a critical mass of heart coherent practitioners to promote a global shift in consciousness toward harmony and balance. Their effort would include helping people to realise the need to become more self-responsible for their own thoughts, emotions, and actions. By taking charge and raising one's heart coherence, the life-supportive energy will in turn generate a mutually beneficial feedback loop between human beings and Earth's energetic systems.

The scientists at GCI postulate that when large numbers of humans respond to a global event with a common emotional feeling, it can affect the activity in Earth's field. In a situation where an event creates negative responses, this could be seen as a planetary stress wave. Conversely, when a positive response is created, a global coherence wave is generated. These scientists hope to gain support from individuals and groups to establish and amplify coherent out-going fields that interact with planetary fields to help generate global coherence. Their goal is to build a critical mass of 350,000 people who practise heart coherence in their meditation, prayer, intention setting, and contemplative living.

With the help of the Internet and advanced communications, the existence of a collective consciousness has been brought to human awareness. Each passing day more and more people are becoming conscious of the inter-influencing connections between humans, animals, the environment, and social systems. According to Deborah Rozman, Ph.D., a member of GCI's steering committee, the increasing awareness of our intimate connections to our world has awakened us to the desire to co-create a new world. It does not take many people to influence the health of the planetary field, i.e., if they are in a heart coherence mode.

The most important question we need to answer now is: "Are we ready to be part of the planetary solution?" If our answer is "Yes!" then let us first recapitulate the 11 global shift presuppositions and incrementally align our thoughts and actions to it daily.

THE ELEVEN GLOBAL SHIFT PRESUPPOSITIONS

1) We are part of the web of life that supports us; nature is constantly interacting and co-evolving with us. It is crucial that we connect, listen, and cooperate with nature.

2) Our outer and inner worlds mirror and entrain each other. If we don't like what we see in our world, we can start changing it by examining our worldviews and aligning our actions with our intentions.

3) A world of compassion, harmony, collaboration, brilliance, and resourcefulness would easily manifest through a united consciousness.

4) Be the change we want to see; lead by example.

5) What we value in our lives shapes our world. Redefine our values to bring forth a life-supportive world.

6) How we use our natural resources determines the sustainability of the abundance we have. Abundance must include how our thoughts and actions nourish others and the environment.

7) In order for a new beginning to materialise, we must end the old. Ending the old world must be accompanied by a clear and vivid vision of the new world.

8) The questions we ask, the intentions we uphold, the things we pay attention to, and the emotions we invoke everyday are ultimately vibrations that shape the kind of world we live in.

9) Making a deliberate choice to be life supportive is our most vital step to ensure our upward evolution.

10)A life-supportive and sustainable world is attained when we appreciate nature's cycle and align ourselves to it.

11)The first step out of our old world is to choose proactively to live simply, be life supportive and aligned to the collective consciousness. Our world is a consequence of who we become.

CHAPTER 16
WHAT ARE THE SIGNS OF OUR NEXT EVOLUTION?

Contrary to popular belief, humans continue to evolve.
Our bodies and brains are not the same as our
ancestors' were—or as our descendants' will be.

—*Peter Ward*

ARE WE STILL EVOLVING?

Revolutionary discoveries have been made by paleoanthropologists in the last century, and they are reshaping our understanding of where we came from and where we are heading. Below are the brief highlights of human evolution.

- Five studies in the *Journal of Science* (September 2011) put forth strong evidence that *Australopithecus Sediba*—a primitive hominin of more than 2 million years old—may be the ancestor of the *Homo genus*. The well-preserved human fossils found in Malapa, Africa, showed some signs of neural organisation in the orbitofrontal region of the brain, which indicates rewiring toward a human-like frontal lobe.
- *Homo habilis* (handy man) lived in East Africa about 2.2 to 1.6 million years ago, with a brain capacity of 800 cc. He used crude hand axes and stone flakes for cutting. Hunting and scavenging required certain behaviours, coordination, and intelligence.
- *Homo erectus* (upright human) evolved 1.9 million years ago in Africa, and lived until 400,000 years ago. Being upright enabled him to be an efficient long-distance runner, chasing down his game. He made more complex tools and weapons. With a brain capacity of 900 to 1,200 cc, he had better hunting skills and the ability to control fire and to cook. Leaving Africa around 1.8 million years ago, he went to colder climates in Europe and Asia.
- A variety of archaic *Homo sapiens* developed around 500,000 years ago. These included *Homo neanderthalensis*, who lived from 200,000 to 30,000 years ago. He was the first human to live in harsh Ice Age environments. He had immense strength, with thick arm and leg bones. He actually had larger brain capacity (1,500 cc) than modern humans, but his speech areas were not as well developed and his forebrain was smaller. His

tools did not develop much over the entire period, but he had cultural development including rituals such as burying the dead.

- About 200,000 years ago, modern humans (*Homo sapiens sapiens*) appeared in Africa. One branch left Africa around 60,000 years ago, sharing Europe and Eurasia with the more archaic Neanderthals from around 45,000 to 30,000 years ago. In late Ice Age environments, it is speculated that modern humans outlived the Neanderthals because of their superior cognitive skills. They created better weapons, fish nets and traps, and needles to sew garments. They had better communication skills, and the ability to cooperate with others and to plan ahead led to their survival. A recent landmark study conducted by Germany's Max Planck Institute (May 2010) found that Neanderthal genes live in us: 1–4% of the Eurasian (Europe, Asia, and Oceania) human genome comes from Neanderthals.

Our evolution has shown that the physical structure, including the skull and the brain's capacity, developed to allow more skills, cognitive abilities, and tool-making to evolve. This enabled better adaptation to various environments. Then, brain size did not continue growing, but the speech areas and thinking forebrain developed, showing the importance of communication, cooperation, abstract ideas, and planning ahead for survival. Recent DNA studies probing genomes, both present and past, have unleashed a great transformation of evolutionary studies.

Dr. Henry C. Harpending of the University of Utah puts forth strong evidence from his human genome studies that disputed the long-held belief that human genetic adaptation stopped 40,000 years ago. In fact humans continue to evolve rapidly in response to the new challenges presented by agriculture and civilization. Together with Dr. John Hawks of University of Wisconsin-Madison, they focused on genetic markers in 270 people from four groups: Han Chinese, Japanese, Yoruba, and Northern Europeans. They found that at least 7% of the human genes have undergone evolution as recently as 5000 years ago. There are many examples of gene evolution to adapt to diseases like Lassa fever, malaria, and HIV. Another discovery is that few people in China and Africa can digest fresh milk in adulthood, whereas almost everyone in Sweden and Denmark can. This ability arose as an adaptation to dairy farming.

Human cultural diversity developed over time with the desire to control resources. With separate territories, numerous groups diverged in beliefs, behavioural practices, and social norms. Studies that used the

International Haplotype Map show how human genetic data confirm that cultural and demographic shifts accelerated transformations in the human genome.

According to Dr. Harpending and Dr. Hawks's team, over the past 10,000 years humans have evolved as much as 100 times faster than at any other time since the split of the earliest hominid from the ancestors of modern chimpanzees. They attributed the quickening pace to the variety of environments humans moved into and the changes in living conditions brought about by agriculture and cities.

Since the Americas were "discovered" by the Europeans, and especially during the last century, people have been moving around the world, sharing genes and homogenising. Many cultures and languages have been assimilated or lost. International travel and global organisations have led to cultures and genes intermingling.

The advent of modern communications technology, especially the mobile phone and Internet, has brought people from all over the world into close and immediate contact. Social media is developing many virtual communities, and knowledge and information is now readily accessible to one and all. News spreads through communities at viral speeds.

In the opinion of Steve Jones of University College London, culture rather than genetic inheritance may now be the deciding factor in whether people live or die. In short, evolution may now be memetic—involving ideas—rather than genetic. A *meme* is a unit of cultural ideas, symbols, or practices which can be transmitted from one mind to another through writing, speech, gestures, rituals, or other imitable phenomena. Memes, according to memeticists, are cultural analogues to genes in that they self-replicate, mutate, and respond to selective pressure. In the words of Malcolm Gladwell, "A meme is an idea that behaves like a virus that moves through a population, taking hold in each person it infects."

The propagation of memes often involves imitation, which implies the copying of an observed behaviour of another individual, but memes can transmit from one individual to another through a copy recorded in an inanimate source, such as books or a musical score. Some proponents of mimetics see the transmission of memes like social contagions, such as fads, hysteria, and copycat crimes.

Author of the book *The Meme Machine*, Dr. Susan Blackmore asserts that the process of mimetics resolves the mystery of the size of the human brain—three times larger than our closest relatives, the great apes. It is resource-intensive to build and maintain the brain, and many mothers and babies die through childbirth complications caused by the size of the head. So why has evolution allowed the brain to grow so hazardously large? As explained by Dr. Blackmore, our early ancestors imitated useful new skills in making fire, hunting, carrying, and preparing food. As these early memes spread, the ability to acquire them became increasingly important for survival. In short, people who were better at imitation thrived, and the genes that gave them the bigger brains required for it consequently spread in the gene pool; behaviours and ideas copied from person to person by imitation—memes—may have forced human genes to make us what we are today.

Examples of memes in our current world are advertising slogans with jingles, gossips, jokes, proverbs and epic poems, nursery rhymes and children's games, fashion, medical and safety advice, religious practices, certain skills that are highly valued, emoticons and message abbreviations, popular principles in the form of cultural values and virtues, and many more. In other words, our individual memes influence our genes and daily living, which in turn produce the paradigm that we are currently living in. Understood in this manner, in order for us to live a better life in the future, we have to start re-examining our memes today, and make definitive changes.

Evolution of Values

With the evolution of technology and industrialisation, new knowledge and practices change people's values and, by extension, their behaviours—a large proportion of which are governed and driven by values. Below are some different ways to look at how values have shifted and are evolving.

World Values Survey: Social Changes Happening

The World Values Survey, which we have previously mentioned, is a worldwide network of social scientists studying changing values and beliefs and their impact on social and political life. It covers 88% of the world's population and shows how social, political, economic, religious, and cultural attitudes differ across societies, and how they have been changing over time since 1981, with technological and economic development. Some of the findings include:

- two dimensions of cross-cultural variation; human development is moving us from human constraint to human choice;
- the large variation in people's values can be distilled to two dimensions: (1) traditional vs. secular-rational values, and (2) survival vs. self-expression values. These two dimensions explain more than 70% of the cross-cultural variation.

Societies near the traditional pole emphasise the importance of religion, parent-child ties, deference to authority and traditional family values, while rejecting divorce, abortion, euthanasia, and suicide. They have high levels of national pride and want stricter limits on selling foreign goods. Societies near the secular-rational pole have opposite preferences. In industrial societies, worldviews have been shifting from traditional toward secular-rational values.

As industrial societies transition toward knowledge societies, an increasing proportion of people have grown up taking survival for granted. Survival values, such as physical and economic security, have been de-emphasized and there is more stress on subjective well-being, self-expression, individual freedom, and quality of life. Self-expression values give higher priority to environmental protection, tolerance of foreigners, gays/lesbians, and gender equality and participation in political and economic decision-making. Child-rearing values shift from emphasising hard work to imagination and tolerance.

Traditional and survival values emphasise human constraint, whereas secular-rational and self-expression values emphasise human choice; this change is a reflection of human development, as it makes people mentally free and motivates them to develop and actualise their inner human potentials. This Human Development Model is also called Maslowian Value Change.

The study entitled *Climatoeconomic Roots of Survival versus Self-expression Cultures*, by Evert Van de Vliert, is one of many which show that societies in more demanding climates endorse survival values at the expense of self-expression ones to the extent that they face economic challenges. In higher-income societies, families spend up to 50% of household income on housing, clothing, food, household energy, appliances, healthcare, and transportation. In lower-income societies, 90% or more of the people endorse survival values, where cases of abject poverty in harsh climates cannot meet survival needs. This has implications for the cultural consequences of global warming and available financing for human development.

Another study called *Development, Freedom, and Rising Happiness*, by Inglehart, Foa, Peterson, and Welzel, shows that from 1981 to 2007, happiness rose in 45 of the 52 countries with substantial time-series data. This was linked to the extent to which a society allows for free choice among its citizens. Economic development, democratisation, and rising social tolerance have increased people's perception that they have free choice, which has led to higher levels of happiness. On the one hand, our civilisation is moving toward more people having more survival comfort, on the other, we have escalating environmental issues that are destroying the livelihood of our near future. What would the evolution of our values look like in the long term? And how would it impact our future livelihood?

Spiral Dynamics—A Leap for Mankind

Don Beck and Christopher Cowan, authors of *Spiral Dynamics*, describe how the current destabilisation of belief and value structures is "like migrating tectonic plates, several core ways of thinking/paradigms are grinding against each other." Based on Dr. Clare W. Graves's quarter-century of research, we are at a breaking point, a shift in psycho-tectonics, a momentous leap for mankind.

Professor Emeritus of Psychology at Union College, New York, Dr. Graves had tested his model on over 50,000 people worldwide. He formulated "emergent, cyclical levels or waves of human existence." Each wave has important tasks and functions, and is included in subsequent waves, where none of them can be bypassed. Although people don't get better or smarter as they move through the levels, they broaden their perspectives and increase their options to act appropriately in a given situation. As Beck put it, "The focus is not on types *of* people, but types *in* people." Thus, the percentages below do not add up to 100%.

To return to Drs. Beck and Cowan, in their work, *Spiral Dynamics*, explained vMemes (values-attracting meta-memes). These involve a worldview, a value system, a level of psychological existence, a value and belief structure, organising principles, a way of thinking, and a mode of living. Various vMemes express both healthy (for better) and unhealthy (for worse) qualities.

In Spiral Dynamics, these levels or vMeme systems are grouped into two tiers. In the first tier, the first six levels are "subsistence levels," where people believe that their values are the only true and correct ones. There is fragmentation and alienation. In the second tier, other levels are integrated via inclusivity and these are the "being levels." There is wholeness and deep meaning. People at the second tier level appreciate the necessary role that all the various levels play.

The first tier (six levels or vMemes):

i. Beige—Basic-Instinctual; survival sense—do what you must just to stay alive. Food, water, warmth, sex, and safety have priority. Use instincts and habits just to survive. Little awareness of self as a distinct being. Best managed through nurturance and tender loving care. [0.1% of the adult population, 0% power]

ii. Purple—Magical-Mystical; tribal order—keep the spirits happy and the tribe's nest warm and safe. Show allegiance to chiefs/ elders in the clan. Preserve sacred objects, places, memories; and observe rites of passage and tribal customs. [10% of the population, 1% of the power]
iii. Red—Powerful-Impulsive; powerful self—be what you are and do what you want; exploitive independence. The world is a jungle full of threats and predators; conquer and dominate other aggressive characters. Egocentric; enjoy self without guilt or remorse. [20% of the population, 5% of the power]
iv. Blue—Purposeful-Saintly; absolute order; life has meaning, direction, and purpose with predetermined outcomes. Absolute belief in one right way and obedience to authority. Laws and regulations build character and moral fibre. Everyone has their proper place, and impulsivity is controlled through guilt. [40% of the population, 30% of the power]
v. Orange—Strategic-Materialist; enterprising self; act in your own self-interest by playing to get ahead and win. Self-reliant, risk-taking, and optimistic people deserve their success. Scientific achievement; manipulate the earth's resources to create and spread the abundant good life. [30% of the population, 50% of the power]
vi. Green—Sensitive-Humanistic; egalitarian order; seek peace within the inner self and explore with others the caring dimensions of community. Feelings, sensitivity, and caring supersede cold rationality. Consensus, reconciliation, and affiliation; refresh spirituality, bring harmony, and enrich human development. [10% of the population, 15% of the power]

The second tier levels or vMemes emerging are:

vii. Yellow—Integrative-Ecological; integrated self—live fully and responsibly as what you are and learn to become. Flexibility, spontaneity, and functionality have the highest priorities; change is the norm. Knowledge and competency supersede rank, power, and status. Systemic and interdependent.
viii. Turquoise—Holistic-Global; global order; experience the wholeness of existence through mind and spirit. Holistic, intuitive thinking and cooperative actions are expected. The world is a single, dynamic organism with a collective consciousness. Eco-consciousness and macro-level actions. The survival of life on earth. [0.1% of the population]
ix. Coral—still unclear; the next neurological capacities.

[Total for the second tier: 1% of the population, 5% of the power]

The 1% of the population in the second tier vision-logic awareness, being in the integral realm, will more creatively generate solutions to our pressing problems. Ken Wilber, founder of the Integral Institute, states that, "Because the health of the entire spectrum of consciousness is paramount, and not any particular level, this means that a genuinely universal integralism would measure more carefully its actual impact." In other words, more work needs to be done to make the lower-level waves more healthy in their own terms. That is, how to feed the starving millions, how to provide water and sanitation, how to house the homeless millions, how to bring healthcare to the millions who do not possess it, how to educate the illiterate, and so forth.

Dr. Clare Graves surmised that the greatest change in human consciousness would take place between levels 6 (Green) and 7 (Yellow). And this is also what several philosophers and Mayan Calendar theorists believe would actually take place in 2012-2017. Growing numbers of people are surfacing in stages 6 and onwards, which means there is a shift from materialism and individual-oriented outlooks to a more community-oriented worldview. As a result, environmental awareness, social justice, and harmony are the main goals of the higher stages in the second tier.

According to Dr. Beck, "Einstein said that the problems we have created cannot be solved with the same thinking that created them. And this is the hope that we have: that in the very dangerous and precarious global situations that we are in today, created by the Blue and Orange levels, we could prepare the breeding ground and the fertile soil and the habitats to generate what the next models of existence will be. We have reached that stage where our successes and our failures have produced problems that we simply cannot solve at the same level that they were created."

In Our Own Words 2000—Eight American Types

A landmark national research and communications programme, *In Our Own Words 2000*, examined the emerging values and transformations of people in the United States and found eight American types.

- Centred in a material world (14%)—materially successful, not interested in personal growth or spirituality, unlikely to be altruistic.
- Disengaged from social concerns (14%)—moderate and socially reserved, general negative outlook, higher incidence of depression and violence, less spiritually inclined, do not like personal growth activities, highest level of Internet access.

- Embracing traditional values (12%)—fairly conservative, materially successful, believe in church and family, sceptical of technology, do not believe in global awakening.
- Cautious and conservative (10%)—strong belief in God and fundamental religious values, less tolerant of different spiritual outlooks, conservative, less trusting.
- Persisting through adversity (10%)—strong positive outlook, experienced most family trauma, value personal growth and creativity, tolerant, try alternative healthcare.
- Connecting through self-exploration (12%)—believe in the connection to people, life, and the earth, interested in personal growth activities, altruistic.
- Seeking community transformation (12%)—strong values connection with people and unity with all life, strong belief in global awakening, give to charity, optimistic, and compassionate.
- Working for a new life of wholeness (16%)—work hard for a strong material foundation, global perspective with traditional values, strong belief in global awakening, somewhat isolated from others.

The first four, forming 50% of the American population, are the conservative, traditional half. The last four are the more creative and change-oriented half. An overwhelming 85% of Americans agree that "underneath it all, we're all connected as one." Nearly as many believe that the earth is a living organism and is fundamentally alive. The last two, the 28% who seek community transformation and work for a new life of wholeness, manifest the values, beliefs, and vision for a holistic civilisation. This is the segment which matches the cultural creatives described below.

Cultural Creatives

Sociologist Paul H. Ray and psychologist Sherry R. Anderson coined the term *cultural creatives* to describe a large segment (around a quarter) of Western society that has more recently developed beyond the modernists, traditionalists, or conservatives. They hunger for a deep change in their lives, with less stress, more health, lower consumption, more spirituality, more respect for the earth and diversity of species. Their values embrace a curiosity and concern for the world's ecosystems and peoples, an awareness and activism for peace and social justice, and self-actualisation through psychotherapy, spirituality, and holistic practices. Based on over 12 years of survey research with more than 100,000 participants, Ray and Anderson found that cultural creatives share attitudes, concerns, and lifestyles such as:

- having a well-developed social consciousness and social optimism;
- being disenchanted with materialism, status display, social inequalities of race, gender, and age;
- being critical of big institutions of modern society, including corporations and governments;
- seeing the big picture and the interconnections; seeking ecological sustainability; being focused on green issues;
- having strong concerns for the well-being of families, children, relationships;
- giving importance to helping people and bringing out their unique gifts;
- loving exotic things, foreigners, and travelling to exotic places, loving books, arts and culture;
- being careful consumers, well-informed shoppers who do not buy on impulse; they read the labels;
- wanting to be involved in creating a new and better way of life.

They include people of all races, ages, and classes, with 60% being women. This coherent subculture is currently lacking an awareness of what a strong force it is in the world. They potentially have the ability to reshape society to be more authentic, compassionate, and engaged. According to Ray, a critical mass of such conscious individuals can shift our evolution away from wanton destruction and toward a positive future for living things on the planet; they can create a mindset shift.

With what we have discussed this far, the meme of Gaia Consciousness, put forth by James Lovelock, has already begun to take hold among millions of people. The important question is, are we able to get sufficient people around the world to make a sustainable large scale change in time?

The LOHAS (Lifestyles of Health and Sustainability)

There is a recently identified business segment that represents a $540 billion global marketplace consisting of educated consumers who make conscientious purchasing and investing decisions. In an interview with the *Natural Business LOHAS Journal*, Paul Ray describes this business segment as being creative with new kinds of businesses, new movements, new ways of life, and new ways of seeing the world. Traditional advertising and marketing strategies often do not work for cultural creatives, who dislike conventional advertising, easily spot lack of logic or consistency, and are very good "bullshit detectors." Direct mail does not work for them, as they want good relationships with people they buy from.

All the above show a world in transition, where old-world traditional values exist together with new-world emerging ones, not unlike when Neanderthals co-existed with the newer modern humans. Robert Ornstein, author of *The Psychology of Consciousness* (1972) and *The Evolution of Consciousness* (1991), said, "There will be no further biological evolution without conscious evolution." He urged humanity, through the evolution of consciousness he describes, to make conscious changes in the way we think and relate to others, to adapt to an unprecedented new world.

Collectively, we are seeing an evolving wave of people with more social consciousness and tolerance, who are more connected and trust each other, broader and systemic perspectives with global awareness and ecological concern, less focused on materialism and consumption and more attention to spirituality—actualising their potential, more utilisation of alternative healthcare, more responsible for creating a new and sustainable way of life. They have had a consciousness transformation.

Consciousness Transformation

> Before we set out to reform the world, we would do well
> to pause and see if we should reform ourselves.
> —*Ervin László, world renowned scientist & humanist*

What is *consciousness transformation*? But first, what is consciousness? We have provided a brief definition in Chapter 1, but let's look at it in more detail now. Scientists look to quantum physics, information theory, or neuropsychology to explain consciousness, and they are only scratching the surface. According to Peter Russell, an explorer of human consciousness, this is because it is being explained within the existing space-time-matter paradigm. He suggests a new model of reality, with consciousness as a fundamental aspect of it, just as space, time, and matter are fundamental.

The Institute of Noetic Sciences (IONS) defines consciousness as "the quality of mind that includes your own internal reality. It includes self-awareness, your relationships to your environment, the people in your life, and your worldview or model of reality." Consciousness transformation is described as "a profound shift in your perspective resulting in long-lasting, life-enhancing changes in the way you experience and relate to yourself, others, and the world. It may occur in ordinary or non-ordinary states of consciousness. It can be gradual or sudden."

Model of Consciousness Transformation

For personal transformation, Director of Research at IONS Cassandra Vieten, Ph.D., introduces the IONS research-based Consciousness Transformation Model in these terms:

- groundwork for change—a series of events (or a single event) termed as "destabilisers" created the groundwork for change. These destabilisers could be the environment, peak experiences, numinous or mystical moments, life transitions, or a combination of some of them.
- Aha! (noetic) experience—a specific episode, period of life, or series of experiences; this moment forces awareness to expand and changes people's perspective or worldview.

She also states that some of the people who embrace the model:

- get obsessed with a desire to repeat the original experience; most find a practice to integrate new insights, including four elements: attention toward greater self-awareness, intention toward healing outcomes, repetition of new behaviours, and guidance from more experienced people in the practice;
- face the challenge to integrate these practices into everyday life, and not having them isolated from the rest of their life;
- face the challenge of going beyond personal quest for self-benefit, moving from "I" to "we," transforming the community;
- face the challenge of altruism; getting caught up in cult mentality or forgetting to take care of self and own well-being;
- live deeply—a balance between self-actualisation and self-transcendence, giving and receiving, formal and informal practice;
- experience equanimity, daily sacredness even in mundane experiences, and presence of being; these amount to a collective transformation, which further triggers individual transformation.

Many people who have experienced moments of consciousness transformation can relate to most of the stages, if not all. The key is to know which stage one is at, and know the steps to take to move on. The heart of this process lies in developing and living the practice, as well as staying and resonating with a group of like-minded people.

Turning Awakening to Practice

After 30 years of ground-breaking research on human consciousness, Marilyn Schlitz, Ph.D., Cassandra Vieten, Ph.D., and Tina Amorok, Ph.D. wrote a book called *Living Deeply—The Art and Science of Transformation in Everyday Life*. They describe how our behaviours, attitudes, and ways of being in the world are changed in life-affirming and lasting ways only when our consciousness transforms and we commit to living deeply in that transformation. They share the good news that opportunities to transform our lives in small and big ways abound in every moment of every day. They highlight that, "Transforming your consciousness may be the most important thing you can do for yourself and the world."

Having read all the preceding chapters, you may well have decided that you too want to be part of the solution, but wonder how to do it. According to research presented in *Living Deeply—The Art and Science of Transformation in Everyday Life*, there are four essential elements in transformative practices that can produce your own consciousness transformation.

i. Intention: You personally choose and are determined to act in a certain manner. This implies being fully aware and in charge of your own path of evolution, and knowing that your personal growth influences the collective evolution of humanity. It also means being in greater alignment with who you really are at your core.
ii. Attention: Changing your normal views brings you greater self-awareness when you focus on your mind and body or meditate. You are able to maintain focus on whatever you choose without getting distracted by other things. You are thus free to elect who you wish to be and able to perceive more clearly the minds, feelings, and intentions of others.
iii. Repetition: Through repetition and behaviours the brain can be consciously shaped and new habits can be formed. It is very much possible to form new neuro-pathways and even new nerve cells, and functions can be transferred across different parts of the brain.
iv. Guidance: In order for your transformative practices to be correct and sustained, it is useful to get guidance from teachers with experience. It is important, however, to harmonise this guidance with your inner wisdom, which appears in the form of symbols and metaphors in your dreams, when you experience altered states of consciousness, synchronous events, when you have visions, or when the voices within speak to you.

Transformative practices include daily prayer, meditation, reflection, periodic fasting, journaling, attending worship services, positive affirmations, walking in nature, yoga, energy healing, rituals, or others. Many practices cultivate insight, which, according to Michael Murphy, can help to identify the roots of problems within oneself, such as faulty assumptions, dysfunctional behaviours, or beliefs that no longer serve us. He talks about ego-transcending agencies for high-level change and loosening the grip of self-centredness.

Constant reminders are needed to integrate transformative practices into everyday actions, new ways of being in everyday life. You choose to use conscious intention to give direction and meaning to your life, to be proactive instead of reactive; to choose life-enhancing actions, for your body, mind, spirit, environment, and society.

Ram Dass, a spiritual teacher who was awarded the Peace Abbey Courage of Conscience Award in 1991, stressed that the most important thing we can do to help the world is systematic inner work that leads to deep psycho-spiritual transformation. Peter Russell says that a person's state of consciousness has an effect on the consciousness of others. A tipping point will be reached when a critical mass of people support planetary life-enhancing solutions. Every individual's effort counts. This is similar to the old *Three Musketeers* saying, "one for all and all for one."

Personality Transformation

The above consciousness transformation can be expressed differently via people's different personalities. The Enneagram Personality Model describes nine distinct categories of personality styles, with their individual worldviews and beliefs, underlying issues and motivators, and ways of being in the world. It is also a predictive model, showing how people change under stress, and how people develop as they progress. Although this segment focuses on an individual's personality transformation, the frame to hold in mind is that when individuals raise their consciousness, they lift the field of consciousness of the collective. This is in alignment with David Hawkins's Map of Consciousness, where each individual's upward climb (above 200) increases their own positive energy logarithmically, which in turn raises the field of collective consciousness.

Enneagram Model

The nine Enneagram styles are described below. No style is better or worse than the others, they are just different. Within each style, there are healthy, average, and unhealthy levels of personality health or development. Therefore, two persons with the same style could look very different if they are at different levels of personality health.

Type 1: People who want to improve things, to make the world a better (or perfect) place. They have a deep sense of what's right and wrong, are very methodical, efficient, and diligent. They are very rational, principled, have high standards, and tend to be critical and rigid. Healthy

Ones have high humanitarian values, integrity, and are purpose-driven. Examples of Type 1 include Mohandas Gandhi and Al Gore.

Type 2: People who want to help, nurture, and care for others. They have a deep sense of being of service, give advice, and are appreciated for their kindness and generosity. They are highly relational, emotional, people-pleasing and can play the martyr. Healthy Twos serve with unconditional love, connected to their inner feelings and convictions. Examples of Type 2 include Mother Teresa and Desmond Tutu.

Type 3: People who want to be successful and who strive hard to be the best. They have a deep sense of goals and outcomes and are determined and work hard to achieve them. They are image-conscious, prone to workaholism and are opportunistic. Healthy Threes enjoy motivating others to greater personal achievements. They are optimistic, authentic, inspiring, and dynamic. Examples of Type 3 include Oprah Winfrey and Barack Obama.

Type 4: People who need to express their authentic selves, who passionately reveal their uniqueness. They have a deep sense of the beauty and meaning of life, are creative, sensitive, and often subject to their moods. They can be dramatic, feel that something is wrong or missing and fall into spells of depression. Healthy Fours are connected to deep meaningful inspirations and idealised themes. They are emotionally intelligent and balanced. Examples of Type 4 include Thomas Merton and Anne Rice.

Type 5: People who need information and data to understand how things work in the world. They have a deep sense of mental constructs, systems, and planning, and enjoy intellectual stimulation and mastery. They need their space and can get emotionally detached and isolated. Healthy Fives bring the gifts of new knowledge and concepts into the world; they are visionaries and profound pioneers. They are perceptive and very focused. Examples of Type 5 include Albert Einstein and Bill Gates.

Type 6: People who can see all the possible dangers ahead and make contingency plans to feel safe. They are vigilant and great trouble-shooters. They have a deep sense of loyalty to their group, are highly responsible and dedicated. They tend to be plagued by doubts, anxiety, and suspicions. Healthy Sixes are committed to working hard for the common good. Examples of Type 6 include Malcolm X and Princess Diana.

Type 7: People who love fun, freedom, options, and new experiences. They are spontaneous, flexible, and entrepreneurial. They can get scattered, have trouble following through, and run a monkey mind. Healthy Sevens are futuristic thinkers, optimistic, and enthusiastic, contributing new ventures to the world. Examples of Type 7 include John F. Kennedy and Richard Branson.

Type 8: People who like to be in charge, take action, and be in control. They have a deep sense for justice, power, and fighting for those under their responsibility. They try to hide their vulnerabilities, can be confrontational and require anger management. Healthy Eights are courageous and magnanimous and have a passionate drive to make substantial positive changes in the world. Examples of Type 8 include Mikhail Gorbachev and Martin Luther King.

Type 9: People who seek peace and harmony around them and within themselves. They have a deep sense for seeing others' points of view, being non-judgemental and having consensus for decisions. They can get complacent, check out, and be stubbornly passive-aggressive. Healthy Nines seek to bring peace, love, and contentment to all around them. Examples of Type 9 include Carl Jung and the 14th Dalai Lama.

Pathways of Integration—Manifestations of Healthy Individuals

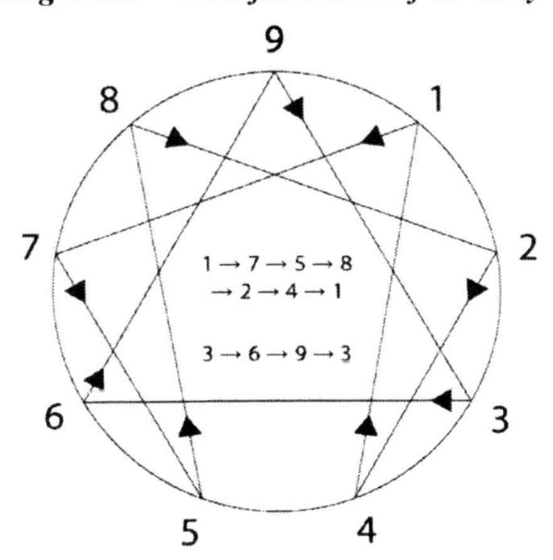

The Enneagram Personality Model
Pathways of Integration

Through the Enneagram symbol above, the Paths of Integration are predictive associations showing how each Enneagram style would manifest the healthy life-supportive qualities of the next arrowed style as they progress.

These are the personality transformations as we move toward healthy progression, indicating the antidotes to the rigidities and fixations of our styles, healing and freeing us from their bondage.

Type 1 moves to 7

The meticulous, rigid, and critical One lightens up, relaxes and enjoys fun and humour at Healthy Seven. Healthy Ones are able to catch themselves being critical or judgemental and learn to see the silver lining in the situation. They are open to making mistakes and appreciate the perfection in imperfection.

Type 7 moves to 5

The scattered, impulsive, and flighty Seven calms down, focuses, and develops mastery at Healthy Five. Healthy Sevens admit that they are running away from pain, and make time to keep still, reflect, or meditate. They intentionally make time to acknowledge the things they are grateful for each day to feel calm and content.

Type 5 moves to 8

The mental, analytical, and insecure Five takes bold action with confidence to make quick decisions at Healthy Eight. Healthy Fives open their emotional channels with others, and learn to assert themselves with other people. They carry out physical activities that connect them back into their bodies and energy.

Type 8 moves to 2

The powerful and well-defended Eight opens the heart to feeling compassion, becoming vulnerable and very helpful at Healthy Two. Healthy Eights trust that others can accept their vulnerabilities, and are able to make heart connections with people. They reconnect with their inner child and bring themselves back to a time of innocence and openness.

Type 2 moves to 4

The people-pleasing, externally attentive Two starts to go inwards, re-connects to personal needs, and develops outlets for creative pursuits. Healthy Twos authentically connect to their own needs, and meet these needs themselves, rather than seeking validation from other people. They connect with their creative passions and make time to bring their creativity to fruition.

Type 4 moves to 1

The emotional and moody Four becomes more objective, rational, methodical, and productive at Healthy One. Healthy Fours disconnect from endless replays of their painful past stories and focus on their wells of creativity. They replace the old emotional juice with the juice of being productively creative.

Type 9 moves to 3

The slow, apathetic, and easy-going Nine becomes motivated, more energetic, and outcome-focused at Healthy Three. Healthy Nines claim their own presence, acknowledge their own strengths, and actively work toward making a difference in the world. They speak up and take the right action, working on and delivering results.

Type 3 moves to 6

The what's-in-it-for-me and self-inflating Three sees a bigger picture and connects with a group or community, working for the greater good of all at Healthy Six. Healthy Threes realise that when the group wins, they win. They discover their inner authentic self behind their self-image, and courageously bring their authentic self out to relate to others.

Type 6 moves to 9

The anxious, suspicious, and hyper-vigilant Six becomes peaceful, trusts the goodness of life, and goes with the flow at Healthy Nine. Sixes see the games they play in their minds and see them for what they are without distorting them. They celebrate the diversity in life, and appreciate the forest from afar, rather than getting caught up in the trees. True courage emerges to embrace life and they lead others with a quiet confidence.

Levels of Health; Transcending the Ego

As people move onto their pathways of integration, they begin to connect to healthier levels of consciousness. In the intricate and dynamic system developed by Don Riso and Russ Hudson, two of the most prolific researchers and writers of the Enneagram today, they presented the world a map called the Levels of Development. The map shows how each Enneagram type would evolve healthily and devolve unhealthily. As people evolve into healthier aspects of their personalities, they become more liberated from their egos. In doing so, they fulfil their real needs and actualise their essential selves. Spiritual capacities emerge from each type as a person moves from personality to essence.

There are nine levels of development; these describe wide ranges of possible attitudes and behaviours, from the healthiest (levels 1–3) aspects to the most pathological ones (levels 7–9), as well as the average in-between (levels 4–6).

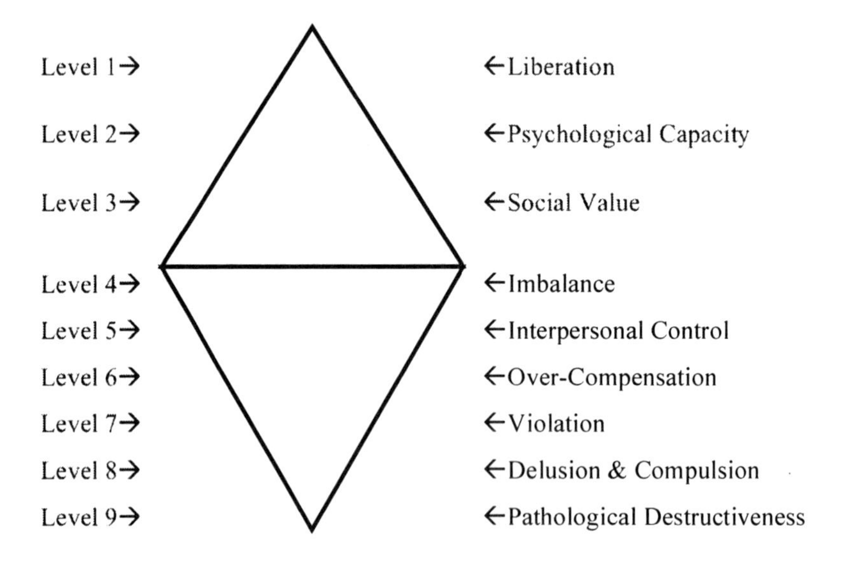

The levels of development show people a way to observe and measure their degree of identification with their personality structures. At level 1, they are free from their ego-based personality structures, whereas level 4 is where people are most identified with them. This system correlates with human experiences very well, as anyone can relate to the fact that when people are high-functioning they are generally open, balanced, stable, and able to handle stress well. On the other hand, when people are more troubled, they are reactive, emotionally stuck, and fragile in stressful situations. Everyone has experienced a wide range of states over the course of their life, from open, joyful, and life-affirming ones to painful, dark, destructive ones.

The basic negative emotion in the Enneagram system that drives people down the levels of disintegration is fear. Fear drives people deeper and deeper into restrictiveness, painful ego states, and further away from their true natures. Rather than resisting the fear, the Enneagram beseeches us to embrace it and let go of our need to hang on to old behaviours and attitudes. Transformation is the process of moving our consciousness up the levels within our type, until we gradually leave the realm of personality and move into the realm of essence. Moving our centre of gravity up even one level is an enormous accomplishment and everything in life changes when we do so. While it is necessary to have experiences of the higher levels, this is not the same thing as living in them. To shift our centre of gravity permanently entails a profound reorganisation of our sense of self, and everything that holds it in place.

In the words of Riso and Hudson, "Level 1 is not the end of the road, but the beginning of another [...]. Once a person has become liberated from the trap of his own personality type, he will begin to experience all kinds of wonderful expressions of themselves, the world, and of life." Most of all—along the same thread of thinking as Ken Wilber (Integral Beings), Sri Aurobindo (Gnostic Beings), and Arthur Koestler (Holon Beings)—Riso and Hudson stated that a person at level 1 would be able to integrate the positive qualities of all nine types, since they are no longer attached to the behaviours and beliefs connected with only one of the types; as people move upwards, toward level 1, they begin to be less attached to differences and move toward similarity with others.

People who move to their essence begin to manifest the spiritual qualities described by Dr. David Hawkins's realm of 540–599: joy and unconditional love. This is not the sudden joy of a pleasurable turn of events; it is a constant accompaniment to all activities. This is often characterised by enormous patience and the persistence of a positive attitude in the face of prolonged adversity. At the high 500s, the world one sees is illuminated by the sheer beauty and perfection of creation. There is a desire to use one's state of consciousness for the benefit of life itself, rather than for particular individuals. This capacity to love many people simultaneously is accompanied by the realisation that the more one loves, the more one can love. The realm of 540–599 is very important: it is the gateway to higher levels of development, such as the 600s (peace and bliss), 700–849 (self-realisation), and 850–1,000 (full enlightenment). Just like what Riso and Hudson have said, reaching level 1 of the levels of development is not the end, but the beginning of the road to higher and subtler realms of consciousness transformation.

Pause: The Next Step in Human Evolution

The state of the world is a consequence
of who or what we become.

—*Dr. David R. Hawkins*

When we daydream about our desired future we tend to envision the wonderful things we are going to have or perhaps possessing a sexy body and mastering new capabilities. We would probably like to have a scientifically-more-advanced home, a hydro-fuel car, and a more intelligent mobile phone. Most of us see the world's technology evolving to meet our needs, but how many of us daydream about *who* or *what* we want to evolve into?

The quote above points to the choice and power we have as individuals to co-create collectively with nature the kind of world we want. The statement is trying to help us realise that the kind of world we build collectively reflects who we are as a species: the dreams, values, and beliefs that drive us shape our world. Therefore, when we change the purpose of our existence, how we see ourselves in relationship to this planet and the cosmos, we change the world.

The study of purpose is called *teleology* and it is highly subjective. It has been extensively studied by philosophers and mystics throughout the ages, and within the confines of these pages, we want to emphasise that how we define the purpose of our existence determines the kind of world we build. Using Jean Gebser's evolution of consciousness as a framework (see Chapter 13), not only do we see the shifts in human dimensional awareness, the values and capabilities through each era, but we also witness the kinds of cultures and environment they created. At the same time, we also see the concretisation of human self-concept and the progressive separation of "I" from nature. Looking at where we are today, would we say we are a successful species on this planet? Maybe. Are we truly happy? Maybe. Are we ecologically wise? Definitely not.

It has been clearly established in this book that our thoughts and emotions influence the field that surrounds us. It is the governing source that stores, connects, dematerialises, and materialises every known entity in the Universe; it influences all the things that we see, hear, and feel in our world. Most people's reality is based on what they have experienced, but science and spirituality are telling us that the field is the governing source that moves, holds, shapes, and transforms everything. Like many other experts on this subject, Eckhart Tolle, the bestselling author of *A New Earth* (2005), clearly states that in raising our own consciousness, we greatly contribute to the evolution of the human race as a whole.

Crystallising what we have discussed this far, the idea of our next evolution is tied to the following emerging qualities and phenomena:

a) the progressive realisation of human collective consciousness and how it is accelerating human co-evolution with the planet;
b) the presuppositions of holism, integralism, and transpersonalism are embodied and applied across many learning disciplines and human endeavours;

c) the invisible and powerful relationship between the field and matter is increasingly being accepted and practised by evolving humans through meditation, contemplation, prayer, and intention-setting;
d) as we evolve we gain access to higher-dimensional awareness (fourth and above); this re-contextualises our experience and relationship with time, space, and energy.

In essence, the qualities described above are crucial for our future evolution, and, according to Timothy Lenton's hypothesis which supports the Gaia theory, evolutionary traits that benefit the system as a whole tend to be reinforced, while those that are incompatible are restrained. Who we are today (*Homo sapiens sapiens*) may be obsolete in the future if we do not evolve in harmony with the planet.

Unlike evolution in the past where we have seen distinctive biological changes, this time, according to leading evolutionary theorists, our evolution is happening at the consciousness level: we are consciously aware of the choices we have.

The vision of what our next evolution would look like has been expounded upon by many scholars. Perhaps when their works are put together, they can help us become future-fit and bring forth the latent and healthier versions of ourselves, and thus heal our world.

► **Vladimir Vernadsky** (Russia, 1863-1945)

Regarded as one of the founders of biogeochemistry and radiogeology, he stated that Earth's evolution began with the geosphere, followed by the biosphere and the noosphere—the sphere of human thought. Just as biological life transformed the geosphere, the emergence of human cognition changed the biosphere. One of the focal points of his work is that the complexification of human cultures would accelerate evolution, where cultural evolution rapidly outpaces biological evolution. This is linked to the recent emphasis on planetary sustainability, which focuses on the harmonisation of cultural and biological evolution. Humans at this stage of evolution would co-evolve or co-create with nature.

► **Sri Aurobindo** (India, 1872-1950)

A world renowned scholar, philosopher, and yogi, he aimed at advancing evolution of life on earth by establishing a high level of spiritual consciousness which he called supramental manifestation. He envisioned a new race of humans called gnostic beings who would usher in a new phase of life on earth. In supramental perception, one understands issues from their many sides, their relation to other things, as well as their essence and wholeness. This supramental consciousness is the intermediary between man and the gnostic being, who is described as living a spiritual existence in an integral way. He discovers that the spirit is everywhere in the world and in every person; this eliminates the separation between himself, others, and life. Gnostic beings can work together, near or far apart, aware or unaware of one another to create a new ecological way of living. A critical mass of such gnostic individuals could create the foundation of a new social order and divine life on earth.

► **Teilhard de Chardin** (France, 1881-1955)

Trained as a palaeontologist and geologist, he strongly asserted the idea that evolution occurs in a directional, goal-driven way and all of creation is pulled toward a converging point, which he called the Omega Point. He posited that human beings represent the layer of consciousness on earth that "folded back in upon itself" and, therefore, have become self-conscious. Based on the law of complexity and evolutionary theory, de Chardin made the startling discovery that evolution is becoming increasingly optional. He proposed that evolution is an ascent toward higher collective consciousness, but this unification can only be voluntary. As opposed to Darwin's concept of natural selection, this phase of human evolution calls for conscious co-creation. De Chardin predicted that the next evolution stage would be led by *Homo progressivus*, who values human progress above all things and has faith in the future of mankind. These new humans would be drawn together into groups, work seamlessly well together, even though their social origins might be quite different.

► **Alberto Villoldo** (Cuba, 1950s-present)

He was a clinical professor of psychology at San Francisco State University. He later resigned and pursued an intense study of shamanic practices of the Incas in South America. According to his studies, when there is great turmoil and upheaval in the world, a new species of humans will give birth to itself, and the physical, emotional, and spiritual traits acquired will be passed on to future generations. Through a process which he developed from ancient traditions, he helps people to perceive the vibrations that make up the physical world (dense) at a much finer level (light). In the study of quantum physics, all matter is densely packed light. In his studies, a new kind of human must come from a new luminous matrix. His focus is to transform the light energy field to usher in a new age of humanity—*Homo luminous*. Broadly speaking, a *Homo luminous* possesses the following characteristics: empathy, fearlessness, tolerance, and the ability to heal.

► **David Hawkins** (US, 1927-2012)

Mentioned several times in this book, from his massive body of research he explained that life began through the movement from the nonphysical to the physical realms, the energy of consciousness (divinity) itself interacted with matter. Life then evolved gradually into higher forms: from bacteria to fish, amphibians to reptiles, then mammals to primates. On the overall, according to Dr. Hawkins, the energy level of mankind is evolving toward spiritual reality. With the rapid development of the pre-frontal cortex in *Homo sapiens* arose the capacity to rise above the sovereignty of ego-based animal instincts and evolve past self-interests to friendship, affection, and care for others. This can then evolve into love and even unconditional love and compassion, which are spiritual qualities. With more and more fields of knowledge pointing to divinity as the ultimate source of existence and creation, a new evolutionary branch of mankind is emerging; Dr. Hawkins called it Homo Spiritus. This is an awakened man who connects the physical to the spiritual, the material to the energy realm, and passes from linear to the nonlinear. Above all, the awakened man realises that consciousness is the core of evolution.

The common themes that run through all these discoveries from different fields of knowledge are: (a) a growing deeply-felt sense of collective consciousness and co-creation, (b) processing information holistically, integrally, and transpersonally, (c) moving from denser to lighter energies, (d) progressively tapping into higher dimensional realities, i.e., fourth dimension and higher, (e) alignment to spirituality. All these open up entirely new possibilities for mankind to relate with and operate in time, space, and energy. This inevitably impacts the geosphere, biosphere, and sociosphere in a totally pervasive and generative manner.

Conscious Evolution

In the light of what we have presented so far, the wisdom of "becoming the change we want to see" is clear. Ever since man came into existence and until now, the notion of evolution has been unconscious. The increasing complexity of biological life and the convergence of the changes in the systems around us, however, together with the rapid advancements of technology and knowledge have enabled us to sketch a picture of the kind of beings we need to become, so we have moved into the conscious realm.

With the image of the better version of our species in sight, the process of acting as if we were the change we wanted to see is immensely useful. The presuppositional frame of "acting as if" is popular in the field of Neuro-Linguistic Programming: a technology for modelling human excellence and promoting desired change. It is a process by which an individual or a group acts as if a desired state had already been achieved. This enables us to drop our current perception of the constraints of reality and mobilise our imagination more fully; that includes releasing the constrictions of our personal history, belief systems, and egos. It helps us turn our ideology of "I" into a function or process, rather than identifying with it as a static noun or frozen entity.

Adjusting our thoughts, emotions, and behaviours to be as we desire, often activates the untapped and unawakened aspects of our potentiality. In quantum psychology, forming in our minds and hearts an energetic template of our desired new being helps us evoke the necessary vibrations to entrain the possibilities—as David Bohm puts it—in the implicate order (unmanifest) to flow into the explicate order (manifest). The enfolding realm represents the relationship of our hearts and minds with the Universe, and the unfolded reality is the experience we confirm in our bodies and environment. In other words, the more we fold inwards, the more we fold outwards the qualities or realities we want to experience. The way out of our current impoverished world is to focus on the desired changes within ourselves, and to express these changes in our daily lives.

Global Mind and Oneness

> What is needed is for man to give attention to his habit
> of fragmentary thought, to be aware of it, and thus bringing it
> to an end. Man's approach to reality may then be truly whole,
> and so the response will be whole.
>
> —*David Bohm, renowned quantum physicist*

> The truth is, there is only one world… People think there
> are two worlds by the activity of their own minds. If they could
> get rid of these false judgements and keep their minds pure
> with the light of wisdom, then they would see only one world
> and that world bathed in the light of wisdom.
>
> —*the Buddha*

For scientists, understanding our world means discovering its universal principles, a single theory that would explain everything. For spiritual people as well, the Holy Grail is to discover the unity behind all things.

Metaphorically speaking, it seems that the earth is divided into two hemispheres of the brain. Asia represents the right hemisphere, where its wise ones turned inwards, and through intuition and meditation searched for truth. Europe represents the left hemisphere; in this part, the search for truth was focused outwards. It became a process of deconstructing and analysing the world. The Asian wisdom sees things more holistically, while the West focuses more on examining differences.

The rapid proliferation of modern communication functions as a connector for the two hemispheres and uncovers a startling coherence about the laws of nature. The integration of the two great ways of knowing is producing a new consciousness that is non-dual: both self-aware and transpersonal, both self-sufficient and supportive, both highly focused and multi-dimensional, and both present and transcendent.

As we now know, a new being is emerging. Perhaps this is what the Hopis, Native American Indians, were referring to when they told us, centuries ago, that: "We are the ones we've been waiting for." The ancient past has prepared us. The time has come for us to welcome the birth of our new selves.

CHAPTER 17
ARE WE READY TO BE PART OF THE SOLUTION?

We have so far been presented with the dire threats in our world, as well as the amazing breakthroughs in scientific discoveries which seem to complement spiritual wisdom around the globe. We have been given entirely new visions of our desired future. At the same time we must constantly remind ourselves of the gravity and urgency of our collective quest to transform our ailing world:

> Avatar [movie] a reality for Indian tribe [Dongria Kondh] fighting mining company [to defend their sacred mountain].
>
> —*CNN, 9 February 2010*

> Doctors in the Iraqi city of Falluja are reporting a high level of birth defects, with some blaming [highly sophisticated] weapons used by the US after the Iraq invasion.
>
> —*BBC, 4 March 2010*

> Web hit by hi-tech crime wave [...] thieves trawl social network sites for information about victims.
>
> —*BBC, 20 April 2010*

> UN report: Eco-systems at tipping point [...] including the Amazon rainforest, freshwater lakes and rivers and coral reefs may never recover.
>
> —*CNN, 10 May 2010*

> Pesticide link to ADHD in kids.
>
> —*The Straits Times, 18 May 2010*

> More US kids on prescriptions.
>
> —*The Straits Times, 20 May 2010*

> The Gulf's silent environmental crisis [...] the massive oil spill in the Gulf of Mexico is harming numerous forms of life.
>
> —*CNN, 28 May 2010*

Is this the world we want to be living in? Is this the world we want to hand over to our children and future generations? It is clear by now that we can make a difference in our world. But the question is: are we ready to do it? Are we willing to look at our similarities rather than differences to solve our problems? Are we prepared to shift our priorities to address the threats to the survival of our world? Are we ready to internalise new life-supportive presuppositions that would help us co-create a new sustainable age?

Media Information Creates Awareness

While we are surrounded by overwhelmingly bad news from almost every facet of our lives, we are also seeing individuals and institutions making efforts to turn the global crisis around. The mainstream media and social media are playing a vital role in bringing life-supportive information to our awareness. While we ask ourselves if we are ready to be part of the solution, let us take a look at some of the many positive possibilities today.

Life-Supportive News from BBC

Material World by BBC Radio 4 reports on developments across scientific disciplines, where scientists convey the excitement they feel for their research work.

- Biodegradable products (20 April 2006)—with new mobile phone models constantly being released, the huge amount of waste created by discarded mobiles is a burgeoning problem. The need for biodegradable electronics is urgent. Later, in 2009: the first degradable and biodegradable bags were released into the market by the Indonesian Company Tirta Marta. After eight years of research by Sugianto Tandio, his environmentally-friendly bags are increasingly being used around the world. His Ecoplas bags are made from tapioca and are 100% biodegradable. The challenge is now to make them more affordable. His other degradable bag is made with a special additive that breaks down the plastic in 2 years.
- Energy consumption (27 July 2006)—scientists are wondering how to meet the ever-increasing demand. Solar power was once publicised as the answer to the planet's power issue, but low efficiency and high costs prevented it from being used on a large population level. Scientists from Southampton and Bath University are attempting to develop solar cells that mimic photosynthesis. Later, in 2011: with the improvement of cell efficiency and manufacturing technology, the cost of solar electricity was coming closer to affordability in the developed world. Installations have been built in different parts of the world such as the United States, Germany, Italy, India, Israel, Japan, South Korea, and many more. Morgan Solar Inc (MSI), a Canadian based company has developed a totally new

optical technology for concentrating sunlight—up to 1400 folds. Additionally, the technology is 100% recyclable and uses only environmentally safe materials.

- Seawater Greenhouse technology (18 September 2008)—with agriculture accounting for 70% of all water used, the shortage of fresh water is largely connected to food production. This technology converts seawater to fresh water, produces food and clean energy in arid regions, and also re-vegetates large areas of desert. The idea is simple: in the natural water cycle, seawater is heated by the sun, evaporates, cools to form clouds, and returns to the earth as refreshing rain. The Seawater Greenhouse technology uses more or less the same process. Hot desert air going into their greenhouse is first cooled and then humidified by seawater. This humid air nourishes crops growing inside and then passes through an evaporator. When it meets a series of tubes containing cool seawater, fresh water condenses and is then collected. And because the greenhouse produces five times the fresh water needed to water the plants, some of it can be released into the local environment to grow other plants. This technology has already been used in Jordan, Australia, Oman, and Abu Dhabi and good results with bountiful fruits and vegetables have been yielded. Countries like Qatar, Egypt, Morocco, and Mexico are actively discussing the viability of this technology with its British inventor, Charlie Paton. One of the ambitious projects that Seawater Greenhouse technology and many other similar companies are looking into is the greening of a 15 kilometres wide and 7775 kilometres long stretch between Senegal in the west to Djibouti in east Africa.

Life-Supportive News from CNN

In *Going Green*, CNN's complete coverage on the environment, many stories from around the world attest to creative ways people, organisations, and nations are dealing with issues affecting us. Some of the stories include:

- "Shoppers weigh up green premiums" (5 May 2010)—a grocery store in central London, called Unpackaged, gets customers to bring their own reusable bags and containers. This reduces carbon emissions related to packaging and food consumption, and waste that goes into landfills. The products are organic or fair trade, from local and small suppliers. The customers are among a growing number of consumers who are willing to pay a premium for green goods and services, which is great for business and the planet.
- "Argentine 'carrotmob' stick up for green business" (5 May 2010)—a global movement based on the carrot-or-stick concept, this group

rewards small businesses for employing sustainable practices by bringing them customers. Founder Brent Schulkin, 29 years old, says:

"As citizens and as consumers we may not be able to wield a whole lot of individual power, but when we do things together, our power is ridiculous. Carrotmob is going to organise our spending, because we are the economy. We decide who gets rich."

Carrotmob started in San Francisco in 2008 and has since had 55 gatherings around the world, from Helsinki to Bangkok.

- "India's farmers profit from organic boom" (5 May 2010)—the Punjab, India's breadbasket, has been using chemicals and pesticides on crops for decades. In 2008, research proved that this has caused many health problems, including cancer, and that drinking water contained heavy metals. Tens of thousands of farmers went back to traditional organic farming (300,000 farms), citing reasons such as lower costs, healthier produce, higher selling prices, and a healthier environment. India's organic farming sector accounts for only a small portion of the global $50 billion market for organic products; but with growing demand, the potential is huge.
- "Building a greener future from bamboo" (8 December 2009)—bamboo, one of the world's fastest growing plants, captures carbon emissions. The Argentinean government launched a programme to teach ways to manage bamboo in a sustainable way, increasing production and profits while absorbing harmful carbon gases and slowing down climate change. Bamboo is used to build houses, fences, boat oars, fishing rods, chairs, lamps, baskets, bowls, etc.

Life-Supportive News from Magazines

Another shining example of converting deserts to farmlands is Carl Hodges's decades of work, reported in *Discovery Channel Magazine* in March 2010. The 72–year-old atmospheric scientist formed Global Seawater Inc., which diverts seawater into inland rivers and uses it first to farm shrimp, shellfish, or fish aquaculture. Then it uses the nutrient-rich water from the farms to grow fields of salicornia, which is edible and whose seeds are processed into edible oil or biofuel. Salicornia grows on desert land watered directly by seawater. Finally, the water is diverted from the salicornia fields to mangrove forests, which filter the water. This method eliminates expensive desalination, and can pump enough ocean water into deserts to counteract the rise in sea level caused by melting polar ice caps. If this innovative and life-supportive idea catches on in our world many large unproductive desert lands can be turned to provide food, water, and wildlife habitat all over the world.

In August 2012 *Living Green Magazine* featured colleges in the United States that are greening up their campus by reducing energy consumption and offering more environmental courses. Buildings are certified by recognised bodies on factors like the sustainability of their building design and practices on water and energy saving. Colleges provide students with food that is bought from local farms to reduce the impact of transportation on food. Some colleges have on-campus farms so students can learn first-hand about farming practices. Many of these farms also contribute to the colleges' cafeterias and farmers' markets. Many colleges are moving to electric and hybrid shuttles to improve fuel efficiency and reduce carbon emissions. For all the colleges, going green isn't just a fad; it's a lifestyle. Top ranking green institutions include University of Washington, Green Mountain College, University of California, San Diego (UCSD), Warren Wilson College, and Stanford University.

From the scientific community, the *Scientific American* magazine presented an important article entitled, "Can We Feed the World and Sustain the Planet?" This article featured the works of Jonathan A. Foley, director of the Institute on the Environment at the University of Minnesota. It clearly articulated the problem every major food producer must face in the near future: (a) it must guarantee that all seven billion people alive today are adequately fed, (b) it must double food production within the next 40 years, and (c) it must achieve both goals while being truly environmentally sustainable.

Agriculture occupies 38% of the earth's land surface and is by far the biggest human use of land on the planet; nothing else comes close. According to Dr. Foley, feeding people would be easier if all the food we grew went into human consumption, but 35-40% of the world's food crops are used to feed animals; thus, meat—coming from the animals that are fed with these crops—is the biggest issue in terms of feeding the human population. Agriculture is also the single largest source of greenhouse gas emission, due to poor practices.

After many months of research, based on the analysis of global agricultural data, Dr. Foley's international team came up with a five-point plan to deal with the food and environmental challenges: (a) halt the expansion of agricultural footprint, (b) close the world's yield gaps, (c) use resources much more efficiently, (d) shift diets away from meat, and (e) reduce food waste. Dr. Foley's plan offers governments and corporations around the world the scientific roadmap to ensuring food availability in a sustainable way.

Life-Supportive News from Youtube

Having investigated factories and dumps around the world for more than 20 years, Annie Leonard is an expert in international sustainability and environmental health issues. Her 20-minute web-film "The Story of Stuff," which tracks the life of the stuff we use every day, such as aluminium cans, laptops, and cotton T-shirts, has been viewed over 10 million times since its release in December 2007. It explores the often hidden environmental and social consequences of consumerism, and can be found on her website, at http://storyofstuff.org. She has since added films like "The Story of Broke," "The Story of Bottled Water," "The Story of Cosmetics," and "The Story of Electronics".

In her book *The Story of Stuff: How Our Obsession with Stuff Is Trashing the Planet, Our Communities, and Our Health—and a Vision for Change*, she further communicates how consumerism and materialism have impacted global economies and international health. Basically, we have too much stuff, and a lot of it is toxic. The book covers the five stages in the materials economy: extraction, production, distribution, consumption, and disposal. She also gives examples of solutions in several lists of policies and practices that any of us could start adopting. In 2008, Annie was named one of *Time* magazine's Heroes of the Environment.

Through Youtube and the Internet, the idea of using clear water bottles as light bulbs is illuminating millions of poor homes in Brazil. The brilliant idea belongs to Alfredo Moser and he had it first during a blackout. He used a simple two-litre bottle, filled it with water and two capfuls of chlorine and created his light bulb. By cutting round holes in his roof and inserting the water-bottle bulb, it could shine the sun's rays into his home and workshop. The water diffracts the light, letting it spread throughout the house instead of focusing on one point. Its luminance was tested and proved to produce as much light as a 50 watt incandescent bulb. Friends, neighbours, and eventually the town started to use his invention and drastically cut down electrical bills.

The idea soon spread to other third world countries and was massively successful in the Philippines—that is where it went viral on Youtube. Millions of homes in the Philippines and around the world are left in the dark because metal roofs block all light and there are no connections to electrical grids. The organisation Isang Litrong Liwanag (A Litre of Light) in the Philippines helps to spread the brilliant invention and assists others to install it. Most of all, it is environmentally friendly—using sunlight and recycling abundant plastic bottles—with zero carbon and heat emissions. Furthermore, it can be installed at almost zero cost. The brilliant idea is now increasingly being used in parts of Africa, South America, India, and other third world countries.

There are many innovations and uprisings happening around the world today, showing us different ways we can turn our situation around. These people from all walks of life, with different passions and callings are creating a win-win-win situation for themselves, others, and the world. These people are not asking themselves whether they can be part of the solution: they are doing it every day. The question is, are there enough people doing these life-supportive activities? Will we be able to do them in time before the point of no return is reached?

What Does It Take to Be Part of the Solution?

A belief is not merely an idea the mind possesses;
it is an idea that possesses the mind.

—*Robert O. Bolton*

What Beliefs Do We Need to Redefine?

Collins Dictionary defines a belief as "a principle, proposition, idea accepted as true," or "opinion; conviction." In the context of the paradigm that is currently predominant, the belief that drives most of our existence is that no matter what happens, business must continue as usual. Business and work have become the absolute centres of our lives, almost a religion for most people. So beliefs surrounding our ideas about work, income, livelihood, spending habits, lifestyle, and business have dramatically shaped our lives and the fate of the planet over the last 200 years.

In the past 15 years, the previously mentioned noteworthy concept of "do good, do well" has been circulating around in business. Do corporations and businessmen who promote this concept truly believe that by doing good they will do well? More often than not, businesses are focused on doing well, and if they do well enough, they might then do some charitable good. But which of these comes first in focus has monumental impact on how one sets priorities and runs the organisation, and largely depends on the values and beliefs of the stakeholders and chief officers. The belief that by doing good one will do well is one of the vital beliefs of our time that we must redefine, communicate, and demonstrate in our society and economy.

In recent years, instead of just focusing on profitability and shareholders' earnings, a growing number of company founders are operating under the belief that the company exists to plough back life-supportive activities into the community and the world. However, it is not just companies that can make a difference in the world; individual citizens have equal or more power to do good! Several examples are given below where company founders and individuals had to overcome limiting beliefs such as:

- "I am a nobody, how can I—one person—possibly make a difference?"—a sense of impossibility and helplessness;
- "The small, poor individuals are of no consequence; they cannot amount to anything."—a lack of connection and empathy;
- "This is their fight or plight; it has nothing to do with me."—a dissociation from others and self-absorption;
- "We must compete to have more, be more successful."—a competitive and status consciousness;
- "I must keep the best to myself, to have a winning edge over others in order to survive."—a scarcity consciousness;
- "The world is for our taking; we must achieve our ideals at all costs."—a worldview disconnected from others, nature, and the planet;
- "Bigger, more, newer, better, and faster are sure ways to building a successful life."—a lack of inner reflection and connection to others, nature, and the planet.

By looking at the story of the people below, we will come to the full realisation that there are many things that we can do to make a positive difference in the world. The important point to keep in mind is that they all started by taking small steps—humble beginnings—even though they had noble grand visions of contributing to a better world. No matter how grand our vision is, we need to begin from where we are now. Ultimately, it must involve taking incremental steps to change our routine and priorities in life and aligning the appropriate actions to our daily activities.

Life-Supportive Activism

The Body Shop

Dame Anita Roddick founded The Body Shop in 1976 with the mission statement: "To dedicate our business to the pursuit of social and environmental change." The company's five core values are: support community fair trade, defend human rights, against animal testing, activate self-esteem, and protect our planet. The Body Shop set up their own Community Fair Trade programme over 20 years ago, and it now operates in over 20 countries, providing essential income to more than 15,000 people. The Body Shop Foundation, a charity launched in 1990, funds projects related to human and civil rights, environmental and animal protection.

It was one of the first businesses to offer recycling and extend support to many global campaigns, such as Stop Violence in the Home and the Campaign for Safe Cosmetics. The company also promotes renewable energy with Greenpeace International. Roddick has received many awards, including Philanthropist of the Year in 1996 and Dame Commander of the British Empire in 2003. By 2005, there were 2,045 stores worldwide. In her autobiography—*Business as Unusual*—Anita Roddick describes her belief that businesses are powerful institutions in our society and they need to assume a moral leadership, with heart, soul, and conscience.

All of the great social and environmental achievements that Roddick has accomplished started because she wanted to create a livelihood for herself and her two daughters while her husband was trekking across the Americas. Without any business experience, she distilled her initial business wisdom down to this: it is all about creating products or services that are of high quality and that people are willing to pay for. Her business philosophy was based on the experience she gained from her early travels around the world, when she interacted with women and understood how people bought and sold in their businesses, and she also discovered the body rituals of women around the world. The other source of her business wisdom came from her mother's frugal practices during the Second World War: "Why waste a container when you can refill it? And why buy more of something than you can use? We reused everything, we refilled everything, and we recycled all we could."

In other words, Roddick's personal practices have been translated in her business. With her passion and the belief that a business has the power to do tremendous good, she continually sought natural healthy products, making sure they are made without harming animals and they are environmentally friendly. She started giving business to struggling natives in third-world countries and fighting for the rights of people that have been violated by corporations, government, and militias. She believes in the power of telling stories; every product she sells tells a cause-worthy story, so that people are not just buying her high quality products, but also supporting a worthy cause. As her products were so loved by her customers that they too wanted to sell them, she created a franchise system which spread all over England first, and then to the rest of the world.

One of her empowering messages to people who have great visions but small budgets and a sense of great doubt is:

If you think you're too small to have an impact,
try going to bed with a mosquito.
—Anita Roddick

The Grameen Foundation

Professor Muhammad Yunus and Grameen Bank, 2009 Nobel Peace Prize winners, are pioneers in offering micro-loans and other financial services to the poor. This enables people to set up small businesses and start moving out of poverty. Professor Yunus and Grameen Bank have made loans to over 7.5 million people in Bangladesh and other countries, 97% of whom are women. No collateral is required, and the loan is usually repaid within six months to a year.

They have a very high repayment rate, averaging 95% to 98%, better than credit card debts in the United States. The Grameen Foundation helps the world's poorest to improve their lives with access to microfinance and technology, to develop or expand a small self-sufficient business and work their way out of poverty. They believe that credit should be a human right. Women are the targeted clients, as they use the profits to send their children to school, improve their living conditions, improve their families' nutrition, and expand their businesses.

By giving poor people the power to help themselves,
Dr. Yunus has offered them something far more valuable
than a plate of food—security in its most fundamental form.
—former President Jimmy Carter

It all started in 1976 when Dr. Yunus visited the poorest household in the village of Jobra (Bangladesh). He found out that 42 women were having a hard time with loan sharks and all their loans amounted to only $27. He used his own pocket money and paid off the loan sharks. He then approached the local bank which refused to lend money to poor women who had no collaterals, so he used himself as guarantor for the women. Each of the women made a healthy profit on the loan and they all paid back the money to the bank by making and selling bamboo furniture. His actions were a result of his beliefs that poor people are entitled to a loan to improve their lives; he sincerely believed that the poor would repay the money they borrow. *Believing in the goodness of people seems to be the common trait among the great humanitarians of our time.*

After several attempts, he finally succeeded to secure a loan from the Bangladesh state-owned bank to lend to the poor in Jobra in December 1976. By 1983, the bank had more than 28,000 members and it was renamed as Grameen Bank. As of 2007, Grameen Bank has issued $6.38 billion to 7.4 million borrowers. To ensure repayment, the bank uses a system of "solidarity groups." These small informal groups apply together for loans and their members act as co-guarantors of repayment and support one another's effort for economic self-advancement.

In over 30 years of arduous work to help the poor, Dr. Yunus has encountered excruciating difficulties and when he was asked what his greatest challenge was, he answered:

> My greatest challenge has been to change the mindset of people. Mindsets play strange tricks on us; we see things the way our minds have instructed our eyes to see.
>
> —*Dr. Muhammad Yunus*

Central Asia Institute

> Changing the world not with bombs, but with schools.
>
> —*Greg Mortenson, author of Three Cups of Tea*

Greg Mortenson, an American nurse, failed to climb K2, the world's second tallest mountain after Mount Everest, in 1993. Very ill, he was sheltered in a poor Pakistani village, Korphe, for seven weeks. He promised to return and build a school to repay the villagers' kindness. He co-founded the non-profit Central Asia Institute and, as of 2009, has built and supported 131 schools in rural Pakistan and Afghanistan, teaching over 58,000 children. This is an incredible humanitarian effort in an Islamic extremist territory—where his life was in danger more than once—in the desire to alleviate poverty and educate children, especially girls.

His adventures are presented in the book *Three Cups of Tea—One Man's Mission to Promote Peace... One School at a Time*, which is a testimony to his belief that education and literacy are the ways to promote peace and save the future. He was awarded Pakistan's highest civil award, the Star of Pakistan, in 2009 and was nominated for the Nobel Peace Prize in 2008 and 2009.

Mortenson was "called" to this mission because he was deeply moved by the kindness and generosity of the poor people in Korphe, who gave a stranger so much when they had so little. The universal bond of human connection that the people of the village demonstrated shows that empathy and kindness go beyond skin colour, religion, and race. Looking at the absence of a school building and teachers, he promised the children of Korphe that he would return and build them a properly equipped school.

Returning to California, he worked long shifts as an emergency room nurse, often living out of his car, to raise $12,000 to build the school. After sending out 580 letters to celebrities and public figures, he received only one small donation. A short article that he published in the newsletter of the American Himalayan Society caught the attention of Dr. Jean Hoerni, a rich millionaire who gave him a donation to help him buy all the materials he needed.

Returning to the village, he first had to build a bridge to transport the building materials over a deep gorge of a river. All the males in the village volunteered to help and he got the involvement from everyone there. The Korphe School was completed in 1996. He hired a local teacher to teach. The second school he built was in Ranga and it took only 10 weeks to build. In the following months of 1996 he built the third and fourth schools. As of July 2012, the Central Asia Institute has established over 300 educational and community initiatives.

He strongly believes that providing education to girls is the most important investment any country can make to create stability, bring socio-economic reform, decrease infant mortality and population explosion, as well as improve health, hygiene, and sanitation standards. Mortenson's view is that fighting terrorism perpetuates a cycle of violence, so instead of it there should be a global priority to promote peace through education and literacy. Greg firmly believes, "You can drop bombs, hand out condoms, build roads, or put in electricity, but unless the girls are educated, a society won't change."

Tumbleweed Tiny House Company

Since 1997 Jay Shafer has chosen to live in tiny houses he has designed and built himself, to lessen his impact on the environment. Though they measure just 89 square feet, his houses have met all his domestic needs and have led to a simpler and slower lifestyle, for which he is very grateful. His Tumbleweed Tiny House Company builds houses from 65 to 837 square feet, and there is no need to own land, for they can be picked up and moved. People can build their own small house for less than $20,000, saving money and the environment. The average American home produces seven tonnes of construction waste and eighteen tonnes of greenhouse gases every year, compared to only one hundred pounds of construction waste and four tonnes of greenhouse gases for a tiny house. Solar-electric panels or wind generators produce electricity, stored in batteries, and utilities bills can plunge to $65 per year. Shafer is leading a movement to change the way people view housing, with his vision of living small. His ideas for efficient living have been featured on Oprah, CNN, and the *Natural Home Magazine*.

Shafer stayed in his 89-square-foot home for almost 15 years. His entire home is smaller than some people's closet. His decision to live in a small home started when he felt that his rights to live the way he wanted were violated by the housing laws in the United States; it was his way of demonstrating civil disobedience non-violently. Another motivator was his desire to fulfil his value of freedom: to be free from the burden of a big house which comes with a high mortgage, free from large areas of cleaning, free from the cost of a long maintenance list, free from worrying about the non-essential stuff, and especially free from managing the things that accumulate in the storeroom that we do not use. Another reason that pushed him to construct his own small mobile home was his awareness of the huge environmental impact a big house creates. According to US environmental statistics, an average American house (2,349 square feet) produces about 18,000 tons of greenhouse gases per year, and 7 tons of construction waste that goes to landfills.

In Shafer's experience in helping other people build their "tumbleweed homes," the toughest challenge is not building the home, but in helping the new house owners get rid of their stuff. Shafer explained in his video interviews that people have great difficulties in separating what they need from what they want. And people live in constant stress, working hard and long hours most of the time because they project their *wants* into their necessary retirement funds. The mainstream economic financial projection does not apply to Shafer. His retirement plan is simply having a small plot of land with a community of tumbleweed homes which are homes of very good friends and family members. Less stress, less things to worry about and manage, smaller footprint on the environment, more mental space, more time for loved ones and to do what you love, more chance for nature to rejuvenate; less is truly more.

> I'm not the kind of person that follows money. For me, I simply follow my bliss. And my bliss is living simply and doing what I love to do, which is designing and building small beautiful homes.
>
> —*Jay Shafer*

Life Supportive Movements Around the World

There are many life-supportive movements around the world that focus on the conditions of humans, animals, and plants, the environment, and our earth. Paul Hawken describes it clearly and succinctly in his book *Blessed Unrest* (2007). An author and environmentalist, Paul Hawken has uncovered two million organisations around the world working toward ecological sustainability and social justice. He describes it as an emerging "global humanitarian movement arising from the bottom up," and concludes that it is the "largest social movement in all of human history."

It is different from previous social movements, having no ideological or religious core of beliefs, and has no strong leader or central leadership. The movement has three basic roots: environmental activism, social justice initiatives, and indigenous cultures' resistance to globalisation, all of which have become intertwined. He shows what is going right in the world, through groups of people who use imagination, conviction, and resilience to heal the planet's wounds with passion and determination. The active groups are listed at *WiserEarth.org*. Some examples of the types of groups and the number of groups in each category as of 2006 include:

- sustainable agriculture (3,349)
- organic farming (670)
- air quality and pollution (1,055)
- animal and plant trafficking (591)
- animal welfare and rights (2,353)
- endangered animal species protection (1,667)
- wildlife habitat conservation (6,149)
- biodiversity conservation (3,048)
- microfinance (1,323)
- child and youth protection (4,645)
- marine ecology and conservation (762)
- community participation (10,053)
- sustainable communities (8,999)
- natural resource conservation (11,393)
- environmental education (11,789)
- sustainable fishing (770)
- sustainable forestry (1,411)
- recycling and reusing (4,346)
- human rights and civil liberties (8,052)
- human trafficking and slavery (573)
- indigenous people and culture (1,341)
- peace and peace-making (7,916)
- poverty alleviation (9,240)
- watershed management (2,638).

Hawken stated that the current turmoil in our world is too complex to be succinctly labelled, but global themes are emerging. These themes include the need for revolutionary social change, the reinvention of market-based economics, the empowerment of women, activism on all levels, and the need for localised economic control. There are constant pleas for autonomy, demands for the re-establishment of cultural primacy over corporate domination, and a rising cry out for transparency in politics and corporate decision making.

More cases of failure than success have been seen in the fight for the environment, and some have said that environmentalism is dead. On the contrary, according to Hawken, everyone on earth will be an environmentalist in the near future, due to the rapidly deteriorating environment. There is a need for a radical global change and that can only happen when our worldview of what it means to be a human being is fully re-examined and redefined. The realisation of our responsibility to a greater whole—which supports all life—must rapidly proliferate and take hold in our consciousness.

Almost everyone today is talking about sustainability, which in simple terms is about stabilising the conflicting relationship between the earth's two most complex systems: human culture and the planet's ecosystems. In Hawken's words, "The way to change the world is to change one's own practices, including one's own home, source of energy, method of agriculture, diet, transport patterns, and communities."

A major structural hold on human culture is the practices that exist in business. Hawken has said many times, "There is no polite way to say this—business is destroying the world." An important observation that is circulating among the myriad of communities in this global life-supportive movement is that goods seem to have become more important and are treated better than people. Making money at all cost seems to be the motto of most corporations. In Hawken's research, 200 of the world's top companies have twice the assets of 80% of the world's people, and that asset base is growing 50 times faster than the income of the world's majority. Revolutionary reforms in the way we do business, as in putting people and the environment first, is going to save the inhabitants of the planet. In support of Hawken's work, Jane Goodall, UN Ambassador for Peace said,

> *Blessed Unrest* [...] describes the growing unrest that
> I encounter around the world, the frustration and courage of
> those who dare to challenge the power of the political and
> corporate world. Paul Hawken states eloquently all that I believe
> so passionately to be true—that there is inherent goodness
> at the heart of our humanity, that collectively we can—
> and we are—changing the world.
>
> —*Jane Goodall*

Earth Hour

Organised by the World Wildlife Fund, Earth Hour began in Sydney in 2007, when over 2.2 million households and businesses switched off their lights for one hour. This was to send the message that it is possible for everyone to make a difference and take action against global warming. In 2008, 370 cities and towns in over 35 countries, with an estimated 100 million people switched off their lights for Earth Hour 2008. Earth Hour 2009 saw hundreds of millions of people in 4,159 cities in 88 countries taking a global vote against climate change. Earth Hour 2010 met with a record 128 countries, over 4,000 cities, and hundreds of millions of people coming together to deliver a powerful action of hope for a better and healthier planet.

Lights went off on heritage sites, ancient and modern marvels, local cafes, schools, royal and presidential palaces, universities, shops and businesses, and backyard dinner parties. Over 1,500 iconic landmarks switched off to show their support, such as Beijing's Forbidden City, Berlin's Brandenburg Gate, London's Tower Bridge, and the Las Vegas Strip.

Andy Ridley, Co-Founder and Executive Director of Earth Hour said of the World Wildlife Fund (WWF), "WWF's Earth Hour, at a personal, local, and global level has become a rallying point for those who want action on climate change and are prepared to be part of the solution." And United Nations Secretary-General, Ban Ki-moon, stated: "The message of Earth Hour is simple: climate change is a concern for each of us. Solutions are within our grasp and are ready to be implemented by individuals, communities, businesses, and governments around the globe."

THE POWER OF ONE

Most of us have seriously underestimated the power of one person to make a positive difference in the world. The key point to realise, after having read about all these examples, is the power of the individual to influence and communicate his ideologies via his own consistent actions. With the fact that we live in an inseparably connected reality and that we cannot not influence each other, the consistency of our actions—small or big—can make a mountain-moving difference. The seed of transformation is within all of us; and so is the desire to make our world a more peaceful, beautiful, abundant, and compassionate place.

Though it may seem that only a few people are adopting a life-supportive lifestyle and promoting this movement, the crux is to remind ourselves that we have the power to influence. Especially with the proliferation of communication technologies, our world is more and more connected every day. News can spread from one small part of a country to millions of people around the world within 3 to 6 hours. This was clearly demonstrated when Farmer Le Dung and his fellow villagers fought local police who were trying to take over their land for a luxury property development in east Hanoi, in Vietnam. With the support of Internet activists, videos of police brutality went viral on the Internet within a few hours via blogs, Facebook, and iPhones. According to Le Dung, "If we hadn't used the Internet, the authorities may have killed us; now they know they have to be careful."

The potential to create a large growing movement via a collective of individuals is not a fantasy, but very possible in our ever-increasing connected world. The most important attitude to embody in our daily lives is to lead by example. We can't tell others to change when we are still part of the problem. Again, we must remember to embody the powerful words of Gandhi, "Be the change you want to see in the world."

Our Collective Choices Shape Our World

The question, "Are we ready to be part of the solution?" would be very difficult to answer had we not taken the journey this book offers. We have come to a point in our evolution where we fully realise the imperative of creating a new world where each living system or species lives in ecological harmony with all other living systems and species.

So, what does it mean for us to be ready to be part of the solution? The seven steps below would help us prepare to take the necessary action.

- Take full responsibility to know what is happening to our world (environmentally, socially, and biologically) and choose to be curious and open. Tune in to our compassion and courage to make a positive difference for ourselves and others.
- Make time to understand how things came to be the way they are today, and be aware of the consequences and options available to improve them.
- In order to influence the world, we must first lead by example.
- Understand and accept the fact that bringing about a paradigm shift creates resistance within oneself and others. The key is to have faith in our visions of a better world.
- Commit to making consistent incremental changes rather than an all-or-nothing big change. See Chapter 18 for eight common vital life-supportive practices to make world-impacting changes.
- Intentionally plan to make changes in our daily and weekly routine.

- Stay connected to the right information by joining like-minded groups. Get involved in life-supportive events and activities. Such involvement helps to increase and sustain one's conviction and motivation.

PAUSE: THE AVATAR MOVIE

Released in December 2009, *Avatar*'s worldwide box office sales were $2,712,115,019 by April 2010. This makes it the world's highest grossing movie of all time, overtaking *Star Wars*, *Titanic*, and *Gone with the Wind*. When Avatar DVDs and Blu-rays were released in the United States on Earth Day in April 2010, 6.7 million of them were sold in four days, bringing in $130 million. What is it about this movie that has attracted people to watch it, many of them more than once?

Avatar, a science-fiction adventure, is set on Pandora, a vibrant planet rich in bio-diversity. Much of the teaming flora and fauna is exquisitely luminescent. Luminescence, or cold light, transforms invisible forms of energy into visible light. What energies are around us that we do not see? All living cells, whether of humans, animals, or plants, emit biophotons (light) which cannot be seen by the naked eye. This dynamic web of light serves as a communication network, regulating all life processes. It is postulated by scientists that the biophoton field's consciousness-like coherence properties indicate its possible interface to the non-physical realms of the mind, psyche, and consciousness.

The indigenous blue-skinned *Na'vi* and other creatures live in a network of energy that flows through all living things, trees, and the entire planet. The Na'vi connect to *Eywa*, this planetary energy that protects the balance of life, at the Tree of Souls, where they can tap into their ancestral memories. Everything is connected. The Na'vi also connect to various creatures through the *heylu* bond of long braided neural interfaces with telepathic communication.

They believe that all energy is only borrowed, and on physical death you have to give it back, while life does not end but continues on in relationship with Eywa. They honour and respect all life forms, only taking what they need and giving thanks for it. This is similar to native cultures, such as the Cherokee, who hunted animals but asked for the animals' forgiveness before killing them. The Na'vi acknowledge each other by a deeply felt, "I see you!" In certain African cultures, the typical greeting is "I see you," symbolising a deeper connection. In fact, until a person is seen and acknowledged, he or she does not exist.

Unfortunately, the Na'vi are under threat from the earth *sky people* (earth humans) who are mining a precious mineral from their land, because the richest veins lie under the enormous Home Tree. The Na'vi stand and fight for their home, their way of life, their sacred places, while the sky people show no remorse in killing the Na'vi and other creatures, destroying their Home Tree, and the Tree of Souls. The significance of Trees can be seen in the Kabbalah's Tree of Life, believed to be a representation of the process by which the Universe came into being out of nothing (a map of creation), and Buddhism's Bodhi Tree, seen as the Tree of Wisdom.

In less than one month of screening, many fans experienced depression and even suicidal thoughts after watching the movie, because they wanted to enjoy the world of Pandora. One fan wrote, "The movie showed [...] something we don't have here on earth. People saw we could be living in a completely different world and that caused them to be depressed." Other fans expressed feelings of disgust with the human race for what we have done to the earth and they wanted to escape reality.

James Cameron—the writer and director of the movie—says that, "One of the themes of the film is symbolised by the fact that it begins and ends with the main character's eyes opening—it's about a change of perception, and about choices that are made once our perceptions change."

On Earth Day 2010 James Cameron was presented the Environmental Hero Award in Santa Barbara where he lives, honouring his help in awakening the environmentalists in us. He described writing Avatar in 1995 to connect the audience to the beauty and glory of nature and he stated that denial is the big disconnect; the urgency of environmental issues is not in the forefront of people's minds as it should be. As a society, we don't believe what scientists tell us, so Cameron urged us all to be science literate in order to have freedom; we can't have freedom if we don't know what's going on.

Besides having created this evocative movie, the director is himself an active contributor to bettering the world's landscape: he has joined 18 indigenous tribes in Brazil to stop the Belo Monte dam project, one of the largest hydroelectric projects in the world. Flooding hundreds of miles of Amazon rain forest would displace 25,000 indigenous people from these 18 tribes and 20,000 non-indigenous residents, and release millions of tonnes of carbon (as methane) from the dead trees. He quoted native North American leaders as saying, "Walk lightly upon the earth, for the faces of future generations gaze up at us." He called upon people to be eco-warriors, to confront "deniers" with the idea that our children have to live in the world we create, asking, "What kind of ancestor do you want to be?"

Apologies to Mother Earth

Forgiveness is a common theme in spirituality and religions, but not apologies. An apology is a written or spoken expression of one's regret, remorse, or sorrow for having insulted, failed, injured, or wronged another. Regret is having an awareness of the pain caused to another, requiring empathy or feeling how our behaviour has affected the other. Apologies recognise responsibility for wrong behaviour and acknowledge that amends need to be made. In *The Power of Apology*, author and therapist Beverly Engel says that meaningful apologies require three *R*s: regret, responsibility, and remedy.

When we can see how we have—individually and collectively—contributed to the global problems we are facing now, we can take ownership of our responsibility and express our heartfelt regret. More importantly, we need to take action to remedy the situation and to be part of the solution!

Let's take a first step and sincerely apologise to Mother Earth:

Dearest Mother Earth,

I am sorry for what I have done to you.

I am sorry for the innumerable thoughtless acts and intentional ill-deeds, for my constant need for more, for my greed and selfishness, for being wasteful and indulgent. I am sorry for being out of touch with nature, for plundering the resources you have so generously shared with me, for my lack of compassion for other living creatures that I have caused suffering to, and for poisoning the environment.

I apologise for my lifestyle, which gobbles fuels and energy and sends greenhouse gas emissions into the atmosphere. I am sorry for my fear and hatred of those not my own kind, for creating and stockpiling nuclear weapons which could destroy you many times over.

And you are the only planet I have to live on!

I am now choosing to be more mindful of the larger picture of life. I am willing to change my ways, even if they inconvenience me. I know how I can make a difference, even if it is in small steps. I truly value being part of the movement to live in a life-sustaining and life-enhancing manner. I am purposefully dedicated to helping you and being a steward for life here, in our beautiful home. I am ready to be part of the solution!

CHAPTER 18
BE THE SOLUTION IN OUR WORLD—TODAY!

What we must do now is increase the proportion of humans
who know that they can cause change.

—*William Drayton, author of* Everyone a Changemaker *(2006)*

THE POWER OF I-WE CONSCIOUSNESS

The current global problems are beyond the capabilities of governments, corporations, and scientists. At the same time, we know that these same huge institutions do wield immense power and influence over human lives and the environment.

Be that as it may, they are also filled with convoluted, thick layers of centuries-old beliefs and practices that are resistant to change. They contain groups of people with different agendas, many stakeholders and decision points. New ideas that challenge their existing paradigm often take months, years, or even decades to be considered and accepted, let alone to be implemented.

Considering the peril of our times, we need to mobilise a much more powerful force, one that the planet has yet to witness: a unified group of people with a common vision of a more sustainable and harmonious world. And today, through the Internet and telecommunication networks, it is fairly easy for large numbers of people from around the world to exchange life-supportive knowledge, practices, and resources.

The more these practices spread and become daily routines, the more the consciousness, the awareness, the appreciation, of the importance of supporting life grows. The more this consciousness grows, the more Gaia consciousness responds. As these two forms of consciousness—human and Gaia—reinforce each other, they further strengthen physical transformations on the planet and enable more individuals to gain access to this consciousness. When a tipping point is reached, a new living paradigm will be installed and a new world will rapidly emerge.

This is the power of the "I-we" consciousness; it begins with individuals like you and me, leading others by our own daily routines, passionately connecting with like-minded groups to spread our message by becoming guides, coaches, and mentors for others.

The focus of this entire book is to awaken and empower individuals to take small but high leveraging steps consistently, and incrementally build a world of cooperation, harmony, and ecological growth.

Living a Life-Supportive Paradigm

Earth's history could one day read: "Man built, man destroyed, man eventually destroyed himself"—*if* we do not change our life-depletive ways—now.

At the same time, because *we* are the ones who have brought ourselves to these dire straits, we are also the ones who are going to change the course of evolution on this planet. As the most "self-aware" and "conscious-choice-making" species on earth, we are in the most advantageous position to tip the scale of evolution toward the proliferation of life. We cannot take Mother Earth for granted anymore; the days of unlimited resources are over.

The idea of greening the earth and the environmental movement was perhaps ignited by Rachel Carson with her book *Silent Spring* in 1962. She sounded the alarm about the pervasive negative effects of using synthetic pesticides on the environment and biodiversity. In Carson's own words, "The question is whether any civilization can wage relentless war on life without destroying itself, and without losing the right to be called civilized."

The concept of life-supportiveness is not just confined to protecting and rejuvenating the environment, but is also deeply connected to our humanity and spirituality. It therefore influences our definition of success and happiness, as well as our daily priorities and focus. All in all, it is about embracing a new paradigm that supports life on this planet, and this requires deep transformation within ourselves and an understanding of our systemic world. We then need to create a clear vision and a viable plan to pace and lead ourselves into a New Living Paradigm, and most of all, to sustain the desired change.

Sustainability, Ecology, and Alignment

In order to develop and implement life-supportive ideas one must look into the concept of sustainability. It is a buzzword and dominant concept in elaborating economic, political, social, and environmental policies. In the United Nation's 1987 *Brundtland Report*, the concept of sustainability involves these two main descriptions:

- sustainable development is people-centred in that it aims to improve the quality of human life and it is conservation-based in that it is conditioned by the need to respect nature's ability to provide resources and life-supportive services; in other words, it means improving the quality of human life while living within the carrying capacity of supporting ecosystems;

- sustainable development is a normative concept that embodies standards of judgement and behaviour to be respected as the human community—society—seeks to satisfy its needs of survival and well-being.

Thus, improving the quality of human life within the capacity of our ecosystems points to "optimising simultaneously the goals of biological, economic, and social systems." In other words, tackling any one of the three critical systems independently, without also satisfying the other two, would virtually guarantee failure. In complexity theory everything is connected: changes in any sub-system affect the other sub-systems and eventually the whole. In short, in order for any healthy change to last and be fully integrated into our world, we must take into consideration the relationships and goals of the social, economic, and environmental systems we live in.

The study of ecology examines the dynamic relationships between multiple living and nonliving systems in an ecosystem. The knowledge of ecology has been applied to human sciences as well—to natural resource management, city planning, community health, economics, biology, and sociology. While its historical roots are in biology and natural history, the principles of ecology are important for any field involving the interaction of elements which form complex systems such as organisations, families, communities, or an individual's emotional and mental life.

We can now affirm that all changes are not equal at every level or subsystem of a large system. In other words, a positive change in one part of a system can be negative in other parts. So the wisdom of understanding whole systems—the concept of holism—is absolutely critical in making lasting change. In holistic healthcare for instance, all aspects of a person's needs (psychological, emotional, physical, and social) are seen as a whole; they are intimately interlinked. To heal a person sustainably, all these different aspects must be taken into consideration to find effective, high-leveraging interventions. On a larger scale as well, sustainability can only be ecologically achieved when we look at the desired change from multiple perspectives and take all of them fully into account.

To ensure that the different levels of a system are harmoniously working together, the concept of alignment is crucial. Alignment is a vital process in effective planning, problem solving, and facilitating personal and group change. In an efficient system the actions and outcomes of individuals within their micro environments are congruent with their strategies and goals. These goals, in turn, are congruent with the system's culture and mission with regards to the macro environment. In other words, to promote sustainable change at an individual and group level, the parties involved must understand the relationship between the various levels of change, and align their activities to fit those dynamics.

The next segment provides a powerful framework for aligning the different levels of the systems involved in making life-supportive changes individually and collectively.

LOGICAL LEVELS OF CHANGE

As mentioned earlier, any system of activity is a subsystem entrenched inside another system, which is entrenched inside another system, and so on. (This reflects Arthur Koestler's concept of a holon, first presented in his 1967 book, *The Ghost in the Machine*.) And the relationship between different systems produces different levels of processes. In other words, a particular level of process exists because of the relationship between two or more systems. Our brain, body, social, and environmental systems are examples of processes that develop natural hierarchies of processes. And without alignment of these different levels of systems, all major desired changes are not sustainable, or worse, can often create undesirable consequences.

One such model that helps to explain and align systemic changes in a multi-systems environment is the Logical Levels of Change. This model was initially developed in the behavioural sciences by Gregory Bateson, a world renowned anthropologist, social scientist, and cyberneticist, and later expanded upon by Robert Dilts in the field of Neuro-Linguistic Programming in the 1980s. This model refers to a hierarchy of levels of processes within an individual or a group. The purpose of each level is to integrate, coordinate, and guide the processes of the level below it. Changes at the higher level have a greater propensity to create cascading and pervasive changes on the lower levels. Changes at the lower levels can affect the levels above, but rarely produce multi-level pervasive changes. The levels, from upper to lower, are: (i) vision; (ii) identity; (iii) values and beliefs; (iv) capabilities; (v) behaviors; and (vi) environment.

In broad definition:

- the ***environment*** level involves the details of the external surrounding in which our behaviours take place. It points to the opportunities and constraints of a given time and place. This corresponds to the "where and when" of change;
- the ability to sense and organise our body movement and expression in a specific external environment is the level of ***behaviours***. This corresponds to the "what to do" and "what's supposed to happen" of change;

- the next level upwards belongs to the process of cognitive and mental ***capabilities***. It points to our mental map, plan, or strategy. At this level we have the capability to choose and modify a category of behaviours to a broader range of external situations. This corresponds to the "how to" and "what's our plan and strategy" of change;
- our capabilities are in turn shaped and influenced by our ***values and beliefs.*** They serve to encourage, inhibit, or generalise particular strategies and plans. They correspond to the "why something is important" of change;
- at the level of ***identity***, whole systems of beliefs and values are integrated into our role and sense of self. These complex identity processes have profound impacts on our beliefs, thoughts, and behaviours. This level corresponds to the "who am I and what's my mission" of change;
- the uppermost level involves our sense of ***spirituality***, going beyond our sense of self, from "I" toward "we." It points to the relationship between our role and our vision of the larger system that we operate in. It corresponds to the "for whom" or "for what" of change, and it can refer to our family, community, humanity, and planet.

It is important to note that as one moves upwards, the ideas become more abstract compared to the details of the behaviours and sensory experience levels. At the same time, every upward movement mobilises more of our nervous system resources, and has a far more pervasive impact on our behaviours and experience. For example, the nervous system resources required at the level of values and beliefs are far more extensive than the ones required at the environment level; at the same time, they have the power to unleash or hold back a wide range of mental capabilities and behaviours.

If we examine change with this model in mind, we can understand that any level which is not aligned with the others can create a conflict in the change process. For instance, an individual may be able to perform something new in a given context, but may not have the know-how to generate creative actions in different situations. Even when one is capable of changing, one might not value the change as a critical life skill and therefore may hardly use it. When one is able to change and believes it *is* a critical life skill, one may not see oneself as an "agent of change." It is therefore crucial for us to recognise and deal with issues that may arise at any one of these levels.

When it comes to making life-supportive changes, the questions below facilitate the alignment of our logical levels.

- Environment—"Where and when do I want to make life-supportive changes?" "In what kind of daily activities or situations do I want to be life supportive?"
- Behaviours—"What actions do I want to carry out in those activities and situations?" "What behaviours do I need to start doing, stop doing, and continue doing?"
- Capabilities—"How can I use my mind to carry out those actions?" "What knowledge or skills do I need to mobilise or develop to perform those actions effectively?"
- Values/Beliefs—"Why do I want to use and develop those life-supportive knowledge and skills?" "Listening to my heart, what core values must I uphold or redefine that will drive my thoughts, priorities, and actions?"
- Identity—"By embodying and mobilising those values and beliefs, what does that say about my role and mission?" "What is a metaphor or symbol that represents my life-supportive identity or mission?"
- Vision/Spiritual—"For whom or for what am I striving?" "As a life-supportive being, what kind of world do I want to co-create with others and Mother Nature?"

With this model, both personal and collective change is created with less struggle and the transition from our old to our new selves becomes much smoother. It is important to realise that the logical levels are wired to different aspects of our neurology. Therefore, to align our logical levels, we need to:

- open and strengthen our senses;
- express and frame our physiology;
- creatively enrichen our mental maps, frameworks, and strategies;
- reframe meanings and create vivid representations and powerful states to redefine our values and beliefs;
- shift our worldviews and redefine our purpose in life; and
- envision a new world that goes beyond our own lifetime.

Addressing the Upper-Logical-Level Issues

If you browse the Internet and bookstores, you will find that there is no shortage of green living ideas. We highly recommend the following books:

- *The Rough Guide to Green Living* (2009), by Duncan Clark;
- *Green, Greener, Greenest: a Practical Guide to Making Eco-Smart Choices a Part of Your Life* (2008), by Lori Bongiorno;
- *Less is More* (2009), by Cecile Andrews and Wanda Urbanska;
- *The Joy of Less* (2010), by Francine Jay;
- *The Plant-Powered Diet* (2012), by Sharon Palmer;
- *Cleanse & Purify Thyself* (2007), by Richard Anderson.

There are more tools & practical ideas at www.choicesofnow.org, where easy-to-get-started tips and well-researched information are available to anyone who wants to begin or further expand their life-supportive living. Gaining access to the various eco-humanitarian support groups and activities can also be done via this blog, as it also provides many "capabilities" in the form of core knowledge, skills, and group support for anyone with the desire to change.

As we can see, the problem is not so much at the level of "how do I go about executing my life-supportive activities?" (capabilities) or "what do I need to do to be life supportive?" (behaviours). There are many simple and effective steps that we can take to be part of the solution. But why don't we see more people doing this around the world? The problem is that we first need to awaken to the severity of the crises we are all facing.

At the heart of our critical planetary situation is that many of us are entrenched in our current vicious cycle of lifestyle-work-money, and live by outdated definitions of success and happiness. Adopting the eleven presuppositions presented in Chapter 15 and aligning our logical levels to embody them is going to empower us to make the transition to the new paradigm more smoothly and ecologically.

At this point in our evolution, we know that we can't be the same cultural species and expect a different life on Earth; stepping into the unknown is our only path to a new life and civilisation. Many psychologists and philosophers are arguing that the biggest crisis we are facing today is one of meaning in our lives; and this corresponds to issues at the level of vision/spirituality.

About 50 years ago, the author of *Man's Search for Meaning*, psychiatrist Victor Frankl, highlighted the three major social problems in our society: aggression, addiction, and depression. According to Dr. Frankl, these social problems can be traced back to an "existential vacuum"—a perception that one's life is meaningless. Up to 1997, before his death, he observed that this existential vacuum was prevalent in many modern countries. Millions of people with this problem couldn't be treated, because psychologists and psychiatrists were still trying to understand what the problem was.

According to Dr. Alex Pattakos, author of *Prisoners of Our Thoughts* (2008), the "crisis of meaning" will not go away through the pursuit of power (correlated to aggression) and pleasure (correlated to addiction). Through his painful experience of surviving the holocaust, Dr. Frankl firmly stated that it is meaning that sustains us throughout our lives, no matter how little or how much power and pleasure we may experience. It is the way we learn and give meaning to our experiences that will help us fill our existential void, which would then effectively address our aggression, addiction, and depression.

In the midst of the human-Gaia crisis, it is up to each of us to find the deeper meaning in our lives and thus raise our level of consciousness to evolve into a new world where:

- all spiritual practices become truly compassionate and inclusive of all spiritual traditions, which pray for each other and manifest miracles;
- we progressively feel a deeper connection to all sentient beings and our daily living demonstrates this love for all life;
- nations come together to solve transnational problems, by sharing resources and actively implementing ecological solutions to environmental and humanitarian problems;
- everyone has sufficient access to nutrition and healthcare;
- fair trade, equal economic opportunities, and equal human rights are practised in every country;
- environmental and biodiversity health become an international priority and nations take full accountability to protect and rejuvenate their natural resources;
- compassionate, joyful, and balanced humans live to contribute their best through their vocation;
- simplicity, a slower pace of life, and spontaneous creative expressions and inventions abound, leading to a new era of scientific-spiritual growth.

It is important to emphasise here that we must turn our heart desires and words into visual (or mental) representations, which we can harness and enrich daily by giving them different colours, brightness, movement, sounds, and emotions.

We can see this inner vision of ours grow with more diverse people involved, with more innovative ideas, and see it transform more areas of our society. And most of all, we need to align our values/beliefs, thoughts, and actions to our daily routine. This is the literal meaning of Gandhi's quote, "Be the change you want to see in the world."

THE SIX LIFE-SUPPORTIVE CONCEPTS SYSTEM

Simplify >

Resonate > **Redefine >**

•

Intention

Embody > **Align >**

Overview

The above Six Life-Supportive Concepts System is a framework to help us understand the vital concepts involved in making sustainable changes in our lives. It corresponds to the *capabilities* of the Logical Levels of Change and it gives us a cognitive map and principles to formulate strategies for manifesting the desired objectives. It is important to remember that the level of capabilities turns the abstract "why" of our values and beliefs into the concrete "what" of our behaviours.

The six concepts are intimately connected and cyclical (clockwise), so the enhancement of one concept boosts, clarifies, and adds momentum to the entire system. At the centre of this system is our moment to moment intention, it is our willingness and focus to manifest these concepts.

> You never change things by fighting the existing reality.
> To change something, build a new model that makes
> the existing model obsolete.
>
> —*R. Buckminster Fuller,*
> *American systems theorist & futurist*

Simplify

One of the greatest and most invisible inhibitors of change in our lives is life itself. We get trapped in fulfilling work aspirations, upholding and achieving a lifestyle deemed successful by society (a house, a car, clothing, luxury goods, etc.), obsessing about financial freedom, trying to meet the unrealistic expectations of important relationships, and keeping up with technological, economic, and social advancements. All these leave us no time or energy to think, decide, and make any life-supportive changes. By deliberately and consistently inserting blocks of time in our schedule to simplify our lives, we gradually open ourselves up for more moments of ease, clarity, and inspiration.

The concept of simplifying is related to the belief in letting go, being sufficient, and trusting life. It leads to a greater sense of freedom and reduction of burden in all aspects of life. The most important point here is to unplug from the world. Without a conscious decision to make time to plan our simplification process, life-supportive changes rarely happen; and if they do, they seldom last. Most of all, these attempts are fraught with deep frustration, feelings of being overwhelmed or totally defeated.

The idea of simplifying is also connected to the presuppositions of choice and abundance mentioned in Chapter 15.

Abundance emerges in our lives when we experience simplicity. It can come in the form of having more time with loved ones, doing things that we love to do, or having a better health as a result of a slower pace in life. Abundance is not about having more things, status, and power without looking at the impact these have on our relationships, health, and environment. The true value of abundance comes from the wisdom of less is more. Choice is about making a conscious decision to select among two or more options; choosing *not* to choose is also a choice, and so is choosing among two or more painful options—i.e., choosing the lesser of two evils. Most people who do not look for options and who do not make their choices in life often end up being forced to make a choice, reactively. It is by generating and choosing our options before crises happen that we are living proactively.

In quantum physics, the quantum vacuum is the source of all matter, although it is "no-thing"; it is in the source of all potentialities. In simplifying our lives, we expand our moments of silence, and it is in silence that answers arise. In the words of Rumi, "Silence is the language of God, all else is poor translation."

Core Actions to Take

(a) Set aside appropriate time in your calendar to plan your simplification process.
(b) Decide on the key areas to take action in. Remember: less is more.
(c) Share your plan with key supportive people in your life. Commit to take action.
(d) Main success factor: commit to taking small steps consistently. Make small, incremental changes over time.

Redefine

The concept of redefining is synonymous to the processes of reframing, recontextualising, and reformulating, which empower us to approach problems differently and solve them by giving them a different meaning.

Each of us has a unique way of framing our experiences in life, and the way we frame them—not the actual experiences—provides their meaning. It is the meaning that produces suffering or joy, limitations or breakthroughs, stagnancy or transformation. So when we change our frames, we change our outlook, feelings, and actions. As history has shown us, there are many people from different backgrounds (Gandhi, Einstein, Mandela, etc.) who have redefined themselves to make the world a better place.

With reference to the logical levels, our values and beliefs are powerful framers of our experience. Thousands of studies on placebos have scientifically proved the power of beliefs; it is not what something actually is, but what we believe it to be that matters. Our values and beliefs hold our motivation and permission that support or inhibit our capabilities and behaviours; they govern our autonomic nervous system and changes at this level have a profound impact on our emotions, thoughts, and physiology.

> If we don't change our beliefs, our lives will be
> like this forever. Is that good news?
>
> —*W. Somerset Maugham,*
> *British playwright*

In the context of making life-supportive changes, we must take a hard look at what we need and what we want. Needs are tied to basic life necessities such as having a roof over our heads, enough good food and water to maintain our health, basic healthcare and hygiene products, enough clothing to be comfortably dressed for all weather conditions and certain social settings, and for some, a simple vehicle to commute if they live in the countryside or have to travel long distances. As we have mentioned in earlier chapters, wants are associated with conspicuous consumption, which means spending money on luxury goods and services to display economic power.

With globalisation and technological advancements, more conspicuous-consumption items are readily affordable to the growing middle-income earners around the world. These products and services take up a huge portion of our income, and, over time, they become our needs instead of what they truly are—our wants.

To free ourselves from the rut of a work-money-stress-shopping lifestyle which destroys our environment and biodiversity, we must clearly separate our needs from our wants, which are linked to our values and beliefs. Redefining our values and beliefs about what really gives us happiness, as opposed to success, is paramount to our current evolution.

The concept of redefining is connected to the presuppositions of metamorphosis and values in Chapter 15. Metamorphosis, in a deep sense, is about transformation: the ending of something old and the beginning of something new. It demands our willingness and faith to step into the unknown. Values determine what we focus on and what we stand up for; they are the key to our motivation. What we value literally shapes our world. Thus, it is perfectly natural that if we want to see a life-supportive world, we must redefine our values.

Core Actions to Take

(a) Engage a life coach or mentor to redefine what is most important in your life.
(b) Clearly separate your needs from your wants. Channel the freed-up time, energy, and money to intensify your life-supportive efforts.
(c) What you initially redefine may lack evidence that it will work or that it is correct. Faith, supported by consistent incremental actions, is the key to solidifying our new vision of the future.

Align

Alignment is the main theme of this chapter, without which change is often a conflicting, frustrating, and self-sabotaging experience.

No living and thriving system in the Universe is closed, so it is important for us to know how to interact with and contribute to others, society, and the world. Alignment is an on-going process to promote congruence, so that all communication and actions be clear and well-received by all systems involved. The sooner we begin the alignment process, the sooner we will feel the positive effects of congruence.

The health of any system depends on how integrated (whole) it is as opposed to how divided or conflicted it is. In the process of alignment, we are guided to harmonise polarities in our lives: family and work, "I" and "we," freedom and security, man and nature, compassion and protection, material and spiritual, and so on. By harmonising these polarities we become more whole and therefore healthier. The presupposition of *Homo chrysalis* in Chapter 15 is vital to the concept of alignment because as we turn inwards to transform our world, we must have faith in facing the dark and challenging process of transformation. Once transformed, the new humans have a greater felt sense of connection to the collective consciousness, are able to process information integrally and transpersonally, to perceive subtler energies and higher dimensional realities; they live a more spiritually active and all-embracing existence, and carry out their daily activities in harmony with themselves, other species, and the environment. Having a clear and vivid vision is crucial in energising our hearts, minds, and bodies to be the change in the world.

Through alignment, in Taoist healing, the flow of *chi* is enhanced as the major accupressure points of the body become aligned, and thus the body can heal itself with more ease.

Core Actions to Take

(a) Make time to do the Logical Levels of Change alignment process. Review it periodically.
(b) Ensure that the specified life-supportive actions are in alignment with your values and vision.
(c) Take up courses to improve your knowledge and skills in identifying and transforming your values and beliefs.

Embody

The concept of embodying turns our vision into actions—this is where the rubber meets the road, where we use our body to give our spiritual vision concrete expression. Many attempts to make major changes in our lives fail because we don't translate our vision and intentions into actions.

To embody our ideal vision, mission, and values, we need to improve our mental capabilities constantly, using cognitive maps and strategies to guide our complex behaviours. Cognitive maps are made up of cognitive packages—groups of visual, verbal, and kinesthetic representations of ideas, concepts, or processes (in the form, for instance, of flow-charts with diagrams and key words or mind-maps). Verbally expressing a metaphor with sensory rich words is a cognitive package. Executing a complex performance through spatial awareness, visual and kinesthetic skills, and timing, is a cognitive package, as well. Using these cognitive packages greatly helps in manifesting our desired actions.

We must pay attention to our daily routine, as it offers us so many opportunities to learn and grow. Each moment presents opportunities for us to receive, send, amplify, and neutralise vital information to and from our collective consciousness. These interactions help to provide ongoing feedback to our actions, thoughts, and feelings about being life supportive. To actualise our life-supportive objectives it is essential for us to be clear of the evidence of our progress.

Bringing forth a new world is ultimately a congruent embodiment of our cherished vision; with this, we influence others, not by telling or forcing them to change, but by inspiring them through our own actions. In short, the concept of embodiment is only powerful when our vision is congruent with our actions.

It is also important for us to embrace the wisdom of nature's cycle (Chapter 15), knowing that we cannot expect linear progress in cyclical systems without producing toxic consequences. There are ebbs and flows, rises and falls, eases and struggles when we operate in any system. It is wise that we recognise the ups and downs of our life-supportive actions and turn every feedback into a useful learning experience.

Core Actions to Take

(a) Create a simple and effective daily life-supportive routine. Make sure it is actionable and measureable.
(b) Set daily intentions to carry out your routine. Be clear about your outcomes, then spend all your energy and focus on the process and surrender the outcome.
(c) At the end of the day, backtrack your routine and acknowledge yourself. Focus on your half-filled glass.
(d) Be the change you want to see in our world.

Resonate

Resonance is the state of vibrational reciprocity between two or more entities, analogous to a violin string in tune with a vibrating tuning-fork, or a radio receiving-circuit in tune with a broadcasting frequency. The process of resonance occurs in nature, in the propagation of sounds, light, electromagnetic waves, etc., as well as in our nervous system where neurons that resonate together get wired together, which ultimately produces learning and behaviour. Through Rupert Sheldrake's process of morphic resonance—the resonance between the morphic units (living things) and the morphic field (collective consciousness)—an act, event, or idea can lead to similar occurrences in the future. According to his theory, this is how knowledge, biological processes, and cultural practices evolve over time.

The concept of resonating is essentially about communication and influence, about connecting, sharing, and spreading our message in the form of ideas, activities, and events to individuals and groups. But first there must be consistency in the type of vibration (message) that we sent out to the field (collective consciousness) every day. That is why the concept of consistent embodiment is so important. When our embodiment is inconsistent we neutralise our vibration, whereas when our embodiment *is* consistent, we strengthen the vibration (message) that we want to resonate in the Field.

Through our daily interactions and activities we create a formalised or indirect network. We need to stay tuned to life-supportive events around the world and actively support worthy causes. Our intentions, emotions, actions, and interactions with others and the environment form the vibrations that we hope to resonate in our world.

By going beyond our five senses, we recognise that our inner and outer worlds mirror each other in a fractalistic manner. If we don't like what we see in our world, we can start changing the world by changing our worldviews and aligning our actions to them. With this global spirit, a more robust, creative, and energy-efficient suprasystem can emerge to tackle the complex multi-systems issues. Electromagnetic, geomagnetic, biomagnetic, and other vibrations have discernible impact on our genes, biology, and consciousness. Our thoughts, emotions, and actions are vibrations that we can resonate to shape the kind of world we want to create.

Core Actions to Take

(a) Through virtual and face-to-face interactions, connect with like-minded groups or create a support group.
(b) Actively share your experiences and resources to help others embark on and sustain their life-supportive practices.
(c) Continue to improve your concepts of simplifying, redefining, aligning, and embodying. Improvement in any of these areas would amplify your life-supportive resonance.
(d) Make time to resonate with the collective consciousness through meditation, prayer, or contemplation. We can include sending our blessings to beings who are suffering from poverty, starvation, sickness, natural disasters, wars, and oppression.

Intention

All outcomes can be manifested if we have the intention to begin and follow through. In our day-to-day living, we are often bombarded with many priorities and distractions and get trapped in a fuzzy and lethargic mind, floating from one purposeless task to another. To move to a focused, clear, and motivated state, we need to direct our consciousness to our purpose, reason, and devotion—also known as intention. Intention is misunderstood today as a weak desire, a fleeting thought without the commitment to follow through. But the original meaning of intention, etymologically speaking, is "to stretch out into" and "to strain, exert, and put one's effort into something." According to Marilyn Schlitz, Ph.D., past president of the Institute of Noetic Sciences, *intention* is "the projection of awareness, with purpose and efficacy, toward some object or outcome." In other words, without intention all new concepts and desires are lost in our daily competing priorities, overwhelming distractions, and disengaged mind.

The power of intention to influence seemingly impossible things and situations has been increasingly substantiated by scientists from different disciplines. As we read in Chapter 14, they have carefully studied the effects of individual and group intentions on the performance of machines, properties of water, behaviours of animals and plants, growth rate of cancer cells and viruses, recovery from chronic diseases, behaviours of human DNA, social and economic trends, collective consciousness, and more. These scientific studies have concluded that the act of giving attention to something changes its physical properties. This has led scientists to ask, "What would our world be like if we gave something our full intention, our deliberate desire to make changes?"

Many articles on the interaction between mind and matter have been published in scientific journals, such as the *American Journal of Physics*, *Scientific American*, *Physics Letters*, *Foundation of Physics*, *Physical Review*, and the *European Journal of Physics*. Our thoughts are energies, vibrations, and information, and so are all unmanifested creative forces and manifested matters in the Universe. Moving out from a space of no-thoughts, unclear or crowded thoughts require a moment of silence, a "pause" in our daily lives. This is the vital beginning of our intention-setting process, which allows us to enter the realm of universal consciousness, of infinite possibilities.

In Particle Physics, the Standard Model describes a set of particles (e.g. photon, quark, lepton, gluon, etc.) that build the entire Universe as we know it. Central to the Standard Model is the presupposition that a Higgs Field—a field of energy present throughout the entire Universe—exists. This field provides virtual particles their mass, giving weight and shape to all matter we see in the Universe. This presupposition was proved with the discovery of the Higgs Boson—a fundamental particle of the Higgs Field—by the European Organisation for Nuclear Research, in July 2012, after months of intense evaluation and examination. Our intentions act like the Higgs Boson: they give particles their mass. The quality of our intention strengthens and accelerates our efforts to manifest the other five concepts, that is, simplify, redefine, align, embody and resonate.

Core Actions to Take

(a) Make time in the morning to set your intention.

(b) Quiet your mind and calm your body by focusing all your attention on your breathing. Focus on the pause between your inhalation and exhalation—expand the pause.

(c) Bring the six qualities of intention into your awareness—simpify, redefine, align, embody, and resonate. Be one with the field of consciousness.

(d) State your intention clearly, Remind yourself that less is more. Be open and flexible.

(e) As vividly as possible, see, hear, and feel yourself carrying out your intentions. Be aligned with them.

(f) Surrender your outcome to the Universe (or God, or whatever name you call this totality). Wait for the mind, heart, and body to be still before you move to the next activity.

(g) Honor your intention. "Be" (clearly experience) your intention as you carry out your daily life-supportive routine.

VITAL LIFE-SUPPORTIVE PRACTICES

Not everything that counts can be counted,
and not everything that can be counted counts.

—*Albert Einstein*

Many things can be done to improve our global situation, but not all things produce the same impact. Identifying the high-leveraging activities and knowing the key drivers to focus on within each activity is a major challenge for most people embarking on "being the solution in our world." The crucial question to ask ourselves is, "What are the changes that would make the most difference?"

The answer lies right before us, in the way we live from day to day and in the ambitions that we pursue. Being the dominant species on earth, making high-leveraging, life-supportive changes in our daily lives has a direct positive impact on our ailing world.

Based on the research gathered in this book and the experiences shared by people in the life-supportive movement, there are eight common Vital Life-Supportive Practices: (i) Spiritual, (ii) Relationship, (iii) Diet, (iv) Lifestyle, (v) Resources, (vi) Humanitarian, (vii) Environment, and

(viii) Vocation. These practices are presented in a descending order, in terms of their pervasiveness in promoting desired changes. They serve to pinpoint the daily activities that would create a widespread, transformative, and sustainable change. The essence of each practice is presented here, and the details and steps are available on www.choicesofnow.org. It is important to highlight that the extent of the results depends on the consistency of one's practices and the incorporation of the Logical Levels of Alignment and of the Six Life-Supportive Concepts System in our lives. The Oxford Dictionary defines *practice* as "the actual application or use of an idea, belief, or method, as opposed to the theories relating to it." So, the emphasis here is on *doing* our life-supportive practices, not just *thinking* about them.

Spiritual Practices

One of the recurring problems that we face is managing the different roles we have in our lives (such as parent, professional, spouse, coach, friend, or business owner). Each role we play requires us to value different things, which creates conflict and fragmentation in our psyche. Developing a spiritual practice helps us to deepen our humanity, expand our wholeness, and connect to the source of creation. It also assists us to develop a unifying centre within ourselves that will promote congruence, power, and simplicity. It is from this centre that we can seek clarity, counsel, and confidence to handle the challenges of the different roles we uphold.

Many of the problems that we have presented in this book stem from our separation from other beings and the environment. Connecting to the source of creation—divinity, sacredness, the unnamed, holiness, etc.—helps us to see that we are all one. This awareness brings out our deeper humanity, and this is where our compassion and reverence for all life becomes a natural state for us.

Almost all spiritual traditions emphasise on the need to meditate. Meditation helps us quiet our minds and hearts so that we can alter our vibrations and tune in to the stillness and silence where all the answers reside. A large part of the meditation process is about surrendering our need for control and our attachments. We enter the realm of pure consciousness, free of rules and regulations, free of fear and desire, of physical distractions. This is also the process that connects us to the collective consciousness, to what is beyond us as individuals.

Our ego is the deterrence to our freedom and new desired world because it is culturally conditioned. Built on fear, pride, and control, what we call our ego creates a worldview that strongly determines how we think, feel, and act in every situation. As a result of that we are literally not able to see anything outside of it. When we can imagine something outside of our worldview—when we create an image of a new world—we develop new connections between our neurons and over time establish new meanings and relationships to the image. Without this new image, neural connections, and meanings we have no energetic and informational means of attuning ourselves to a new paradigm.

> We are living through one of the most fundamental shifts in history—a change in the actual belief structure of society. No economic, political, or military power can compare with the power of a changed mind. By deliberately changing their images of reality, people are changing the world.
>
> —*Willis Harman, Ph.D., American social scientist and futurist*

Through meditation we can:

- progressively surrender our fear, pride and control;
- soften our "I" and strengthen our "we" consciousness;
- develop mindfulness, compassion, divine courage, and faith;
- create a new world vibration and resonate with others;
- use this new world consciousness to guide our daily actions.

Many people detach their meditation experience from day-to-day living, but it is important that we bring this practice into our daily routine; the deeper meaning of spiritual development is to bring the wisdom of heaven to earth. More and more people in spiritual practice believe that heaven is not just a realm outside of us, but also a place we live in and cherish within ourselves. Again, the important point is to bring what we experience during meditation into our moment-to-moment living; if we experience compassion in our meditation, we must also be compassionate in other activities that we engage in.

If there is one thing that can improve everything in our lives if we do it consistently, that is meditation. This mind- and body-emptying, centring, or focusing process, is sometimes called praying and contemplation; it increases our compassion and understanding in all relationships and helps us to see the relevance and importance of eating responsibly and compassionately; it redefines our priorities and simplifies our lifestyle consumption by promoting self-sufficiency; it helps us to examine our usage of resources and reduce wastage; it raises our awareness of suffering and willingness to help the less fortunate; it expands our consciousness to appreciate the beauty, interdependence, and divinity in all living and non-living things.

Meditation is an easy process to learn, but it takes a lifetime to master. It requires the right attitude of openness, acceptance, trusting, and patience. For the beginner, it is useful to accept that meditation is not about the quantity, but the quality of the experience. Here are the basic guidelines:

(1) find a quiet space; stretch or scratch to make yourself comfortable;
(2) sit on a chair or on the floor; close your eyes, progressively relax your muscles from the top of your head to your feet;
(3) focus all your attention on your breathing; notice the pauses before and after each exhalation and inhalation;
(4) visualize your surroundings and expand your peripheral vision; immerse yourself into the environment;
(5) be aware of the presence of your consciousness; tune in to the silence and emptiness;
(6) mentally, emotionally and physically, hold the silence and stillness as long as you can;
(7) after a period of silence, post your question or intention gently; it is alright if the answer is not clear—it may present itself to you after your meditation;
(8) to end the practice, bring your awareness into your body; wait for a sense of certainty and calmness, then open your eyes.

Relationships Practices

Our sense of who we are and what we value is formed in our childhood; how we relate with our family, relatives, and friends in that period forges our identity and shapes our governing values and beliefs, our priorities later in life. Our family is often the primary motivator for our achievements and also one of the main criteria for happiness. Collectively, our family dynamics shape society and our world.

We can say that a country is a collective of families. With happy families, a country is much more able to respond to challenges. Therefore, in order to change our community, country, and world, we must begin with our own family. Developing a relationship practice helps us to stay connected to the people who are most important in our lives, and helps us to strengthen the bond between individuals, groups, and nations.

Our definitions of self and others are largely influenced by our relationships. According to many metaphysical studies, our relationships serve as mirrors to who we are and therefore, by examining our relationships, we have the chance to bring out a better version of ourselves.

In a world where everything is so intimately connected, what we do to others, we do to ourselves.

A global shift in consciousness implies the understanding that all of us are of the same humanity and that we are all in this together. As the living conditions on the planet become increasingly inhospitable, the need for each of us to create peace in our relationships is ever-more imperative. And congruence—which is the source of effortless power—can only come when we are able to build peace, true care, and love in our own homes. For many of us, the healing and the revival of our authentic humanity begin at home.

The understanding and support from our family and loved ones is a very important foundation that we need to strengthen in order to make life-supportive changes quickly and smoothly, in an atmosphere of encouragement; otherwise, these can often be misunderstood and looked at with scepticism.

Here are the guidelines to improve relationships, starting with the people in your home:

(1) make a conscious choice to progressively de-anchor yourself from work commitments; move from a work-centred to a family/relationship-centred life;
(2) develop and uphold compassion when dealing with others; daily meditation and intention-setting help greatly;
(3) take a powerful life-supportive view that everyone has a positive intention behind their behaviour; look beyond the overt behaviour and attempt to understand the hidden underlying unexpressed need of the other person; then work toward helping that person achieve that need ecologically (i.e. without negatively affecting other areas/relationships in their life);
(4) make time to eat with your family or the people close to you;
(5) be genuinely interested in them and demonstrate your appreciation; make time to help them in meaningful ways;
(6) help family or friends to practise "paying it forward": the recipient of a good deed passes it on to another person, instead of returning it to the giver; this is one way in which we can actively resonate the positive effects of relationship caring.

Diet Practices

As we have clearly shown in this book, the largest contributor to the global greenhouse effect is our meat-based diet. In the last 150 years, the rapid growth of the meat and dairy industries has cleared more forests, and consumed more grains and natural resources (especially water) than any other industry. The agricultural methods used to grow animals have caused a long list of meat- and dairy-related health problems that affect billions of people around the world. Animal farms and animal markets are the most dangerous incubators of deadly viruses, with more than five lethal flu pandemic outbreaks since 1918. The improper treatment and disposal of animal waste contributes to the release of methane and other greenhouse gases, which, combined with other animal farming factors, contributes to 18–20% of the planet's greenhouse emissions.

Thus, adopting a plant-based diet has a positive wide scale impact on our planet's geosphere, biosphere, and sociosphere. On a personal and individual level, the quality of our energy determines the quality of our lives. And what we eat on a daily basis has direct impact on the quality of our energy.

Below are some guidelines to adopting a life-supportive diet:

(1) Slowly cut out meat. Start with "meatless Mondays" (or another day) and then, over time, expand to other days of the week. For some, it's easy to focus on having one or two plant-based meals every day. The key is to keep it consistent over one or two months and then gradually increase the frequency of meatless days. Make time to look for good vegan/vegetarian restaurants or food outlets.

(2) Cut out dairy and eggs. This can be done progressively. Remember that dairy foods are "liquid animal protein," equally harmful to health (refer to Chapter 6); they are not environmentally friendly and cruelty-free.

(3) Eat more whole, unprocessed, plant-based food. Dark leafy vegetables and protein-rich foods like tempeh, tofu, mushrooms, grains, beans, nuts, and fruits.

(4) Organic soy and wheat "mock meats" can be included in your weekly diet. This meat alternative has a remarkable texture, similar to meat and it is made 100% from plants. Organic mock meats are a great alternative for heavy meat eaters who are on their way to becoming vegetarians.

(5) Vegan and vegetarian dishes are very tasty. Invest your time in getting new and simple recipes to cook. When your food is enjoyable, it's obviously easy to sustain the diet.

(6) Cut out preservatives and highly processed food. Eat in a way that supports your own health. Invest in an alkaline water machine. Use only natural salt, raw sugar, and high quality coconut or olive oil at home. Make time to look for or make simple healthy snacks.

Lifestyle Practices

By looking closely at someone's lifestyle, we discover their attitudes, values, and worldview. A person who values success and recognition would typically have a lifestyle that displays material wealth and status. One who values green-living would focus on simplifying their life, consume less natural resources, and purchase environmentally friendly products.

It has been thoroughly explained in the book that defining success in our lives through our current paradigm of material wealth and fame will not get us the happiness we truly want. As a case in point, the 2012 Happy Planet Index, which measures the extent to which countries deliver long, happy, and sustainable living to their citizens, ranked Singapore 90th out of 151 countries. On the other hand, *The Straits Times* reported in August 2012 that Singapore is number one in the world for GDP per capita, according to the Wealth Report released by Knight Frank and Citi Private Bank. *The belief that more money equates more happiness is thus clearly disproved.*

The intention of this segment is for us to examine the things we spend our money on, which are associated with our current lifestyle. By bringing our awareness to what we buy, we can make conscious life-supportive decisions before buying. We can focus on reducing, re-using, and recycling; yet there are certain things we can even cut off our shopping lists and lives.

Whenever we buy lifestyle goods, we need to ask ourselves several questions. One of them is, "Do I *need* this or do I *want* it?" As many products are toxic to our bodies and the environment, another question would be, "How is this product made, and what is it made of?" It is likely that when we buy something of great value at a very low price, it comes at the expense of someone's livelihood. So, yet another question to ask ourselves is, "How is this product traded and how are the people (farmers, craftsmen, factory workers, etc.) involved in the manufacturing process being treated?" (To read up more on this topic, we recommend the book *Ecological Intelligence*, by Daniel Goleman.)

Here is a short list of websites you can visit to get some basic knowledge on ethical production, manufacturing, and trading:

- www.goodguide.com—for evaluations of health, environmental, and social impacts of products;
- www.fairtraderesource.org—an information hub to support the fair-trade movement;
- www.cosmeticsdatabase.com—the database of ingredients in cosmetics and personal care products;

- www.echa.europa.eu—for chemicals legislation regarding human and environmental health.

The things we collectively buy determine many aspects of our society: the kinds of business practices, professions, and studies that are in demand, our industries and their impact on economy; also, the human resources we mobilise and use, the natural resources we extract from the earth, and the animal and plant species we support or kill. To a large extent, the kind of world we live in is determined by the things we choose to buy. Naturally, when we collectively stop buying or investing in certain things, the industries that supply those things will have to reinvent themselves.

Based on the 2007/2008 Monthly Household Expenditure Survey in Singapore, here are the top 8 biggest ticket items:

- housing and utilities (22.4%): the bigger the house, the higher the cost of maintenance and utilities;
- transport (15.8%): motor car, petrol, maintenance, and other vehicle expenses (e.g. parking fees, road taxes, etc.) take up 69% of this expenditure;
- food serving services (13.5%): hawker centres and food courts, restaurants and fast foods take up 97% of this expenditure;
- recreation, telephone, and Internet services (12.3%): audio and visual equipments, games and toys, paid television and mobile-phone services take up 43% of this expenditure;
- food (7.6%): meat, seafood, and dairy/eggs take up 45% of this expenditure;
- educational and tuition services (7.2%): private tuition takes up 26% of this expenditure;
- clothing, personal care and personal effects (6.6%);
- health cost (5.3%): medical products, outpatient services take up 70% of this expenditure.

Looking at the list above, the size of the house we choose to live in, how we travel and commute, what and where we eat, how we entertain and communicate, how we educate our children, how we take care of our health, and how we dress and groom ourselves take up 91% of our monthly expenditure.

Here are some suggestions that will help to drastically reduce stress and pressure in life, to facilitate our transition from a work centred to a relationship-spiritual-health centred life. These suggestions will also reduce our impact on the environment:

(1) downsize and simplify your home;

(2) downsize or sell your car;
(3) cook at home and progressively switch to a plant-based diet; it is also a good way to increase your bonding and caring time with loved ones;
(4) stop chasing new electronic, communication, and computing devices; spend more time physically connecting with people, artistic performances, and nature;
(5) increase your coaching time on your children's studies;
(6) strengthen your overall health by focusing on whole plant-based food; do more meditation and exercises; aim to be free from doctors and pharmaceutical drugs;
(7) focus on developing your inner beauty and confidence; de-anchor yourself from cosmetics, clothing, and accessories; these things focus on your external beauty;
(8) take the challenge to free yourself from the addiction of shopping for 90 days; develop a meaningful hobby or interest;
(9) clear out your storeroom; let your unused stuff be of great joy and use to poor people.

Resources Practices

This section refers to the effective management of the following resources: water, electricity, petrol, and any other recyclable and reusable materials such as paper, bamboo, glass, aluminium, etc. This also includes reducing the usage of plastics and other non-biodegradable materials. How we use and discard our resources is going to determine the habitability of our planet within the next 20 years.

According to the United Nations Environmental Programme (UNEP), global consumption of natural resources could triple to 140 billion tons a year by 2040 with a huge human population of 9.3 billion. The panel of scientists at UNEP say that we need to completely rethink the use of resources and massively invest in technological, financial, and social innovations to at least freeze consumption levels in wealthy countries.

The Energy Report produced by the WWF (World Wildlife Fund) asserts that a radical near-total global shift to clean fuels within the next 40 years could yield savings of €4 trillion annually and definitively tackle climate change. The report also states the need to halve meat consumption in rich countries to reduce methane emissions from livestock. While the governments and scientists around the world are stepping up to create wide-scale, sustainable, clean energy, we, as collective individuals must develop a resource management practice to reduce, reuse, and recycle our resources and waste.

Below are the key suggestions for the two major categories of resource consumption management; details can be found on www.choicesofnow.org.

Electricity, Water, and Petrol

- Turn off all non-productive electrical appliances and equipments. Be conscious to turn the electrical socket switches off, do not leave equipments on stand-by mode, they are still consuming energy and generating heat. A few major areas to focus:
  (a) air-cons (air-conditioners), washing machines and dryers, television and home-theatre equipments; if comfortable, switch to fans rather than using air-cons; use energy-efficient refrigerators;
  (b) computers, printers, PDAs, digital cameras, and mobile phones;
  (c) turn off the lights, fans, and air-con when you leave the room.
- Clean water and clean air will be a valuable resource in the future; now, most people in modern cities take them for granted. The main idea is to see and feel the water you use, to develop a "relationship" with it. A few major areas to focus:
  (a) install low-flush toilets and water restrictors on faucets and shower heads;
  (b) find creative ways to reuse water; e.g. water used for washing food can be used for watering plants or washing heavily soiled linens or cooking utensils;
  (c) fill the basin up before washing your face, turn off the water while lathering in your bath, brushing your teeth, shaving, cleaning the dishes.
- Reduce your petrol consumption by:
  (a) changing to a hybrid car; CNG is much cheaper than petroleum, and it emits 30% less carbon dioxide;
  (b) switching to a smaller and fuel-efficient car; a clutch-shift car is less petrol intensive than an automatic car;
  (c) switching off your car engine when not in use; maintain the correct tyre pressure and clean your air filter regularly;
  (d) reducing usage of your car: consolidate car trips, carpool, ride a bike, walk for short trips, take public transport.

Reducing, Reusing, and Recycling Waste

- Cut down plastic usage. Use paper bags or bring your own green bags for shopping. Use biodegradable plastic bags for garbage disposal. Cut down the purchasing of anything that is made of plastic.
- Use recycled paper for printing. Print on both sides of the paper. Find ways to reuse unwanted one-sided printed paper.
- Convert to electronic billing to cut down paper billing. Unsubscribe from magazines that you do not read or use.
- Ask for recycling bins in your neighbourhood. Things to focus on are metal cans, glass, and paper.
- Be mindful of not throwing away but donating the following items: furniture, handphones, computers, electrical appliances and equipments, clothing, and footwear.

Humanitarian Practices

The dictionary defines a humanitarian as someone who is devoted to the promotion of human welfare and the advancement of social reforms. Being a humanitarian does not necessarily involve getting on the next plane to a third-world country, hugging children, and following the typical western "do-gooder" stereotype. In fact, humanitarianism encompasses a variety of political, social, and economic activities appealing to the full spectrum of humanity. Right in our own backyard, we have many causes that we can stand up for to make a positive difference.

Being a humanitarian is rooted in our compassionate nature. Professor Dacher Keltner, author of *The Compassionate Instinct* (2010) and Director of Greater Good Science Centre, together with other scientists have provided scientific evidence that we are compassionate by nature, and this helps us to survive and evolve successfully as a species. It is the ability to step authentically into the shoes, skin, heart, and eyes of others, and to care enough to do something helpful. It is the ability to look deeper, beyond the differences that separate us, to find the common needs, emotions, and values that make us human. To be compassionate, to look at our internal and external differences as resources, and to choose to focus on the commonality that binds us all are the values and skills that we need to develop to unite the world and solve problems that cannot be solved at a national level.

A humanitarian understands that while countries, religions, politics, and races have boundaries, our humanity transcends them; we are in this world together. At the same time, while knowing that the focus is on "we," a humanitarian also acknowledges that every single person (every "I") has the power to change, to care, to inspire others, and to generate a pervasive chain of changes that could make the world a better place for everyone.

Our humanitarian quest is not only confined to humans, but also encompasses all sentient beings (as explained in Chapter 6). To further support this, in July 2012, a group of eminent neuroscientists gathered at the University of Cambridge to discuss the question, "Are animals conscious?" At the end of this meeting, which was called the Cambridge Declaration on Consciousness, David Edelman of the Neurosciences Institute of La Jolla, California, Philip Low of Stanford University, and Christof Koch of the California Institute of Technology publicly proclaimed that:

> [...] nonhuman animals have the neuroanatomical, neurochemical, and neurophysical substrates of conscious states along with the capacity to exhibit intentional behaviors. Consequently, the weight of evidence indicates that humans are not unique in possessing the neurological substrates that generate consciousness. Nonhuman animals, including all mammals and birds, and many other creatures, including octopuses, also possess these neurological substrates.

In short, science is stating that animals, just like humans, are self-aware, and able to feel complex emotions and make intentional choices. Our humanitarian quests stem from our humanity, and as Dr. Albert Schweitzer said, our humanity must include all living creatures. Besides, animals and plants are the parts of the ecosystem that support our existence.

There are many ways in which we can embark on our humanitarian practice; here are some ideas to get us started:

(1) select your field of passion and learn about it; humanitarianism is a wide area and the Internet is filled with humanitarian causes;
(2) take action on your field of choice; give your passion a voice, choose an outlet to express it: (a) choose to join Internet petitions on animal and human rights; (b) start a virtual discussion with like-minded groups to launch a social project; (c) use social media—blogs, Facebook, Twitter, & Youtube—to promote important causes; (d) donate and raise money to support worthwhile projects;
(3) volunteer in local social projects; help out at: old folk homes, animal shelters, poor shelters, centres for the learning disabled, troubled youth centres, prison rehabilitation centres, etc.;

(4) take meaningful vacations to poor countries to offer your help; there are non-profit organisations that can assist you; the help you offer may be aligned to your existing skill-sets, such as your ability to teach, cook, manage projects, build, etc.;
(5) volunteer to help in challenging places that are hit by natural disasters, famine and diseases, war and oppression that would require more information and preparation; there are relief and specialised organisations supported by the United Nations that you can consult with;
(6) practise random acts of kindness (RAK); notice how your behaviours affect others, and hold yourself accountable for your actions; kindness is contagious—one good deed deserves another; buy a poor elderly person a meal for a week, offer to clean the home of someone in need, help a child to learn every week, help someone whose car is broken down, clear your storeroom and donate useful items to a poor family; these are just some of the RAK that you can perform weekly or daily.

Environmental Practices

Taking care of the environment is a personal responsibility. But first, we need to establish an authentic connection with it. Making time to go to the park, nearby forest, river, lake, or seaside is a very good start. Focus on the concrete sensory experience: it is important to relate with the environment with our body and heart, rather than just our mind. Create metaphors for the things that we see, hear, and feel in nature. Be open and willing to learn from what we observe, as the cycles and events in nature are mirrored in our daily lives.

There is a deep healing connection between humans and nature. According to Professor Yoshifumi Miyazaki, a leading scholar in forest medicine, a 20-30 minute walk in the forest can produce tangible health benefits. His study of 288 volunteers in 24 different forest sites showed the following endpoint results compared to an urban control group: 13% decrease in cortisol, 18% decrease in sympathetic nerve activity, 2% decrease in blood pressure, 6% decrease in heart rate, and 56% increase in parasympathetic nerve activity, indicating heightened relaxation in the body.

Taking the dog for a quiet walk in the woodlands, going hiking in the forest, enjoying outdoor gardening, or having a picnic in the park are great simple ways to connect with nature. This helps us recognise our oneness with the process of life. Nature has its own stories to tell and by listening and learning from them, we can broaden and deepen our eyes, ears, and minds. In the words of Lao Tzu,

> When you know nature as part of yourself,
> you will act in harmony.
> When you feel yourself as a part of nature,
> you will live in harmony.

It is from this authentic reconnection (99.99% of our evolutionary history we have lived in nature) that we make meaningful decisions as to how we can protect the environment. The main thrust of environmental activism, campaigns, or lifestyle changes is to get the green message to spread through the social media. Such environmental changes do not happen overnight; they require the efforts of the majority to be sustained.

Here are some key suggestions:

- plant a tree or create a garden; during photosynthesis, trees and other plants absorb carbon dioxide and give off oxygen. A single tree can absorb approximately one ton of carbon dioxide in its lifetime. Richard Reynolds's book, *On Guerrilla Gardening*, is a good read on how to turn small patches of unproductive land to beautiful life-giving gardens;
- clean up the environment. Start up your own group or get involved in environmental clean-up projects. These can be focused on cleaning up beaches, streams, or parks. Start in small areas, promote public awareness, and work your way to bigger areas with the help of volunteers;
- protect forests, woodlands, and wildlife. Get involved with governmental or non-governmental organisations to protect your local and national forests and wildlife. Support the creation and protection of animal habitats. Contact the WWF forest conservation programme and understand the importance of forests and wildlife and their contribution to our ecosystems. Petition and write letters to policy-makers: boycotts and non-violent civil disobedience are tools for protecting forests and wildlife. Stand up for what we believe is the right thing to do;
- support and encourage renewable energy development. The burning of fossil fuels has significantly contributed to global warming. Supporting governments and corporations to speed up innovations and their implementation is a powerful way to phase-out inefficient energy industries and promote a cleaner environment in the future. Learn about local developments. Find

ways to educate and unite people to push for such developments;

- promote reducing, reusing, and recycling (3Rs) programmes. This concept is about reducing consumption and the amount of waste produced by reusing materials whenever possible and recycling them. This helps to reduce pile ups in landfills and natural spaces as well as to cut down the pollution in drainage systems and waterways. There are many means of incorporating the 3Rs in our lives and the key, again, is to create social awareness through our own example;
- use environmentally friendly products. From personal care to household and gardening products; many people can greatly help in the healing of the environment through education and intelligent purchasing. In many cases, although it is initially more costly to use these products, the long term savings and benefits are huge. Getting used to reading product labels to know what the products are made of and how they are made is important. When we buy environmentally friendly products, we directly support companies that take care of our planet.

Vocation Practices

How we define our life purpose, what we choose to dedicate our lives to, has a profound impact on the way we lead our lives. At a collective level, our quest for purpose and happiness in life shapes our civilisation, environment, and evolution. Happiness and wealth studies gathered from tens of thousands of people from more than 150 countries, conducted by leading professors such as Nobel Laureate Daniel Kahneman, Edward Diener, Ruut Veenhoven, and many others, state that money ceases to be a "giver or creator" of happiness after a certain income level is achieved. Beyond that, more money does not equate more happiness.

What these studies also made clear is that, regardless of the society we live in, what makes us happy is a meaningful life that comprises: loving and supportive relationships, the joy produced by the simple things, volunteering and helping others, keeping a physically and spiritually healthy life, pursuing one's life purpose—which is often expressed as our vocation—, and taking pleasure in the process.

Interestingly, the Merriam-Webster dictionary defines vocation as "a divine call to an individual or group to perform important duties or a particular function in life." Profession is a paid occupation that involves extensive training and a formal qualification, whereas a vocation is an often-voluntary work that is carried out more for its altruistic benefits. A vocation fulfils a spiritual need for the worker and is often linked to their divine gifts or talents.

Vocations are for those who are really dedicated not just to work, but more so to serve people. In the words of best-selling author and influential spiritual leader Eckhart Tolle, "It comes as no surprise that those people who work without ego are extraordinarily successful at what they do. Anybody who is ONE with what he or she does is building the new earth." Many may think that carrying out one's vocation means leading a life of poverty. Tolle's words reassure us that by following our vocation, our path is abundantly blessed.

Metaphysically speaking, our vocation is an all-important activity or service that our soul has come into the world to accomplish. If we search deeply within ourselves, expressing our purpose through our vocation is something that we need to do in order to feel complete and fulfilled. The more we fixate on the gratification of our bodily and ego needs, the more we feel something is incomplete or missing.

Our vocation may not be publicly adored and it may not have a pompous name or status. It may be about empowering our family, facilitating support groups (divorce, addiction, phobia, etc.), teaching or coaching children on subjects we are passionate about. For many, it could be contributing to a social, political, or environmental cause, or sheltering and healing injured animals. It could also be honing a particular life skill, developing a body of scientific work, creating and inventing a software application or machine that can alleviate or solve certain social or environmental needs. Still for some it might be entertaining or spreading a life-supportive message with their singing or acting talents, educating and empowering others with certain knowledge and skills, or creating beautiful gardens and plantations. The possibilities are endless.

When we uncover our vocation and act on it, we have a sense of doing something that feels right and motivates us to grow by mobilising our known and unknown capabilities, and it is emotionally rewarding. When we are in tune with our vocation, we seem to be at the right place, at the right time; synchronicities of meeting the right people, receiving the right resources, and being in the right situation happen more frequently.

Often, our vocation becomes clearer after we have worked on some of our major life challenges. Sorting out significant relationship issues, solving chronic financial debts, overcoming a crippling fear or limiting belief, rising above social resistance or cultural expectations, surrendering our need for control, may all be part of the "foundational work" that needs to be done before we can embark on our vocation.

In the context of our current world, exercising the practices we have mentioned so far greatly facilitates the discovery and expression of our vocation.

Below are some additional pieces of advice on realising your vocation.

- Listen to your heart: what do you love to do or be the most? Do not be swayed by the suggestions of others.
- Engage a career coach: answering and going through well-crafted questions and exercises greatly helps you gain more clarity and confidence about what steps need to be taken to uncover or reconfirm your vocation.
- Be highly explorative: often discovering your vocation requires you to jump in and try something out without knowing ahead how things are going to work out.
- Make time to develop yourself: take up personal development courses and read to expand your worldview; learn new ways to learn.
- If realising your vocation requires career change, designing a well-paced training plan to arm yourself with the necessary knowledge and skills is essential.
- Pay attention to synchronicities: when you are moving toward your vocation, wonderful coincidences will happen to give you the assurance that you are on the right path.
- Understand and embrace the fact that fulfilling your vocation may require taking risks and giving up some of the comforts in your current life. Thinking about ways to get support for taking such risks is greatly beneficial.
- Visualise your desired vocation as vividly as you can. Be sure to include details such as the possible work activities, lifestyle, the people you are with, the environment you are in, and the time of the day (day time or night time). A multi-sensory representation would serve as a daily reference to sort out your priorities and help you to materialise your desired vocation.

A Thought Experiment for Singapore

Singapore has achieved many accolades ever since its independence in 1965. It has raised itself from a poor third-world country to a first-world, wealthy nation in one generation (35 years). Some of its achievements include:

- ranked first out of 34 countries in 2011 for Net International Reserves per capita (international reserves minus external debt) by the Central Intelligence Agency;
- ranked third among 185 countries for the highest GDP purchasing power parity per capita by the International Monetary Fund (2011) at 59, 710 dollars;
- ranked fourth in terms of Asia's Most Livable City (Standard of Living) by the Economist Intelligent Unit;
- ranked first in 2010 by Gallup's Potential Net Migration Index (most desired country for migration);
- ranked third out of 55 countries on Health Infrastructure by the International Institute for Management Development's (IMD) World Competitiveness Yearbook (2010);
- first in Asia in 2009 for Social Achievement Capital (ethical business practice) by Caux Round Table;
- first together with New Zealand and Denmark (2010) in the Corruption Perception Index by Transparency International;
- the most wired country in the world (2008) at 99.9 percent broadband penetration according to Infocomm Development Authority of Singapore.

Having achieved so much, with the resources that Singapore has gathered over the last 47 years, what can the nation do to improve the global ecological crisis?

First of all, Singapore needs to come to terms with the part it plays in the global warming problem.

According to the Global Footprint Network, most developed countries in the world are turning resources into waste faster than waste can be turned back into resources, which puts the world in a global ecological overshoot. Conservative UN projections suggest that if current trends do not change, humanity will be in the impossible position of needing the equivalent of two Earths by the 2030s. As mentioned in Chapter 5, if everyone alive today lived like a Singaporean, we would need 4 planets!

We must begin, as individuals and as a nation, to make ecological limits central to our decision-making in all aspects of our daily living. We need to invest in technology and infrastructure as this will enable us to function in a resource-constrained world.

What would it be like for Singapore to take up the challenge of showing the world that modern city living can be made ecologically self-sustainable?

Let us explore this possibility with a thought experiment. We fast forward and project Singapore in 2035. A life-supportive nation has already been established. What is the country like now (in 2035)? Based on what kind of lifestyle, values, and consciousness does Singapore think and act? What kind of impact does Singapore have on the region and on the world?

VISION OF SINGAPORE IN 2035

It is late in the afternoon and people are walking about in a leisurely pace, talking to neighbours and even to strangers at the town centre. They are open and at ease with each other. Young and old are taking time to notice the trees, flowers, and birds. They are enjoying the ponds and the fresh air in parks. Some are taking long walks, some are cycling with family members and friends. They are also making time to buy fresh local produce for their evening dinner.

Morning and evening traffic jams are rare occurrences. The major business districts are now decentralised. Smaller offices are operating in residential town areas. People walk, take short bus rides, or cycle to work.

Seen from above, the bulk of the island's activity revolves around five mega structures known as Eco-Farming Towers (EFTs), strategically placed in the North, South, East, West, and Centre of Singapore. These 30 to 50 storeys high mega-structures hold leading-edge technologies on vertical farming that produce 55% of the republic's basic food requirements. These EFTs grow more than 80 species of organic vegetables such as chye sim, pak choy, kai lan, eng chye, bean sprouts, wheat grass, as well as 30 to 40 varieties of mushrooms and fruits. Tofu, tempeh, and different versions of soybean products are massively produced in the EFTs for local consumption.

Due to the fact that more resources are needed to grow seafood and chickens, a smaller area of the EFTs are allotted for such farming. In any case, the change in local and global consciousness around healthy, humane, and sustainable food has resulted in more than 60% of the population being vegetarian or vegan. In the early 2020s, in the desperate global movement to secure sustainable food sources, the practice of slaughtering hundreds of millions of cows, pigs, and goats for meat was abandoned. A major restructuring of the meat industries has taken place around the world. Most people now view such practices as morally wrong and irresponsible in terms of wasting water, energy, and productive lands that could grow food for people instead of feeding animals.

EFTs have become the business and social centres of Singapore, with offices, shopping, entertainment, and transport facilities built around them.

Here, people meet, recreate, shop, and dine, find out about eco-humanitarian news and get to know how their food is grown and made. These towers are built using renewable technologies and therefore they are completely self-sustained; they use primarily solar power and biofuels—organic waste from the tower itself. The "sweat" from plants (evapotranspiration) is collected, and, using condensation and filtering technologies, it is recycled to be used for farming or bottled for drinking. The crops are grown with the most advanced technologies in hydroponics and aeroponics—farming methods that do not use soil to grow food.

The idea of vertical farming began to mushroom rapidly around the world in 2015 to 2020. Singapore led the way in South East Asia, and was followed by China, Qatar, United Arab Emirates, the United States, and France. As more and more EFTs have been erected, more agricultural lands have been returned to nature, and forests around the world have started to be repopulated. The ecological and energy-efficient EFTs have put an end to harvests being damaged by bad weather and have drastically slowed down the rampage of converting forests into farmlands. The growth of forests around the world helps to reduce global warming and revitalise biodiversity, halting the mass extinction of species.

The phenomenal success of this radical economic and cultural change in Singapore has not been in the assertion of the political power of the government, but more so in the willingness of citizens to voice out and take action on their collective vision. It started with a small group of about 50,000 eco-humanitarian-conscious people meeting in health-environmental-social-spiritual related conferences and seminars, and through the social media. As the sharing of information about climate change ideas has become more publicly noticeable, more people have started to voice their thoughts to the government. As younger generations started taking interest in grassroots activities, the idea of securing the nation's food source has become entrenched in the minds and hearts of most citizens.

The government engaged foreign vertical-farming experts who provided the infrastructure and means of controlling the EFTs; the first one was built in 2017 and subsequent ones rapidly emerged, as the government had the full support of the citizens.

People now actively participate in the decision-making and operation of the EFTs. The selection of the type of crops to farm and how they should be made available to the public were decided by volunteers who took key appointments in managing the EFTs. With active and consistent public campaigns, more and more people see and experience the benefits of a plant-based diet. Public health statistics gathered from the last decade show that the nation saved trillions of dollars in terms of healthcare costs due to the healthier diet and lifestyle changes.

After the major shift in consciousness of 2012–2013, confirmed by scientific institutions tracking global consciousness, many priorities of nations around the world began to shift toward social happiness and ecological living. Work hours in Singapore have been revised to 30 hours per week after a major shift in national priorities garnered by the government's bold decision to move away from GDP to a new national progress index called Genuine Progress Indicator (GPI). Rather than working longer hours, people started to focus on working efficiently and having more time with their family, enhancing their health and spirituality, and developing their own gifts to contribute to society.

The government has long known that how a nation defines and measures progress has a profound impact on the quality and meaning of the life of its citizens, and the world at large. With the support of the public, the government gradually phased out GDP and phased in GPI over a three-year period. GPI is a comprehensive measurement system that takes into account the social and environmental assets rather than just the economic value of consumption—which is what GDP does. GDP gives no indication of sustainability because it fails to account for depletion of either human or natural capital. Furthermore, it doesn't register the cost of pollution and the non-market benefits associated with volunteer work, parenting, and the ecosystem services provided by nature.

In reaction to the bleak and harsh realities of the converging crises of the early 21st century, the scientists, governments, and citizens of many countries were desperately looking for an economic system that measured progress by the improvement in well-being rather than the expansion of economic consumption activities. The more evolved nations were looking for sustainable frameworks that:

(1) promoted genuine progress built on several dimensions of human well-being;
(2) encouraged a quick transition to renewable energy technologies;
(3) fairly distributed both resources and opportunities;
(4) safeguarded and restored natural capital; and
(5) economic localisation.

GPI is designed with the above in mind. It adjusts a nation's personal consumption expenditures upwards to account for the benefits of non-market activities (e.g. volunteering and parenting) and it adjusts downwards to account for costs associated with income inequality, environmental degradation, and international debt.

With the establishment of GPI in Singapore in 2017, nation-wide surveys conducted every three years indicated that citizens reported a higher life satisfaction. They appreciated a slower-paced and balanced life, caring for family members, a helpful and inclusive community, a growing and healthy environment, and the freedom to pursue a meaningful vocation. The size of our ecological footprint has achieved a sustainable earth-share thanks to the establishment of EFTs, the installation and utilisation of solar energy island-wide, the advancement of technologies for desalinating seawater and waste-water recycling, the reduction and recycling of waste, the utilisation of clean renewable energies (liquid hydrogen, electric, and other hybrid technologies) for all transportation, and a simplified lifestyle.

More nature theme parks and forests, as well as lakes and reservoirs were created and slowly introduced with wildlife that is native to the habitat of Singapore. People enjoy tree-top and lakeside walks and they can dine and drink in nature as health-food kiosks are available. Tree planting in government approved lands, indoor gardening, rooftop gardening, and having regular picnics have become a way of life for a significant portion of the citizens. Latest satellite footage of the country's green cover grew from 46% in 2012 to 61% in 2035.

Equal and fair opportunities for all races to advance in the Singaporean society have long been secured and reinforced with special education for certain races to bring out the best in them. Education is seen as a major framework to shape future generations toward living a life-supportive existence. The main topics that are reinforced and enriched throughout the different educational levels are ecological intelligence, public health and welfare, cross cultural communication and compassion, and local and global well-being through sustainable utilisation of planetary resources.

Healthcare and education are easily available to everyone as costs have been kept low. This is largely due to the national shift toward a healthier diet and lifestyle and the deconstruction of the military forces. 20% of the original full-time military personnel have been assigned to the United Nations Peacekeeping Force to carry out regional peacekeeping work. The rest were retrained and assimilated into the business, social, education, and other sectors. What used to take up a large chunk of the nations' financial resources is now freed up and channelled to accelerating healthcare technologies and Research and Development projects in universities, focusing on food production technologies, genetics and healthcare, and renewable and quantum energy. The government invests in the research and development of energy medicine which uses light, vibrations, and quantum information approaches to help the human body repair itself.

Instead of overreliance on the global economy, Singapore is investing in its own resources to produce a significant portion of the goods and services, food, and energy it consumes from its local endowment of financial, natural, and human capital. In the government's economic localisation plans, major corporations are incentivised to use local goods and services, therefore new business sectors are created and they are growing sustainably. Business is no longer a separate "private" sector; companies have become a functional part of the community. Production and marketing decisions are made not only with the sole intention of achieving success in the marketplace, but also to minimise the impact on the environment, to reach employee satisfaction, and to maximise human and social usefulness. The fundamental purpose is not just to increase value for stakeholders by exploiting available resources, but to assume social and ecological responsibility for them.

Rapid but well-paced life-supportive changes made Singapore the first developed nation to be ecologically self-sustaining. Singapore has become the "model eco-nation" for the world to see that ecological living in modern cities is possible. In partnership with green technology universities from the USA and UK, Singapore has become the regional expert in helping other nations and cities to attain ecological sustainability. Entirely new green industries and jobs emerged with this positive social, economic, and environmental transformation. New eco-disciplines and degrees in universities were created and enriched over the last two decades. Singapore naturally became the regional hub for eco-living and eco-business leadership and management.

From an individual perspective, Singaporeans are able to slow down through a less material-driven and simpler lifestyle. Their lives abound in spiritual experiences like epiphanies, spontaneous insights, or moments of serendipity because they are able to tune in to a higher consciousness. These occurrences were once thought to be exclusive to healers, spiritual masters, and uniquely gifted people, but this is no longer the case. More and more Singaporeans are using them to guide their decisions and activities in life.

These spiritual or transpersonal experiences are open, all-inclusive, connected to nature, humanity, and the intelligence of the Universe. They are helping people to intuit the connections between economics, humanity, environment, and spirituality, to make ecological decisions, and to act in ways that truly support life.

Back from the Future

This thought experiment gives us a glimpse of what is possible for Singapore. It helps us realise that each country has a critical role to play in creating a brighter future for all of humanity and every inhabitant on earth. Given the financial, human, economic, and technological resources we have, we are in a privileged position to show the world how a developed nation can achieve ecological sustainability. So far, in our history, we have strived for our own livelihood and we have done exceptionally well. We have always put our own interests in all our economic and social endeavours. *It is about time we saw ourselves not as Singaporeans, but as citizens of this beautiful planet.* It is about time we summoned all the courage and faith we can to mobilise all our resources and lead the way by showing the world an ecological way out of the crises we are in.

The important thing to remind ourselves over and over again is that the sooner we begin, the sooner we are able to manifest the new world we hope to live in. By making the change today, individually and collectively, we are giving these new values and new consciousness ample time (to 2035) to manifest this new desired world. It is also important for us to keep in mind that we must strive to do this by cooperating with other nations in a spirit of openness, humility, and willingness to learn. To turn the world around let us change ourselves—you and me— first, then invite others to join us.

So let us begin by making a pledge to be the solution in our world—today!

I pledge to be a responsible global citizen, to choose a lifestyle that is simple and sustainable by developing the gifts I have been offered, and to serve the community for the betterment of all.

I'm starting with my body. I will consume food that is good for my health, the environment, and that supports the life of other species.

I will be conscious of unnecessary consumption, by being mindful of waste, and grateful for what I already have.

I will support individuals and organisations that promote life-supportive messages and activities. I will be mindful of using eco-human friendly products and services.

I will make time to connect with myself and others and be compassionate, and to develop meaningful relationships.

My actions will serve the common good and not the satisfaction of my ego. I will always be aware that we are all connected, as one big family living on the same planet that we want to take care of.

PAUSE: THE WAY OUT IS IN

The message for this chapter—and perhaps for the entire book—is that the way out of our current crises is to go inwards, into ourselves. The consciousness expansion that we must make happen now and the work we need to do to shift our values are both inward-looking processes.

To add further richness to the concept of consciousness, we can say that if all our experiences, our various states of consciousness, were weather patterns—clouds, rain, rainbows, hurricanes, tornados, or cooling breezes—, our consciousness would be the sky in which they take place. This general consciousness is the context in which all of our experiences, perceptions, thoughts, and feelings converge. It is this ever-present aspect of consciousness that we are most focused on when we speak about consciousness transformation, because a change in our general consciousness will have a profound impact on the way we think, behave, and feel in our daily lives.

Bearing in mind that our consciousness is transpersonal in nature, the only question is, what is the degree of permeability for each individual? For some it is a more ego-bounded and brain-limited consciousness; for others, it is beyond the limits of personal identity, open to people and the world. We have learned that it takes an evolved consciousness to access the all-unifying energy field; when we do that, we can grow and be happier without creating conflict between people, nature, and the environment.

The expression "the way out is in" resonates with most ancient spiritual traditions. *The Way*, spoken of by the world's greatest teachers, is not one single path, yet all paths lead to the same destination—the wisdom within.

> Value only those things in life which are eternal.
> Search for these things within yourself. [...] For I tell you,
> the living God is within you, and you are in Him.
>
> —Dialogue of the Saviour
> *(attributed to Jesus: the Coptic text, in the Nag Hammadi library)*

> Brahman (God) allowed man to look outward, but in seeing the outer,
> the inner is ignored. Those who see the eternal, turn their gaze within.
>
> —the Katha Upanishad *(Krishna)*

> Seeking within, you will find stillness. Here there is
> no more fear or attachment—only joy.
>
> —the Dhammapada *(the Buddha)*

> The Way is empty, the Way is full. There is no way to
> describe what it is. Find it within yourselves.
>
> —the Tao Te Ching *(Lao Tzu)*

True wisdom is ultimately the result of the intuitive mind, not the rational mind. Gnostic Christians called the opening of the intuitive mind *gnosis*, which is not intellectual knowledge, but a deep understanding of how reality works. Such moments are flashes of insights that come to us in an instant, without effort. Zen Buddhists call them *satori*, while Christians call them *revelations*. Gnostic Christians spoke of gnosis as being "secret knowledge," yet it is secret only to the degree that most of us don't have spiritual eyes to capture these powerful and meaningful moments in our daily lives. To use a Christian metaphor, we miss seeing the kingdom of God both within and around us because we are caught up with the kingdom of earth.

So how can we raise our consciousness to see the ecological solutions we need at this point? This question must be handled in two parts: first we need to de-anchor ourselves from the vicious lifestyle-work-money cycle (kingdom of earth) we are in; secondly, we need to cultivate practices in our lives to strengthen the values that would raise our consciousness. The notion of de-anchoring ourselves from the cycle we are in implies creating more time, space, and energy to re-invent our world. A good way to start the de-anchoring process is to apply the Six Life-Supportive Concept System. Then, naturally begin cultivating the Vital Life Supportive Practices in our lives. The key is to incrementally pace and lead ourselves into these practices, and to be consistent in doing them.

The four great teachers mentioned above emphasise the importance of devotion and practice to attain higher wisdom:

> These things that you have learned and received and heard and
> seen in me, do these. And the God of peace will be with you.
>
> —Philippians 4:9

> Once you have understood the importance of Self,
> seek the wisdom of yoga [spiritual practice].
> In this way you will free yourself from karma.
>
> —the Bhagavad Gita *(Krishna)*

> Wisdom is achieved by those who overcome
> their natural resistance to diligent practice.
>
> —the Dhammapada *(the Buddha)*

Like glass, polish your inner vision until it becomes spotless.

—the Tao Te Ching *(Lao Tzu)*

Turning inwards for answers has been practiced by spiritualists in all traditions for thousands of years and it is also what we currently need in order to solve our problems on earth. We must now summon the faith that the revolutionary change we are hoping for can only come into reality when there's a definitive shift in our inner subjective world. And this inner shift cannot happen unless we make time to turn away from the material realm, do the inner work to transform our consciousness and values, and align them to our daily routine. This will help us transform conflicts and limitations so that the differences between the old and new worldviews can be reconciled. In doing so, we reform our outer world through the alignment of our spirit, mind, heart, and body.

When we are aligned to the new worldview and when we have a different perception of what is possible, we begin to see sporadic flickers and then, gradually, clear signs of the sacred shining through in our seemingly most ordinary daily experiences. And when this happens on a collective level, when individual and group *satori* start to cross-pollinate all over the world, the appropriate solutions will be abundant and we will witness unprecedented innovations and breakthroughs across the planet.

Perhaps it is the divine plan for all of us to come to this point of our evolution and face odds that are insurmountable by one group or one nation. And in order to save ourselves and the other inhabitants of the planet, we have to come together as one big family and co-create our world with Mother Earth. Having come to the end of the book, we now know why we need to act and the steps to take. It does not matter which area in our lives we start changing and we needn't worry whether the changes are significant enough. What does matter is that we start changing something *today*, and build on it everyday. We, the authors, have faith that by having made clear the urgency and seriousness of the situation we are all in, you too will choose to be life-supportive—as of now. Let our collective efforts and intentions manifest a new sustainable world of harmony, happiness, and abundance for everyone!

When you are inspired by some great purpose,
some extraordinary project,
all your thoughts break their bonds:
your mind transcends limitations, your consciousness
expands in every direction, and you find yourself
in a new, great, and wonderful world.

Dormant forces, faculties and talents become alive,
and you discover yourself to be a greater person
by far than you ever dreamed yourself to be.

—*Patañjali, Indian teacher, creator of the* Yoga Sutras

Simple Acts to Change the World

Nothing of any real positive social value has ever been achieved
from the top down; it's always been achieved from the bottom up.
Everything starts with personal choice,
and realizing that you can make choices.
—Howard Zinn, American historian and political scientist

OVERVIEW

The main message that we've emphasised throughout the chapters of this book is that the world will not become a better place if we continue living the way we have been in the past; and it is very clear that simply talking or dreaming about solutions to our global problems will not solve anything.

Perhaps there has been a time in your life when you were excited about a certain project, but soon after you became absorbed by your old routines, distracted by seemingly more important things, overwhelmed by too much information, or in doubt regarding the project's feasibility. So, with this bonus section, we want to help you gain a different and positive experience; we want to make your ignition easier by suggesting a simple plan with simple actions, which—when acted upon consistently—can produce waves of positive changes in your life, and most importantly, which you can start doing *today*, to change our world for the better.

Now that you have the intention to be the solution in our world and that you are ready to act, you will find this segment a very useful gateway to your life-supportive journey.

THE PROJECT

Here we present to you five life-supportive actions that are easy to implement, and you can start as soon as you finish reading this section. What you want to do is:

(a) choose **any three** actions in the list below;
(b) do the three chosen actions consistently, **three months**;
(c) once you have started to do them, encourage **three other people** you know to join you; these three people will each reach out to three other people, and so on;
(d) **help them** with their chosen actions if possible.

Five Simple Life-Supportive Acts

(1) Clear out your storeroom and give the things you no longer use to people in need.
(2) Have at least one or two vegetarian meals every day.
(3) Cut down air-conditioning, electricity, and water usage as much as possible and grow more plants/trees at home.
(4) Practice random acts of kindness every day (smile, help someone by carrying their bags, offer your seat, share your umbrella on a rainy day, buy a less fortunate person a meal, tell someone you appreciate them; the possibilities are endless).
(5) Stop shopping for "wants." Buy only basic necessities. Remember, what you choose to buy is your vote for what you want to see in our world.

Clearly all the actions above are very simple and anyone can do them. The reasons why you can make an immediate positive difference with these acts will not be explained here, for we have done so extensively in this book. However, let us take a closer look at the impact of your participation in this project: if you reach out to three people, and they each reach out to three people, who help three others, and so on, billions of people can be actively involved in making our world a better place within 20 months! Life-supportiveness can ripple out quickly and pervasively, and we could experience what it is like to be part of a single global community, with a common vision to heal our planet.

> Small acts, when multiplied by millions of people,
> can transform the world.
> —*Howard Zinn, American historian & political scientist*

The Science Behind It

This project is based on the scientifically validated idea that a big change can come from a high leveraging of small acts, if we carry them out consistently (refer to chaos theory and the butterfly effect in Chapter 14, and to the spiritual science of tipping points in Chapter 15). So it is not necessary for us to make herculean efforts individually, nor does every person on earth need to be part of this project, because when a critical mass is reached, our collective consciousness will spontaneously reorganise our multi-systems to create a new world order.

What's in It for Me?

By doing these small and simple acts consistently for three months, you are not only contributing positive change to the world, you are also improving your own world because "our world" includes you: we are inseparably united as the web of life connects us all. You create more space in your life, you free up the time and energy that kept you in the trance of consumerism, you build a stronger connection with your loved ones, you improve your health and overall state of mind, and you reduce the toxic impact of food and other products on your body, home, and environment. Furthermore, you significantly cut down on your monthly expenditure, which enables you to use money in more enriching ways; your faith in humanity is restored and you feel a true sense of connection to the world. The actions you choose will lead to a shift in your daily routine and priorities, which positively change your relationship with others and the world; ultimately, your life is brimming with life-supportiveness.

We hope that after three months of life-supportive activities, you will notice how much your life has improved and you will want to continue sustaining your actions. In alignment with the Six Life-Supportive Concepts System, we strongly encourage you to share your discoveries and strategies at **www.choicesofnow.org**; it is an act of resonating with the growing life-supportive community around the world. We urge you to add on more of the life-supportive practices suggested in Chapter 18, in an incremental way, so that together with an increasing like-minded community, we can realise our common dream of leading a sustainable life, in harmony with nature and all sentient beings, on a thriving, green, and beautiful planet.

BE the change you want to see in our world—today!

Epilogue

I've learned so much and on so many levels from the experience of writing this book that I could write another one just about that. And I trust that the lessons that have yet to be revealed to me, will come when the time is right.

One of the most profound realisations I had was that I only sought God when I was in trouble; He was not really part of my life, and until recently, I didn't fully understand the beauty and power of just "being with Him." It was through my moments of stillness, through the simplification of my lifestyle, and the shift to a relationship-centred life that I had this deep understanding.

One day, not long ago, as I succeeded to let go of structures and rules and to allow myself to be fully present and still, a presence within emerged. I had a feeling of peace like never before and with it came a sense that I and everyone is going to be alright, as long as we feel the connection with God (or the Universe, or sacred reality, or the divine dimension of existence, and so forth; we each talk of this in our own terms). Most of us yearn for an intimate bond with family, friends, or partners, in a desire to avoid loneliness, to fill an inherent void, but all our attempts to fill this void are temporal. This presence I felt has an eternal feeling and it is complete on its own. It is from this union that I sense that in spite of the challenges which will arise, this presence will guide us.

Now, I know that daily events can be turned into divine moments when we choose to treat them with acceptance, love, compassion, faith, and joy; and when we are collectively able to do this at will and witness the presence of God (or sacred reality, etc.) in ordinary instances, we manifest a new world before us. The crises that we are facing now are mostly trying to help us go back to our source and bring a holistic and spiritual way of being into our ailing world.

Today, it is with great joy and ease that my family and I are leading an existence filled with life-supportive routines; we are on a plant-based diet 98 % of the week, we have drastically cut down our grocery budget with meat and dairy out of our purchases, and even our dog, a street mongrel that we rescued from an accident, is now a healthy and calm vegetarian. Our health has improved and our temperaments have de-escalated; water and electric bills have been reduced by 30%; we recycle partially-used paper for most of our office needs, which significantly lowers our stationary budget. We spend a little more money on quality household products that are friendly to the environment and promote fair trade. We've converted our petrol car to a hybrid that uses Compressed Natural Gas, which greatly reduces emissions.

Our business has taken on a new life: 100% of our marketing, selling, and customer servicing is happening virtually. The interactions with our prospects and customers have increased by 50%, and the most satisfying fact of all is that our business is driven to do well so that we can do good—standing by other life-supportive organisations and groups—socially and environmentally, with 3–5% of our company revenue being directed to them. So far we have financially supported animal shelters, tree-replanting initiatives, eco-living Buddhist centres, schools for autistics, the Life-Skill Centre for Children in Danger, and many more. We also use our marketing resources to spread social and environmental awareness and we firmly believe that by doing good, we can do well. Our training and coaching business has achieved breakthroughs in revenue, even though we are running fewer classes every year.

We are part of the global-shift in consciousness by bringing life into everything we touch. As I write these pages, my team is devising simple activities to enrol people into life-supportive living, and we have so far reached out to hundreds. Our message is that through consistent simple acts, we can collectively gain unstoppable momentum to change the world.

Externally, life is pretty much the same, with people and businesses still going about their usual ways. But my internal world is totally changed. For me, time has slowed down. I have been abundantly blessed and I treasure the people in my life, everything that I have and experience. By having less "wants," I have become much wealthier.

This book is the result of thousands of hours of reading, researching, and experimenting; the volumes of scientific journals, more than 50 books, the hundreds of websites, and the life-supportive changes that we've made are all testament to the fact that we needn't be scientists, rich, nor powerful to make a paradigm-shift in our lives.

Through understanding, compassion, and our desire to promote peace, my team and I are a living proof that it is possible. I sense that my mission is to reach out to others and help them awaken to the new life and world that they want to be part of, to help them walk through their discoveries and challenges with the knowledge and experience I have gathered in the last four years. The greatest challenge of our time is perhaps to awaken from our separateness and to see the invisible and all-important connection that binds our seemingly divided existences.

I end one journey and begin a new one with the wisdom of Anwar Sadat's words, "In the power to change oneself, is the same power to change the world."

References and Resources

Chapter 1: What's Happening to Our Time?

1. *In Search of Time* (2008), by Dan Falk, Thomas Dunne Books
2. *Cosmic time*—presented by Michio Kaku, http://www.bbc.co.uk/bbcfour/documentaries/features/time4.shtml
3. "Law of time: artificial time", by Jose Arguelles, http://www.lawoftime.org/law/law.html
4. "A singularity in time", http://www.peterrussell.com/Odds/SoundsTrue2012.php
5. "How long have we been here?", http://www.nhm.ac.uk/nature-online/life/human-origins/modern-human-evolution/when/index.html
6. Timewave Zero, http://www.fractal-timewave.com/derivation.htm
7. NIST Cesium Atomic Clock is world's most precise, http://www.beaglesoft.com/mainfaqtime.htm
8. NIST Time Scale data archive (leap seconds), http://tf.nist.gov/pubs/bulletin/ leapsecond.htm
9. The Earth's slowing rotation, http://pages.prodigy.com/suna/earth.htm
10. Leap seconds, http://tycho.usno.navy.mil/leapsec.html
11. "The expanding universe: from slowdown to speed up", http://www.scientificamerican.com/article.cfm?id=expanding-universe-slows-then-speeds
12. Dark energy and matter, http://nasascience.nasa.gov/astrophysics/what-is-dark-energy
13. Dark matter "beach ball" unveiled, http://news.bbc.co.uk/2/hi/science/nature/8444038.stm
14. Ukrainian scientists provide a new look at time, http://www.physorg.com/news64.html
15. Calendar systems, http://en.wikipedia.org/wiki/Calendar#Calendar_systems
16. "An error in time and the Gregorian calendar", Jose Arguelles, http://www.13moon.com/time-is-art.htm
17. "Mayan calendar: The classic Maya", Jose Arguelles, http://www.13moon.com/time-is-art.htm
18. *Breaking the Maya Code*, by Michael D. Coe (1992), London Thames and Hudson
19. "Prophetic closing date", by Jose Arguelles, http://www.13moon.com/time-is-art.htm
20. Armageddon: Mayan prophecies, http://www.history.com/video.do?name=armageddon&bcpid= 3887230001&bclid =5983807001&bctid=6085856001
21. *Galactic Alignment* (2002), by John M. Jenkins, Inner Traditions International
22. "What is the galactic alignment?", by John M. Jenkins, http://alignment2012.com/whatisga.htm
23. Cusp of a Great Cycle, http://www.adishakti.org/mayan_end_times_prophecy_12-21-2012.htm
24. Entering the Mayan year of the Caban, http://www.kachina.net/~alunajoy/97march.html
25. The risk of believing Mayan end date as Dec 21, 2012, http://www.calleman.com/ content/articles/risk_of_2012.htm
26. *The Mayan Calendar and the Transformation of Consciousness* (2004) by Carl J. Calleman, Bear & Company
27. Precession and the dawning of the Age of Aquarius, http://www.iki.rssi.ru/mirrors/stern/ stargaze/Sprecess.htm
28. "Age of Aquarius", http://www.adishakti.org/age_of_aquarius.htm

29. Astrological meaning of the Age of Aquarius, http://en.wikipedia.org/wiki/ Age_of_Aquarius
30. Earth's precessional cycle and sacred geometry, http://www.lunarplanner.com/HolyCross.html
31. *Fingerprints of the Gods* (1995), by Graham Hancock, Three Rivers Press
32. Review of the Orion Prophecy, http://www.diagnosis2012.co.uk/orp.htm
33. Characteristics of the Kali Yuga, http://www.hinduism.co.za/kaliyuga.htm
34. "The topology of time", by Jay Weidner, http://www.jayweidner.com/2012 Topology2.html
35. High energy lurks at the galactic centre, http://www.eurekalert.org/pub_releases/2004-09/ppa-hem092204.php
36. *The Mystery of 2012* (2007), pg 46, by John Major Jenkins, Sounds True Inc.
37. Solar Cycles and the Earth's weakening magnetic field, http://portland.indymedia.org/en/2009/06/392087.shtml
38. "Solar super flare: the Carrington event", http://science.nasa.gov/headlines/y2008/06may_carringtonflare.htm
39. Watching solar activity muddle Earth's magnetic field, http://www.esa.int/esaSC/SEMF75BNJTF_index_0.html
40. *Fractal Time* (2009), by Gregg Braden, Hay House Inc.
41. "Giant hole in Earth's magnetosphere", http://www.examiner.com/x-2383-Honolulu-Exopolitics-Examiner~y2009m2d24-How-Obama-can-take-bold-action-on-hole-in-Earths-magnetic-field
42. "Vostok ice cores" http://www.ncdc.noaa.gov/paleo/icecore/antarctica/vostok/vostok.html
43. "Rapid changes in the Earth's core discovered", http://www.physorg.com/news1346 42306.html
44. Magnetism triggers a brain response, http://www.sciencemag.org/cgi/pdf_extract/260/5114/1590
45. "Magnetic analysis of human brain tissue", http://www.springerlink.com/content/x4223w2110201620/
46. IPCC targets lead to 54% odds of catastrophic climate change, http://www.securegreen future.org/content/ipcc-targets-lead-least-54-odds-catastrophic-climate-change
47. More record warmth as scientists warn of global tipping point, http://edition.cnn.com/2012/06/08/us/record-warmth
48. "Scientists uncover evidence of impending tipping point for Earth", http://newscenter.berkeley.edu/ 2012/06/06/scientists-uncover-evidence-of-impending-tipping-point-for-earth/
49. Mayans protest 'twisting of truth' over 2012 doomsday predictions, http://www.huffingtonpost.com/2012/10/31/ mayans-protest-2012-doomsday-predictions_ n_ 2050519.html
50. What the Mayan elders are saying about 2012 by Carlos Barrios, http://www.seri-worldwide.org/id435.html
51. 2012 in retrospect by John M. Jenkins, http://21stcenturyblues.wordpress.com/2013/02/07/2012-in-retrospect-by-john-major-jenkins/
52. Some new reflections on the Mayan calendar end date by Carl Calleman, http://calleman.com/content/articles/SomeNewReflections.htm

53. The true meaning of the 2012 by John Perkins, http://www.greatmystery.org/nl/vancouver2012perkins.html
54. "Hopi prophecies", http://everything2.com/title/Hopi+prophecies
55. Hopi prophecy rock, http://www.viewzone.com/hopi.prophecy.html
56. *The Rocks Begin to Speak* (1976), by Lavan Martineau, K.C. Publications
57. "The Earth will shake three times" (Robert Bossiere). http://www.articlesbase.com/mysticism-articles/the-earth-will-shake-three-times-855448.html
58. Hopi prophecy; Hopi village of Oraibi, http://www.youtube.com/watch?v=TnoxByonalc&feature=related
59. Definition of crisis (Greek), http://dictionary.reference.com/browse/crisis
60. "Transforming crisis into opportunity" (Jewish Liturgy), http://www.inner.org/torah_and_science/mathematics/gematria/examples/transforming-crisis-opportunity.php
61. Sanskrit dictionary, http://spokensanskrit.de/index.php?tinput=vimarza&direction=SE&script=HK&link=yes&beginning= [for vimarśa]
62. The translation of the Great Isaiah Scroll, http://www.ao.net/~fmoeller/qa-tran.htm
63. *The Isaiah Effect* (2000), by Gregg Braden, Three River Press
64. Who were the Essenes?, http://www.essenespirit.com/who.html

Chapter 2: Scenarios and Tipping Points

1. *Worldshift 2012* (2009), by Irvin Laszlo, Inner Traditions
2. "Can a collapse of global civilization be avoided?", by Paul R. Ehrlich and Anne H. Ehrlich, http://www.campaignforrealfarming.org/wp-content/uploads/2013/01/Can-a-collapse-of-global-civilisation-be-avoided.pdf
3. Edward Lorenz's butterfly effect, http://www.exploratorium.edu/complexity/CompLexicon/lorenz.html
4. Insights on debt, derivatives, government guarantees etc., http://www.financialarmageddon.com/
5. "The dollar is heading for collapse". http://www.thedollarcollapse.com/
6. "U.S. dollar collapse could devastate economy", http://www.reuters.com/article/idUSTRE5AN5AP20091124
7. *Guide to Investing in Gold and Silver* (2008), by Michael Maloney, Business Plus
8. *The modern survival manual: surviving the economic collapse* (2009), by Fernando F. Aguirre
9. Top 10 infectious diseases, http://news.softpedia.com/news/Top-10-Infectious-Diseases-That-Have-Killed-Most-People-70741.shtml
10. The influenza pandemic of 1918, http://virus.stanford.edu/uda/
11. "10 Genes, furiously evolving", new swine flu, http://www.nytimes.com/2009/05/05/health/05virus.html?_r=1
12. "Swine flu ancestor born on U.S. factory farms", http://www.wired.com/wiredscience/2009/05/swineflufarm/
13. "Swine flu is evolution in action", http://www.livescience.com/health/090428-swine-flu-viral-evolution.html

14. "Pandemic Flu—Communicating the Risks", http://www.who.int/bulletin/volumes/84/1/interview0106/en/
15. Russia, US, plan radical cuts in weapons, http://www.chinadaily.com.cn/world/2009-12/22/content_9215862.htm
16. "Mahmoud Ahmadinejad says U.S. fabricated nuclear documents", http://www.guardian.co.uk/world/2009/dec/22/mahmoud-ahmadinejad-us-nuclear-documents
17. "Six escalation scenarios to world nuclear war", http://www.carolmoore.net/nuclearwar/alternatescenarios.html#assumptions
18. Lugar survey on proliferation (nuclear) threats and responses, http://lugar.senate.gov/reports/NPSurvey.pdf
19. "Reducing the nuclear threat", http://www.thebulletin.org/web-edition/op-eds/reducing-the-nuclear-threat-the-argument-public-safety
20. "Nuclear weapons: how many are there?", http://www.guardian.co.uk/news/datablog/2009/sep/06/nuclear-weapons-world-us-north-korea-russia-iran
21. Facts about known nuclear nations, http://www.carolmoore.net/nuclearwar/
22. BBC News—Struggle to stabilize Japan's Fukushima nuclear plant, http://www.bbc.co.uk/news/science-environment-12726628
23. "The twenty thousand year poison: nuclear safety and our long term future", http://akiomatsumura.com/2011/05/the-ten-thousand-year-poison-nuclear-safety-and-our-long-future.html
24. Statement of the World Future Council about the situation in Japan (Fukushima), http://www.worldfuturecouncil.org/3877.html
25. "Fukushima nuclear disaster (are you or your friends at risk?)", http://www.greenpeace.org/international/en/campaigns/nuclear/safety/accidents/Fukushima-nuclear-disaster/?accept=b657abbca5cecc9cc5c482cc0866c1b0
26. "U.S. scrambles to contain global WikiLeaks fallout", http://www.thejakartaglobe.com/home/us-scrambles-to-contain-global-wikileaks-fallout/409243
27. WikiLeaks stirs debate on information revolution, http://uk.reuters.com/article/idUKLNE6B504120101206
28. List of facilities vital to U.S. security leaked, http://www.bbc.co.uk/news/world-us-canada-11923766
29. What is WikiLeaks?, http://www.bbc.co.uk/news/technology-10757263
30. US Geological Survey; sea levels rise in 2100 will likely exceed IPCC projections, http://climateprogress.org/2008/12/16/us-geological-survey-stunner-sea-level-rise-in-2100-will-likely-substantially-exceed-ipcc-projections-sw-faces-permanent-drying-by-2050/
31. "Giant iceberg heading toward Australia", http://www.cnn.com/2009/TECH/science/12/09/australia.iceberg/index.html
32. "Kiribati says New Zealand, Australia not doing enough", http://tvnz.co.nz/world-news/kiribati-says-nz-aust-not-doing-enough-3249608
33. Earth's slowing rotation, http://pages.prodigy.com/suna/earth.htm

34. Geological survey of Canada; Geomagnetism, http://gsc.nrcan.gc.ca/geomag/index_e.php
35. Earth's magnetic field and climate variability, http://www.appinsys.com/Global Warming/EarthMagneticField.htm
36. *Scientific American Magazine*, "Faster than Expected", Nov 2012 Issue.
37. "2012 Galactic Alignment—of energetic significance or a non-event?" http://www.ancient-world-mysteries.com/2012-alignment-mayan-long-count.html
38. "It's not the end of the world", http://www.famsi.org/research/vanstone/2012/ index.html
39. The 2012 winter solstice non-event, http://www.astunit.com/astrocrud/2012.htm
40. NASA scientist speaks about 2012, http://www.youtube.com/watch?v=dHGaZMC8E0U&feature=related
41. "Your world in 2015; winning for all; science & spirituality". *Ode Magazine*, December 2005 issue, http://www.odemagazine.com/doc/29/your_world_in_2015/
42. *Einstein and Buddha—The Parallel Sayings*, 2002, by Thomas J. McFarlane, Ulysses Press
43. "Newton the Alchemist", http://www.alchemylab.com/isaac_newton.htm
44. Alchemy; Carl Jung's formulation of collective unconscious, http://metaphysics.suite101.com/article.cfm/what_is_alchemy_today
45. "A singularity in time", http://www.peterrussell.com/Odds/SoundsTrue2012.php
46. *Waking up in time*, 1992, by Peter Russell, Origin Press
47. The Golden Age, http://en.wikipedia.org/wiki/Golden_Age
48. "Hopi Prophecy Stone (Heading towards Golden Age?)", http://www.wicca-spirituality.com/hopi-prophecy.html
49. Butterfly symbology, http://insects.org/ced4/symbol_list2.html#36
50. The holisticshop dictionary: butterfly, http://www.holisticshop.co.uk/dictionary/butterfly.html
51. "Metals are metamorphosed through alchemical work", http://www.alchemywebsite.com/bookshop/metamorphosis_course.html

Chapter 3: What's Happening to Our Weather?

1. Intergovernmental Panel on Climate Change, "2007 Climate Change Synthesis Report", http://www.ipcc.ch/publications_and_data/publications_ipcc_fourth_assessment_report_synthesis_report.htm
2. International Energy Agency, "World Energy Outlook Report", http://www.worldenergy outlook.org/2008.asp
3. Earth's atmosphere, http://www.nasa.gov/audience/forstudents/9-12/features/912_liftoff _atm.html
4. Greenhouse effect, http://www.gsfc.nasa.gov/gsfc/service/gallery/fact_sheets/earthsci/green.htm
5. China's economic growth will have a devastating environmental impact, http://www.guardian.co.uk/commentisfree/2007/dec/09/comment.china
6. Effects of acid rain, http://www.epa.gov/acidrain/effects/

7. Acid rain in Europe, http://www.ace.mmu.ac.uk/eae/Acid_Rain/Older/Europe.html
8. China suffers severe acid rain contamination, 22 Sep 2006". http://www.china-embassy.org/eng/xw/t273190.htm
9. "Acid rain: downpour in Asia", http://earthtrends.wri.org/text/climate-atmosphere/feature-27.html
10. Melting glaciers will speed up sea level rise, http://news.nationalgeographic.com/news/2007/07/070719-warming-glacier.html
11. "Global warming: scientists reveal timetable", http://www.commondreams.org/headlines 05/0203-04.htm
12. Sixty percent of the ecosystems are at risk, http://www.millenniumassessment.org/en/index.aspx
13. Heat related deaths, illnesses and diseases, http://www.euro.who.int/globalchange/Assessment/20070403_2#
14. Dr. Tony McMichael, health risk of climate change, http://nceph.anu.edu.au/Staff_Students/staff_pages/mcmichael.php
15. Climate change amplifying animal diseases, http://www.physorg.com/news 162486984.html
16. W.H.O. air quality guidelines, http://www.who.int/mediacentre/factsheets/fs313/en/index.html
17. China water and air pollution, http://www.wri.org/publication/content/7833
18. "Global oxygen level falling, warn scientists", http://www.bjreview.com.cn/science/txt/2008-08/15/content_142842.htm
19. "The Millennium Ecosystem Assessment Synthesis Report (2005)", http://matagalatlante.org/nobre/down/MAgeneralSynthesisFinalDraft.pdf
20. "The 5 elements and the cycle of life", http://www.fiveelementtraining.com/article_1.html
21. "The Ying & Yang in medical theory", http://academic.brooklyn.cuny.edu/core9/phalsall/texts/yinyang.html
22. "The Yellow Emperor's Classic of Medicine" excerpt, http://www.friesian.com/ yinyang.htm
23. *Tao Te Ching*, by Lao-Tzu, Translation by S. Mitchell, http://academic.brooklyn.cuny.edu/core9/phalsall/texts/taote-v3.html
24. *Living the Wisdom of the Tao*, by Dr Wayne Dyer, 2007, Hay House Inc.

Chapter 4: What's Happening to Our Water?

1. The water resources of Earth, http://www.globalchange.umich.edu/globalchange2/current/lectures/freshwater_supply/freshwater.html
2. Water in the body; water and health, http://www.chemcraft.net/wbody.html
3. Global water crisis, U.N. water reports, http://www.unwater.org/mediaclim.html
4. "Asia's Next Challenge: Securing the Region's Water Future", http://www.wilsoncenter.org/index.cfm?topic_id=1421&fuseaction=topics.event_summary&event_id=515137
5. World water assessment programme, http://www.unesco.org/water/wwap/wwdr/index.shtml
6. "Water for life", World Water Day 2005, http://www.un.org/waterforlifedecade/ index.html

7. Global climate change and agricultural production, http://www.fao.org/docrep/w5183e/w5183e00.HTM
8. BP oil spill damages ecosystems, http://environment.about.com/od/petroleum/a/oil_spills_and_environment.htm
9. "2 years after the BP oil spill", http://www.washingtonsblog.com/2012/04/2-years-after-the-bp-oil-spill-is-the-gulf-ecosystem-collapsing.html
10. IPCC Technical paper on climate change and water, www.ipcc.ch/pdf/technical-papers/climate-change-water-en.pdf
11. *National Geographic Magazine*, April 2009 Issue, "Australia's dry run", "Outlook: Extreme"
12. Dartmouth Flood Observatory, http://www.dartmouth.edu/~floods/Archives/index.html
13. "U.N. Warning: Global flood threat is rising", http://radio.weblogs.com/0105910/2004/06/15.html
14. W.H.O. Flooding and communicable diseases, http://www.who.int/hac/techguidance/ems/flood_cds/en/
15. Fifty percent of the world's major rivers are polluted, http://news.bbc.co.uk/2/hi/science/nature/538457.stm?storyLink=%23
16. Pollution of the Ganges River, http://www.youtube.com/watch?v=wb_yDBmRgmU
17. Global warming and rising sea levels, http://www.science.org.au/nova/082/082key.htm
18. Melting ice sheets and glaciers, http://www.windows.ucar.edu/tour/link=/earth/climate/sea_level_rise.html
19. "Antarctica losing ice faster than expected", http://www.straitstimes.com/World/Story/STIStory_458198.html
20. "The first refugees of climate change", *Discovery Channel Magazine*, Dec 2009 Issue
21. Water footprint and virtual water, http://www.waterfootprint.org/?page=files/home
22. "Britain is the world's sixth largest importer of water", http://www.dailymail.co.uk/news/article-1047158/Britain-sixth-largest-water-importer-world.html
23. "Water science", by West Marrin Ph.D., http://www.watersciences.org/id25.htm
24. *Universal Water*, by West Marrin, Ph.D., 2002, Inner Ocean Publishing
25. "Water is it living or dead?", by Theodor Schwenk, http://www.waterflow.net/article2.htm
26. "Activated Water", by Igor Smirnov on Albert Szent Gyorgyi's work, http://www.bioline.org.br/pdf?ej03016
27. Institute of Heartmath, research on water, http://www.heartmath.org/faqs/research/research-faqs.html
28. *The Healing Power of Water*, by Masaru Emoto, Hay House Inc.

Chapter 5: What's Happening to Our Land?

1. Area of Earth's Land Surface, Science Desk Reference American Scientific, New York: Wiley, 1999
2. UNEP, Degraded soils—Maps and Graphics, http://maps.grida.no/go/graphic/degraded-soils
3. Severity of human induced soil degradation, http://www.fao.org/landandwater/agll/glasod/glasodmaps.jsp?country=PNG&search=Display+map+%21

4. Causes of soil degradation, http://www.acsgarden.com/articles/other-gardening/soil-degradation.aspx
5. Land degradation, Global environment outlook 3, http://www.grida.no/publications/other/geo3/?src=/geo/geo3/english/141.htm
6. Land degradation on the rise (2008), http://www.fao.org/newsroom/en/news/2008/1000874/index.html
7. "2008 Living Planet Report", http://www.grida.no/publications/other/geo3/?src=/geo/geo3/english/141.htm
8. Human ecological footprint, http://www.panda.org/about_our_earth/all_publications/living_planet_report/footprint/
9. "The Silent Spring", by Rachel Carson, 1962, http://en.wikipedia.org/wiki/Silent_Spring
10. International Union for Conservation of Nature, http://www.iucn.org/?4105/Our-Planet-Reviewed
11. World Resource Institute, measuring nature's benefits, http://www.wri.org/publication/measuring-natures-benefits
12. UNEP, mapping human impacts on the biosphere, http://www.unep-wcmc.org/ GLOBIO/index.htm
13. *National Geographic Magazine*, November 2008 Issue, "Borneo's Moment of Truth"
14. Borneo peat swamp forest, http://en.allexperts.com/e/b/bo/borneo_peat_swamp_forests.htm
15. Scientists find new global warming "Time Bomb", http://www.thewe.cc/weplanet/news/arctic/permafrost_melting.htm
16. *Discovery Channel Magazine*, December–January 2011 Issue, "Tundra Burning"
17. *50 Facts that Should Change the World*, by Jessica Williams, 2007, Icon Books Ltd.
18. Waste generation in selected ASEAN countries (2001), http://www.unep.or.jp/letc/Publications/spc/State_of_waste_Management/2.asp
19. World's biggest landfill in the Pacific Ocean, http://science.howstuffworks.com/great-pacific-garbage-patch.htm
20. World's biggest problems, http://www.arlingtoninstitute.org/wbp/peak-oil/161
21. 2008 oil reserves, http://www.bp.com/sectiongenericarticle.do?categoryId=9023769&contentId=7044915
22. *National Geographic Magazine*, March 2009 Issue, "Canada Oil Boom: Scraping Bottom"
23. Number of cars produced in 2007, http://www.worldometers.info/cars/
24. World Energy Outlook—Renewable Energy, http://www.iea.org/weo/docs/weo2008/WEO2008_es_English.pdf
25. Hydrogen the perfect fuel, http://www.commutercars.com/h2/
26. "Frozen power", *Discovery Channel Magazine*, November 2009
27. The Global Footprint Network, http://www.footprintnetwork.org/en/index.php/GFN/page/basics_introduction/
28. *Cradle to Cradle*, by McDonough and Braungart, 2002, North Point Press
29. Singapore's bushfires hit nearly decade high in January 2009, http://www.alertnet.org/thenews/newsdesk/SIN393826.htm
30. Native American totems and their meanings, http://www.legendsofamerica.com/NA-Totems.html

31. Native American proverbs, http://www.legendsofamerica.com/NA-Proverbs.html
32. "Bee Gone", *Discovery Channel Magazine*, February 2009 Issue
33. "Race to Save the Frogs", *National Geographic Magazine*, April 2009 Issue
34. "Which Species Will Live?", *Scientific American*, August 2012
35. "Has the Earth's sixth mass extinction already arrived?" http://www.nature.com/nature/journal/v471/n7336/full/nature09678.html
36. Animal totems, http://www.sayahda.com/cycle.htm
37. Animal totems, http://www.linsdomain.com/totems/pages/frog.htm
38. *Animal Speak*, by Ted Andrews, 1996, Llewellyn Publications

CHAPTER 6: What's Happening to Our Food?

1. World Summit on Food Security 2009, United Nations' Food and Agriculture Organization, http://www.fao.org/wsfs/world-summit/en/
2. APEC Japan 2010 Ministerial Meeting on Food Security in Niigata, http://www.niigata2010 apec.com/e/
3. Hunger and Poverty, Global Issues, http://www.globalissues.org/article/240/beef
4. As Prices Rise, Farmers Spurn Conservation Program, http://www.nytimes.com/2008/04/09/business/09conserve.html?_r=1&scp=1&sq=farmers%209%20April%202020088&st=cse
5. U.S. could feed 800 million people with grain that livestock eat, Livestock Production: Energy Inputs and the Environment, http://www.news.cornell.edu/releases/Aug97/livestock.hrs.html
6. Institute for Natural Resources in Africa, United Nations University, http://www.inra.unu.edu/index.cfm
7. "Genetically Engineered Crops—A Threat to Soil Fertility?" http://www.psrast.org/ soilecolart.htm
8. Food Crisis, The World Bank, http://www.worldbank.org/foodcrisis/
9. Causes of Hunger are related to Poverty, Global Issues, 3 October 2010, http://www.globalissues.org/article/7/causes-of-hunger-are-related-to-poverty
10. "Green Revolution: Curse or Blessing?" International Food Policy Research Institute, http://www.ifpri.org/sites/default/files/pubs/ib/ib11.pdf
11. Biodiversity International, http://www.bioversityinternational.org/
12. United Nations Environment Programme, http://www.unep.org/
13. World Health Organization Pesticides Evaluation Scheme, http://www.who.int/ whopes/en/
14. Pesticides Exposure Associated with Parkinson's Disease, http://www.hsph.harvard.edu/news/press-releases/2006-releases/press06262006.html
15. Misuse of Antibiotics in Food Animals, Union of Concerned Scientists, http://www.ucsusa.org/news/press_release/bill-addresses-antibiotics-misuse-in-ag-0206.html
16. Consumer Concerns About Hormones in Food, Cornell University, http://envirocancer.cornell .edu/Factsheet/Diet/fs37.hormones.cfm
17. Mad Cow Disease, http://www.mad-cow.org/

18. Avian Flu, Centers for Disease Control and Prevention, U.S. Department of Health and Human Services, http://www.cdc.gov/flu/avian/
19. "Bird Flu Timeline", *New Scientist*, http://www.newscientist.com/article/dn9977-timeline-bird-flu.html
20. GMOs in the EU Member States, GMO Compass, http://www.gmo-compass.org/eng/news/country_reports/
21. "Is GM Salmon Swimming Against the Tide?", BBC, 27 September 2010, http://www.bbc.co.uk/news/business-11368657
22. "World's First Flu-Resistant GM Chickens 'Created'", BBC, 13 January 2011, http://www.bbc.co.uk/news/science-environment-12181382
23. Food Additives, Center for Science in the Public Interest, http://www.cspinet.org/reports/chemcuisine.htm#Food additive
24. Outbreaks and Recalls, Food Safety Outbreak Alert, http://www.cspinet.org/foodsafety/outbreak_report.html
25. A Study in Suppression of Information: The Toxicity/Safety of MSG, The Truth About MSG, http://www.truthinlabeling.org/l-manuscript.html
26. Ribo Rash from Flavour Enhancer 635, Food Intolerance, http://www.nogw.com/download/2006_ribo_rash_fe635.pdf
27. Carbon Footprint of Food Flown Around the World, http://voiceireland.org/food-matters/food-matters-distribution-a-complex-issue/
28. Fast Food Restaurants, http://www.wikinvest.com/wiki/Fast_Food_Restaurants_(QSR)
29. "Fast Food Awash with 'Trans' Fats", *New Scientist*, http://www.newscientist.com/article/dn8989
30. *Fast Food Nation: The Dark Side of the All-American Meal*, by Eric Schlosser, 2001, Houghton Mifflin
31. *Super Size Me: A Film of Epic Portions*, 2004, by Morgan Spurlock
32. *Black Gold: A Film about Coffee Trade*, 2006, by Nick and Marc Francis
33. Livestock's Long Shadow: Environmental Issues and Options, The Livestock, Environment and Development Initiative, http://www.fao.org/docrep/010/a0701e/ a0701e00.HTM
34. "Can We Feed the World and Sustain the Planet?", *Scientific American*, November 2011 issue
35. The State of World Fisheries and Aquaculture 2008, United Nations' Food and Agriculture Organization, http://www.fao.org/docrep/011/i0250e/i0250e00.htm
36. Fish In—Fish Out, *Aquaculture Europe*, Vol. 34 (3), September 2009, http://envirofinfish.org/files/Fish-In-Fish-Out-Ratios-Explained1.pdf
37. "Cooking Up a Storm", September 2008, Food Climate Research Network, http://www.fcrn.org.uk/
38. *The China Study: Startling Implications for Diet, Weight Loss and Long-term Health*, by T. Colin Campbell, Ph.D., and Thomas M. Campbell II, 2006, First BenBella Books
39. *Diet for a New America*, by John Robbins, 1998, HJ Kramer, 2nd edition
40. *The Food Revolution: How Your Diet Can Help Save Your Life and Our World*, by John Robbins and Dean Ornish M.D., 2001, Conari Press
41. *Eating Animals*, by Jonathan Safran Foer, 2009, Back Bay Books/Little, Brown and Company

42. "Belgian city, Ghent, Plans 'Veggie' Days", BBC, 12 May 2009, http://news.bbc.co.uk/2/hi/europe/8046970.stm
43. "Food Poisoning in the U.K.", BBC, 26 March 2007, http://news.bbc.co.uk/2/hi/health/6483541.stm
44. "Food Companies Are Placing the Onus for Safety on Consumers", The New York Times, 14 May 2009, http://community.nytimes.com/comments/www.nytimes.com/2009/05/15/business/15ingredients.html?scp=2&sq=14%20May%202009%20processed%20food&st=cse
45. "Tainted Milk Repackaged", *The Straits Times*, 8 February 2010, http://www.straitstimes. com/BreakingNews/Asia/Story/STIStory_487923.html
46. One more victim dies in Geylang food poisoning incident, Channel News Asia, 8 April 2009, http://www.channelnewsasia.com/stories/singaporelocalnews/view/420957/1/.html
47. The truth about milk, http://rense.com/general26/milk.htm
48. The dangers of milk, http://www.notmilk.com/
49. Dioxins and their effects on human health, http://www.who.int/mediacentre/factsheets/fs225/en/index.html
50. *Your Life in Your Hands*, 2000, Jane Plant Ph.D., Thomas Dunne Books
51. Marketing milk and disease, http://www.nealhendrickson.com/mcdougall/030500pudairyanddisease.htm
52. "Sweet Drinks Up Cancer Risk," *The Straits Times*, 8 February 2010, http://www.straitstimes.com/BreakingNews/TechandScience/Story/ STIStory_487935.html
53. *The Omnivore's Dilemma*, 2006, Michael Pollan Ph.D., The Penguin Press
54. About Michael Pollan, http://en.wikipedia.org/wiki/Michael_Pollan
55. Healthcare spending in the U.S. and selected OECD countries, http://www.kff.org/insurance/snapshot/oecd042111.cfm
56. California Healthcare Foundation, http://www.chcf.org/~/media/MEDIA%20LIBRARY%20Files/PDF/H/PDF%20HealthCareCosts10.pdf
57. "An Unhealthy America: The Economic Burden of Chronic Disease", http://www.milkeninstitute.org/healthreform/pdf/AnUnhealthyAmericaExecSumm.pdf
58. Mounting medical care spending could be harmful to the G-20's, http://www.standardandpoors.com/ratings/articles/en/eu/?articleType=HTML&assetID=1245328578642
59. "Reverence for Life", by Albert Schweitzer, http://en.wikipedia.org/wiki/Albert_Schweitzer#Reverence_for_life
60. "1,000 Professors Demand Animal Welfare", http://www.albertschweitzerfoundation.org/news/1000-professors-demand-animal-welfare
61. We are wired to be good, kind, and compassionate, http://greatergood.berkeley.edu/article/item/the_compassionate_instinct/
62. *Six Reasons for Expanding Our Compassion Footprint,* by Marc Bekoff, *The Animal Manifesto,* by Marc Bekoff, 2010, New World Library
63. "Ethics, Food and Spirituality", http://www.ru.org/spirituality/ethics-food-and-spirituality.html
64. "Take a Hint from the Hunzas", http://www.byregion.net/articles-healers/Hunza_Diet.html

65. "Nutrition and the Bible", http://tccsa.tc/articles/nutrition0001.htm
66. *The Lost Religion of Jesus: Simple Living and Non-violence in Early Christianity*, 2000, Keith Akers, Lantern Books
67. Spiritual benefits of a vegetarian diet, http://www.atmajyoti.org/sw_spir_benefits_veg_diet.asp
68. How animal welfare leads to better meat, http://www.theatlantic.com/health/archive/2011/08/how-animal-welfare-leads-to-better-meat-a-lesson-from-spain/244127/
69. Animal stress results in meat causing disease, http://www.scn.org/~bk269/fear.html
70. "What's Wrong with Your T-Bone Steak?", by Dr. Alvin E. Adams, http://www.rawfoodexplained.com/why-we-should-not-eat-meat/whats-wrong-with-your-t-bone-steak.html
71. "Hormones in Animals and Meat Quality", http://www.steadyhealth.com/articles/ Hormones_In_Animals_And_Meat_Quality_a2217.html?show_all=1
72. Consequences of meat protein on human behavior", http://www.ivu.org/congress/euro97/consequences.html
73. Effects of stress and injury on meat, http://www.fao.org/docrep/003/X6909E/x6909e04.htm
74. *The World Peace Diet*, 2005, by Will Tuttle, Ph.D., Lantern Books
75. *Healthy at 100*, 2007, by John Robbins, Ballantine Books

CHAPTER 7: What's Happening to Our Bodies?

1. How women's bodies have been transformed in the past 60 years, http://www.dailymail.co.uk/health/article-1213475/Whats-happened-bodies-Womens-figures-transformed-past-60-years--huge-implications-health.html
2. "The Changing Body: Technology and the Evolution of the Human Physique", http://www.taipeitimes.com/News/editorials/archives/2011/04/30/2003502027
3. The New York Review of Books, http://www.nybooks.com/articles/archives/2011/oct/27/body-and-human-progress/?pagination=false
4. Technology advances; human supersizes, http://www.nytimes.com/2011/04/27/books/robert-w-fogel-investigates-human-evolution.html?pagewanted=all&_r=0
5. Why are women's breasts getting bigger, http://www.thirdage.com/women-s-health/why-are-womens-breasts-getting-bigger?page=2
6. Study: "More U.S. Girls Starting Puberty Early", http://edition.cnn.com/2010/HEALTH/08/09/girls.starting.puberty.early/index.html
7. "Abdominal Fat and What to Do about It", http://www.health.harvard.edu/newsweek/Abdominal-fat-and-what-to-do-about-it.*htm*
8. *The Changing Body: Health, Nutrition, and Human Development in the Western World since 1700*, R. Floud, R. Fogel, B. Harris and Sok C. H., 2011, Cambridge University Press
9. "Global Health Risks 2009", World Health Organization, http://www.who.int/healthinfo/global_burden_disease/global_health_risks/en/index.html
10. "Obesity, Halting the Epidemic by Making Health Easier—At A Glance 2009", http://www.cdc.gov/nccdphp/publications/AAG/pdf/obesity.pdf

11. "Obesity Driving Rising U.S. Health Costs," U.S. News, 10 February 2007, Psychological impact of obesity on school-aged children, http://www.dshs.state.tx.us/ obesity/pdf/psychological%20impact%20of%20obesity.pdf
12. National Centre for Eating Disorders, U.K., http://www.eating-disorders.org.uk/
13. Karen Carpenter, http://www.richardandkarencarpenter.com/index.html
14. The American Society for Aesthetic Plastic Surgery, http://www.surgery.org/
15. "Heidi Montag: Addicted to Plastic Surgery", *People*, 13 January 2010, http://www.people.com/people/article/0,,20336472,00.html
16. "Model's Death Highlights Plastic Surgery Risks," CNN, 7 January 2010, http://edition.cnn.com/2009/HEALTH/12/02/model.death.surgery.risk/index.html
17. "The Young and Plastic Surgery Hungry," Time.com, 7 May 2008, http://www.time.com/time/business/article/0,8599,1738111,00.html
18. "Florida Teen Dies After Complications During Breast Surgery", ABC News, 25 March 2008, http://abcnews.go.com/GMA/Parenting/story?id=4520099
19. "Increased Risk of Suicide Among Patients With Breast Implants: http://psy.psychiatry online.org/cgi/content/full/45/4/277
20. "CEO Dies After Liposuction", *The Straits Times*, 13 February 2010, http://www.straitstimes.com/BreakingNews/Singapore/Story/STIStory_490321.html
21. How body dysmorphic disorder works, http://health.howstuffworks.com/body-dysmorphic-disorder.htm
22. "The Epidemic of Junk Food Sex," by Lisa Love, suite101.com, 26 March 2008, http://improving-relationships.suite101.com/article.cfm/spirituality_relationships
23. Sexually Transmitted Diseases, World Health Organization, http://www.who.int/vaccine_ research/diseases/soa_std/en/
24. Uniting the World Against AIDS, United Nations, June 2008, http://www.un.org/ga/aids meeting2008/PR%20GA%20HLM%20final%20print.pdf
25. Cervical Cancer and the Human Papilloma Virus (HPV), World Health Organization, http://www.who.int/reproductivehealth/topics/cancers/en/
26. "Global HPV Vaccination Called by Physicians", AllBusiness.com, 18 July 2007, http://www.allbusiness.com/services/business-services/4551348-1.html
27. "20 Shocking Smoking Facts: The Dangers We Face From Tobacco Use", About.com, 27 May 2008, http://quitsmoking.about.com/od/tobaccostatistics/a/tobaccofacts.htm
28. 2008 National Survey on Drug Use and Health, SAMHSA's Office of Applied Studies, U.S. Department of Health and Human Services, http://www.oas.samhsa.gov/ NSDUHlatest.htm
29. Alcohol Abuse Facts and Statistics, Drug-Aware.com, http://www.drug-aware.com/alcohol-abuse-facts-statistics.htm
30. Understanding Alcohol Abuse, http://www.webmd.com/ mental-health/alcohol-abuse/understanding-alcohol-abuse-basics
31. National Institute on Drug Abuse, http://www.drugabuse.gov/
32. Prevalent prescription drug abuse, http://www.selfmedicating.info/prescription.htm

33. Safe Use Initiative: Collaborating to Reduce Preventable Harm from Medications, U.S. FDA, http://www.fda.gov/downloads/Drugs/DrugSafety/ UCM188961.pdf
34. Safe Use Initiative Fact Sheet, U.S. Food and Drug Administration, http://www.fda.gov/Drugs/DrugSafety/ucm188760.htm
35. "The fourth leading cause of death in the U.S. is from reactions to FDA-approved drugs," http://www.lightparty.com/Health/HealingRegeneration/html/PoliticsEconomics OfHealth.html
36. "Big Pharma Fraud: Jail Time Could Be the Cure", by Andy Ho, *The Straits Times*, 14 July 2012
37. Sodium Laureth Sulfate (SLES), http://www.natural-health-information-centre.com/ sodium-laureth-sulfate.html
38. Guide to less toxic products, http://www.lesstoxicguide.ca/index.asp?fetch=household
39. "Disinfectant Could Give Rise to Antibiotic-resistant Superbugs", http://www.cnn.com.sg/2009/HEALTH/12/30/disinfectant.superbugs/index.html
40. Emissions of PAHS, dioxins and PCBs, http://www.norden.org/en/publications/ publikationer/2011-549
41. "UW Scientist Henry Lai makes waves in the cell phone industry", http://www.seattlemag.com/article/nerd-report/nerd-report
42. United Nations World Focus on Autism, http://www.autismspeaks.org/about-us/press-releases/united-nations-world-focus-autism
43. *The Body Ecology Diet*, 2007, by Donna Gates, B.E.D. Publications
44. "Our Stolen Future", the Basics of, http://www.ourstolenfuture.org/basics/bookbasics.htm
45. The study of endocrine-disrupting compounds, http://intl-icb.oxfordjournals.org/content/ 45/1/194.full
46. The impact of environmental factors, body weight, exercise on fertility http://www.apothecarybydesign.com/userfiles/files/EnvironmentalFactors.pdf
47. "The miracle of fasting", http://www.evolutionhealth.com/bragg_miracle_fasting.html
48. "The key to excellent health: cleansing the colon", http://curezone.com/cleanse/ bowel/bowel_dr_anderson.html
49. "The history and development of body-psychotherapy", http://www.courtenay-young.co.uk/courtenay/articles/History_of_B-P_article_2.pdf
50. Healing yourself—quantum physics, how the body works, http://tracymalone.webs.com/ Healing%20Yourself.pdf
51. "Work Stress Changes Your Body", http://news.bbc.co.uk/2/hi/health/7203088.stm
52. "Are the Japanese worked to death?", http://www.research.vt.edu/resmag/sciencecol/ AFDC95.html
53. Work stress and risk cardiovascular mortality, http://www.bmj.com/content/325/7369/857
54. "History of Kinesiology and Muscle Testing", http://www.healing-with-eft.com/history-of-kinesiology.html
55. Bioenergetic analysis —Alexander Lowen, http://www.noanxiety.com/psychotherapies/ bioenergetic-therapy-alexander-lowen.html
56. "A Spirituality of the Body", http://www.cslewisinstitute.org/webfm_send/543

57. "Judaism and the Body", http://www.bje.org.au/learning/judaism/ethics/bioethics/body.html
58. The Spiritual Path of Islam: Body-Soul Conflict, http://www.islam101.com/sociology/spiritualPath.htm
59. Spiritual healing in the Sufi tradition, http://nurmuhammad.com/Meditation/EnergyHealing/harvardhealinglecture.htm
60. Sufi treatment methods and philosophy behind them, http://www.ishim.net/ishimj/ 2/02.pdf
61. *You Can Heal Your Life,* 1984, by Louis L. Hay, Hay House Inc.
62. *The Body Is the Barometer of the Soul,* 1994, by Annette Noontil, Gemcraft
63. *The Secret Language of Your Body*, 2010, by Inna Segal, Atria Paperback
64. *Your Body Speaks Your Mind*, 1996, by Debbie Shapiro, Piatkus

Chapter 8: What's Happening to Our Neighbours?

1. "Political Sex Scandals Rock Modest Malaysia", http://www.nytimes.com/2008/08/04/world/asia/04malaysia.html
2. Malaysian politics—Elizabeth Wong, http://archives.thestar.com.my/last365days/
3. "Racism Alive and Well in Malaysia", http://www.amren.com/mtnews/archives/2006/03/racism_alive_an_1.php
4. "Racism is allowed, protest against racism isn't", http://www.achrweb.org/Review/2007/195-07.html
5. Survey: "Most Young Malaysians Want Meritocracy", http://www.themalaysianinsider.com/index.php/malaysia/25354-survey-most-young-malaysians-want-meritocracy
6. Robbery cases in Malaysia, http://archives.thestar.com.my/last365days/
7. Easy for illegal immigrants to enter Malaysia", http://globalnation.inquirer.net/news/breakingnews/view/20070722-78107/Easy_for_illegal_immigrants_to_enter_Malaysia--deputy_PM
8. "Get Out! Malaysia Tells Migrants, Again", http://www.asiasentinel.com/index.php?option=com_content&task=view&id=1019&Itemid=31
9. "Rape Cases Up 300 Percent from 2007", http://thestar.com.my/news/story.asp?file=/2009/4/3/parliament/3619849&sec=parliament
10. 2008 Economic Report on Indonesia, http://www.bi.go.id/web/en/ Publikasi/Laporan+Tahunan/Laporan+Perekonomian+Indonesia/lpi_08.htm
11. Indonesia economy overview—CIA World Factbook, https://www.cia.gov/library/ publications/the-world-factbook/geos/id.html#top
12. "Beware Indonesia's 'Road Pirates'", http://www.globalpost.com/dispatch/indonesia/090305/beware-indonesias-road-pirates
13. "Traditional and Emergent Sex Work in Urban Indonesia", http://intersections.anu.edu.au/issue10/surtees.html
14. "Indonesia's 'Love Shacks'", http://www.globalpost.com/dispatch/indonesia/090213/ indonesias-love-shacks
15. Indonesia 2009 Crime and Safety Report, https://www.osac.gov/Reports/report.cfm?contentID=95858

16. "Indonesians Struggling After Devastating Quake", http://www.latimes.com/news/nationworld/world/la-fg-indonesia-quake5-2009oct05,0,6500927.story
17. Indonesia quake toll could soar, http://www.cnn.com/2009/WORLD/asiapcf/09/30/indonesia.earthquake/index.html
18. Core values in Thailand, http://www2.unescobkk.org/elib/publications/sourcebook.../07THAILA.pdf
19. Thailand: end of a year of political troubles, http://suan84.blogspot.com:80/2009/01/thailand-end-of-year-of-political.html
20. Transparency International: global coalition against corruption, http://www.transparency.org/news_room/in_focus/2008/cpi2008/cpi_2008_table
21. Corruption in Thailand, http://aceproject.org/ero-en/regions/asia/TH/Corruption_in_Thailand.pdf
22. Thai authorities close second airport, http://www.foxnews.com:80/story/0,2933,458266,00.html
23. "Investors look to overcome Thailand fears", http://news.bbc.co.uk/2/hi/business/6260357.stm
24. "Thai floods edge closer to central Bangkok", http://www.businessweek.com/news/2011-11-06/thai-floods-edge-closer-to-central-bangkok-may-last-3-weeks.html
25. Why Filipinos choose to work abroad, http://www.filipinooverseas.org/why-filipinos-choose-to-work-abroad
26. Number of overseas Filipino workers, http://www.census.gov.ph/data/pressrelease/2009/of08tx.html
27. Traffic in Manila, http://www.pinoycars.ph/index.php?option=com_content&view=article&id=83&Itemid=110
28. Poor suffer most from corruption in Philippines, http://ipsnews.net/news.asp?idnews=45652
29. 25 provinces under state of calamity, http://www.webcitation.org/5k6pPDbtd
30. 140 die in Philippine storm; toll expected to rise, http://www.washingtontimes.com/news/2009/sep/28/death-toll-philippine-storm-reaches-100/
31. "China to supplant U.S. as world's largest manufacturer", http://www.moneymorning.com/2008/08/11/china-manufacturing/
32. "China: major challenges ahead", http://www.politicalaffairs.net/article/view/2905/1/155/
33. Seven social problems hinder China, http://english.people.com.cn/200501/24/eng20050124_171731.html
34. "As China roars, pollution reaches deadly extremes", http://www.nytimes.com/2007/08/26/world/asia/26china.html
35. "Excuses for China's high housing prices", http://www.upiasia.com/Economics/2009/04/27/excuses_for_chinas_high_housing_prices/6979/
36. China's crime rate, http://uk.reuters.com/article/idUKPEK3065420080130
37. China's widening income gap, http://www.businessweek.com/globalbiz/content/feb2007/gb20070216_056285.htm

38. Children's materialism in urban and rural China, http://www.allacademic.com//meta/p_mla_apa_research_citation/1/6/9/3/3/pages169333/p169333-1.php
39. New global poverty estimates, http://www.worldbank.org.in/WBSITE/EXTERNAL/COUNTRIES/SOUTHASIAEXT/INDIAEXTN/0,,contentMDK:21880725~pagePK:141137~piPK:141127~theSitePK:295584,00.html
40. India development key issues, http://web.worldbank.org/WBSITE/EXTERNAL/COUNTRIES/SOUTH ASIAEXT/0,,contentMDK:21053816~pagePK:146736~piPK:146830~theSitePK:223547,00.html
41. India has the most billionaires after the U.S., http://techbizlawblog.wordpress.com/2007/08/14/india-has-most-number-of-billionaires-after-the-us/
42. The top 10 challenges of India, http://www.knowledgecommission.gov.in/downloads/news/news218.pdf
43. "Will $500 billion close India's infrastructure deficit?" http://www.worldbank.org.in/WBSITE/EXTERNAL/COUNTRIES/SOUTHASIAEXT/INDIAEXTN/0,contentMDK:21544993~menuPK:3940241~pagePK:64027988~piPK:64027986~theSitePK:295584,00.html
44. India's environmental challenges, http://www.allbusiness.com/professional-scientific/scientific-research/779967-1.html
45. India's economic challenges, http://www.worldbank.org.in/WBSITE/EXTERNAL/COUNTRIES/SOUTHASIAEXT/INDIAEXTN/0,,contentMDK:21214455~pagePK:141137~piPK:141127~theSitePK:295584,00.html
46. Ministry of Trade and Industry (Singapore), http://app.mti.gov.sg/default.asp?id=148&articleID=19101
47. Future of Singapore, http://www.littlespeck.com/informed/2006/CInformed-061012.htm
48. "The unsustainable Singapore model", http://theonlinecitizen.com/2009/09/the-unsustainable-singapore-model/
49. "Relentless rising cost of living in Singapore", http://theonlinecitizen.com/2007/08/the-relentless-rising-cost-of-living-in-singapore/
50. PM on cost concerns, rising expectations, http://www.straitstimes.com/Prime%2BNews/Story/STIStory_449241.html
51. "What happened to Singapore—the land of plenty?" http://theonlinecitizen.com/2007/01/what-happened-to-singapore-the-land-of-plenty/
52. No need to be so hurried in life, http://theonlinecitizen.com/2009/01/%e2%80%9cno-need-to-be-so-hurried-in-life-and-also-no-need-to-earn-so-much-money%e2%80%9d/
53. Singapore's green drive: Is enough being done?, http://www.wildsingapore.com/news/20070506/070512-1.htm
54. Climate change, a hot and stuffy issue, http://www.youth.sg/content/view/7480/59/
55. Significant hike in level of concern in global warming, http://www.acnielsen.com.sg/news/20070604.shtml
56. Singapore rejects emission cuts, http://www.straitstimes.com/Breaking%2BNews/Singapore/Story/STIStory_448079.html

57. More teens hit by sexually transmitted infections, http://www.straitstimes.com/Breaking%2BNews/Singapore/Story/STIStory_446405.html
58. Spike in female underage sex, http://www.straitstimes.com/Breaking%2BNews/Singapore/Story/STIStory_335834.html
59. Net blamed for teen sex, http://www.straitstimes.com/Breaking%2BNews/Singapore/Story/STIStory_426155.html
60. "Internet-initiated Sex Crimes against Minors", http://www.unh.edu/ccrc/pdf/CV71.pdf
61. Mutualism (biology), http://en.wikipedia.org/wiki/Mutualism_(biology)
62. Per capita GDP, rich and poor disparity, CIA, the World Factbook, GDP Per Capita (PPP), https://www.cia.gov/library/publications/the-world-factbook/rankorder/ 2004rank.html
63. Social unrest rising; growing wealth gap, http://www.straitstimes.com/Asia/China/Story/STIStory_469623.html
64. Wealth gap is creating a social time bomb, http://www.guardian.co.uk/world/2008/oct/23/population-egalitarian-cities-urban-growth
65. Cultural intelligence, http:// *www.bkconnection.com/static/culturalintelligence.pdf*
66. *Leading with Cultural Intelligence* (2009), By David Livermore Ph.D., Amacom
67. *The Chaos Point*; *The World at the Crossroads*, by Ervin Laszlo, 2006, Hampton Roads Publishing
68. Poverty facts and statistics, http://www.globalissues.org/article/26/poverty-facts-and-stats
69. Worldwatch Institute, vital signs 2003, http://www.worldwatch.org/
70. "Water in conflict", http://www.globalpolicy.org/security-council/dark-side-of-natural-resources/water-in-conflict.html
71. Third of the world's population in war of conflict, http://www.ploughshares.ca/
72. Morphogenic field, Rupert Sheldrake, http://www.sheldrake.org/papers/ Morphic/part2.html
73. Club of Budapest International Foundation, http://www.clubofbudapest.org/
74. End poverty 2015 millennium campaign, http://www.endpoverty2015.org/english/news/millions-mobilize-worldwide-and-web-demanding-world-leaders-eradicate-poverty/12/oct/09
75. Rainbow mythology, http://www.zianet.com/rainbow/frrelig.htm
76. The nature's mind, http://www.edmitchellapollo14.com/naturearticle.htm
77. *The Intention Experiment*, by Lynn McTaggart, 2007, Free Press
78. *Walking between the Worlds*, by Gregg Braden, 1997, Radio Bookstore Press
79. *Jesus, Buddha, Krishna and Lao Tzu—The Parallel Sayings*, 2007, by Richard Hooper, Sanctuary Publications
80. "The Compassion Instinct", http://odewire.com/118151/the-compassion-instinct.html
81. Ultraprevention—excerpts, http://www.ultraprevention.com/book/excerpts.htm
82. Open hearts build lives: Positive emotions, induced through loving-kindness meditation, build consequential personal resources, http://www.unc.edu/peplab/publications/Fredrickson%20et%20al%202008.pdf
83. Is compassion meditation the key to better care-giving? http://www.huffingtonpost.com/matthieu-ricard/could-compassion-meditati_b_751566.html
84. Loving-kindness and compassion meditation: Potential for psychological interventions http://www.pnei-it.com/1/upload/loving_kindness_and_compassion_meditation_prew.pdf

85. *Resurfacing*, 1997, by Harry Palmer, Star Edge
86. *The 2007 Shift Report: New Physics*, by The IONS
87. Superordinate goals in the reduction of intergroup conflict, http://www.brocku.ca /MeadProject/Sherif/Sherif_1958a.html
88. *The Bond*, 2011, by Lynne McTaggart, Free Press
89. *Great Peacemakers* (Riane Eisler), 2008, by Ken Beller and Heather Chase, LTS Press
90. Pulling together increases your pain threshold, http://www.ox.ac.uk/media/news_stories/ 2009/090916_3.html

CHAPTER 9: What's Happening to Our Work and Money?

1. International Labour Organization (ILO), www.ilo.org
2. Population Bulletin, Sep 2008, Vol. 63, No. 3, Population Reference Bureau Migration
3. Mega City Task Force of the International Geographical Union, http://www.megacities. uni-koeln.de/
4. Population Bulletin: World Population Highlights 2008, Sep 2008, Vol. 63, No. 3, Population Reference Bureau, http://www.prb.org/pdf08/63.3highlights.pdf
5. Key Indicators of the Labour Market (KILM) 5th Edition, International Labour Organization (ILO)
6. Offshoring—What is Offshoring?, SourcingMag.com, http://www.sourcingmag.com/ content/what_is_offshoring.asp
7. "Change in employment by occupation, industry, and earnings quartile, 2000-05," by Randy Ilg, Monthly Labor Review, December 2006, http://www.bls.gov/opub/mlr/ 2006/12/art2full.pdf
8. Global Employment Trends—Update, May 2009, ILO's Employment Trends Team, http://www.ilo.org/wcmsp5/groups/public/---dgreports/---dcomm/documents/ publication/wcms_106504.pdf
9. World of Work Report 2008: Income inequalities in the age of financial globalization, International Institute for Labour Studies, ISBN 978-92-9014-868-5, International Labour Office, Geneva, 2008
10. World of Work Report 2008—Global income inequality gap is vast and growing, ILO, http://www.ilo.org/global/About_the_ILO/Media_and_public_information/Press_releases/lang--en/WCMS_099406/index.htm
11. "How shifting occupational composition has affected the real average wage," by Rebecca Keller, Monthly Labor Review, June 2009, http://stats.bls.gov/opub/mlr/2009/06/ art2full.pdf
12. "Gates: Bankers should give more", CNNMoney, 29 January 2010, http://money.cnn.com/video/news/2010/01/29/n_bill_gates_bankers_davos.cnnmoney/
13. "Global Employment Trends for Women, March 2009," ILO, http://www.ilo.org/ wcmsp5/groups/public/---dgreports/---dcomm/documents/publication/wcms_103456.pdf
14. Part-time work, Information Sheet No. WT-4, June 2004, ILO, http://www.ilo.org/public/ english/protection/condtrav/pdf/infosheets/wt-4.pdf
15. Shift work, Information Sheet No. WT-8, May 2004, ILO, http://www.ilo.org/public/ english/protection/condtrav/pdf/infosheets/wt-8.pdf
16. "Self-employment in the United States: an update," by Steven Hippie, Monthly Labor Review, July 2004, http://www.bls.gov/opub/mlr/2004/07/art2full.pdf

17. International Labour Standards on Migrant Workers' Rights, Bangkok 2007, ILO, http://www.ilo.org/wcmsp5/groups/public/---asia/---ro-bangkok/documents/publication/wcms_bk_pb_184_en.pdf
18. "Migrants shape globalised world", BBC, 18 Dec 2006, http://news.bbc.co.uk/2/hi/6183803.stm
19. "Gangmastsers continue to exploit", BBC, 31 Jul 2009, http://news.bbc.co.uk/2/hi/uk_news/8177468.stm
20. Indonesia maid killed in Saudi Arabia, BBC, 19 November 2010, http://www.bbc.co.uk/news/world-asia-pacific-11795356
21. "Saudi woman jailed for abusing Indonesian maid Sumiati", BBC, 10 January 2011, http://www.bbc.co.uk/news/world-asia-pacific-12151454
22. "Serangoon Gardens Dormitory Saga", The Straits Times, 17 October 2008, http://cjcpig.wordpress.com/2009/07/19/serangoon-gardens-dormitory-saga/
23. "India tops migrant workers table", BBC, 19 March 2008, http://news.bbc.co.uk/2/hi/business/7305667.stm
24. "Laid-Off Foreigners Flee as Dubai Spirals Down", The New York Times, 12 February 2009, http://www.nytimes.com/2009/02/12/world/middleeast/12dubai.html? _r=2&ref=world
25. "Egypt braces for further day of protests", BBC, 27 January 2011, http://www.bbc.co.uk/news/world-africa-12294804
26. "Egypt unrest: ElBaradei returns as protests build", BBC, 27 January 2011, http://www.bbc.co.uk/news/world-africa-12300164
27. Child Labour Web Movie, ILO, http://www.ilo.org/public/english/bureau/inf/wdacl/english.htm
28. Questions and Answers on child trafficking: training tools for ILO constituents, ILO, http://www.ilo.org/ipec/areas/Traffickingofchildren/lang--en/WCMS_113318/index.htm
29. Child onion pickers highlights exploitation problem, BBC, 25 October 2010, http://www.bbc.co.uk/news/uk-11617664
30. "Telecommuting Trends in the 2009 Economy," by Victoria E, Bright Hub, 11 August 2009, http://www.brighthub.com/office/home/articles/22829.aspx
31. "Japan's ambitious digital future", BBC, 16 June 2009, http://news.bbc.co.uk/2/hi/technology/8102854.stm
32. "Cyber-crime rising, report warns", BBC, 31 March 2009, http://news.bbc.co.uk/2/hi/americas/7973886.stm
33. "The Internet—on the Verge of Exploding," *Discovery Channel Magazine*, June/July 2009
34. "Is the U.S. Government losing the battle against hackers?", *Discover Magazine*, 9 December 2008, http://blogs.discovermagazine.com/80beats/2008/12/09/us-government-is-losing-the-battle-against-hackers-report-says/
35. "Hackers Infiltrate Pentagon's $300 Billion Fighter Jet Project," *Discover Magazine*, 21 April 2009, http://blogs.discovermagazine.com/80beats/2009/04/21/hackers-infiltrate-pentagons-300-billion-fighter-jet-project/
36. Representative Money, Answers.com, http://www.answers.com/topic/representative-money

37. *Guide to Investing in Gold and Silver: Protect Your Financial Future*, by Michael Maloney, 2008, Business Plus
38. Social Security, http://socialsecurity.com/
39. International Social Security Association, http://www.issa.int/aiss/Observatory/Social-Security-Databases
40. "The Rise in Occupation Changing Rates in the United States 1979–2007," by Matissa Hollister, Dartmouth College, http://www.dartmouth.edu/~socy/pdfs/HollisterOccupationChange.pdf
41. "Follow the money", BBC, 10 September 2009, http://news.bbc.co.uk/2/hi/business/8249411.stm
42. "Huge crisis, huge causes", BBC, 17 Sep 2009, http://news.bbc.co.uk/2/hi/business/8260447.stm
43. Financial Crisis, http://financialcrisis.org/
44. "CNNMoney.com's bailout tracker", http://money.cnn.com/news/storysupplement/economy/bailouttracker/index.html
45. "Waking up to reality in Iceland", BBC, 26 January 2009, http://news.bbc.co.uk/2/hi/europe/7852275.stm
46. Data and Statistics, International Monetary Fund, http://www.imf.org/external/data.htm
47. Brief review of world socio-demographic trends, http://gsociology.icaap.org/report/socsum.html
48. "Is testosterone to blame for the financial crisis?," by Jordan Lite, Scientific American, 30 September 2008, http://www.scientificamerican.com/blog/60-second-science/post.cfm?id=is-testosterone-to-blame-for-the-fi-2008-09-30
49. "Sweeping US financial reform passed by Senate", BBC, 15 July 2010, http://www.bbc.co.uk/news/business-10654128
50. "U.S. Spending on Mental Health Care Soaring", *Discovery Health*, 6 August 2009, http://health.discovery.com /news/healthscout/article. html?article=629759& category=17&year=2009
51. "Work-related stress: Scientific evidence-base of risk factors, prevention and costs", Prof. Jean-Pierre Brun, http://www.who.int/occupational_health/topics/brunpres0307.pdf
52. "Is the way we work now making us sick?" BBC, 22 September 2009, http://www.bbc.co.uk/worldservice/business/2009/09/090922_mentalhealthwork.shtml
53. "Depression looms as global crisis", BBC, 2 September 2009, http://news.bbc.co.uk/2/hi/health/8230549.stm
54. "Suicides, mental illness up in Korea", AsiaOneHealth, 28 October 2008, http://health.asiaone.com/Health/News/Story/A1Story20081028-96710.html
55. "Mental illness soars as global crisis hits", ABC News, 4 May 2009, http://www.abc.net.au/news/stories/2009/05/04/2559793.htm
56. "Financial crisis increasing violence against women", Uganda People News, 8 September 2009, http://www.ugpulse.com/articles/daily/news.asp?about=Financial+crisis+increasing+violence+against+women&ID=11873
57. Abraham Maslow's Hierarchy of Needs, http://www.businessballs.com/maslow.htm

58. "What is Self Transcendence?", http://www.selftranscendence.org/self_transcendence/what_is_self_transcendence/
59. Bill O'Hanlon, Connection, Compassion and Contribution, http://www.che.org/members/mission/docs/1014/SpiritualityWork%20101907%20JKnuerr.ppt#258,13,Slide 13
60. "Being, Having and Doing Modes of Existence", by Erik H. Cohen et al, Bulletin of Sociological Methodology, 87/2005 July, http://bms.revues.org/index851.html
61. "The five waves of social transformation", Leland Kaiser, Keynote Speakers, Inc., http://www.keynotespeakers.com/speaker_detail.php?speakerid=4724
62. *Business for the Common Good: A Christian Vision for the Marketplace*, 2011, by Kenman L. Wong and Scott B. Rae, InterVarsity Press
63. The Goodwork Project, an overview, http://www.stemcareer.com/richfeller/pages/library/Documents/The%20GoodWork%20Project%20-%20An%20Overview.pdf
64. "Spirituality at work", by Seth Wax, https://docs.google.com/viewer?a=v&q=cache:UupVDiuLB3QJ:pzweb.harvard.edu/ebookstore/pdfs/goodwork41.pdf+seth+wax+spirituality+at+work
65. "Finding God in Our Work", http://kensmessage.blogspot/2011/09/finding-god-in-our-work2011sep11.html
66. The four paths of Yoga, http://www.sivananda.org/teachings/fourpaths.html
67. Karma Yoga: the way to God through work, http://www.netplaces.com/hinduism/paths-to-uniting-with-god/karma-yoga-the-way-to-god-through-work.htm
68. Karma Yoga: the Yoga of awareness in action, http://www.mandalayoga.net/index-what-en-karma_yoga.html
69. 10 big companies that promote employee meditation, http://www.onlinemba.com/blog/10-big-companies-that-promote-employee-meditation/
70. Toward a theory of spirituality in the workplace, http://www.emeraldinsight.com/journals.htm?articleid=1669090
71. Spirituality in the workplace: finding the holistic happy medium. http://www.emeraldinsight.com/bibliographic_databases.htm?id=1306577&PHPSESSID=u39jh8j83mor5ju3n0qaf7hpa6
72. Impact of yoga way of life on organisational performance, http://www.ncbi.nlm.nih.gov/pmc/articles/PMC2997233/
73. Boost performance by tapping employee's altruism, http://blogs.hbr.org/hbr/hewlett/2009/07/general_electric_and_pfizer_am.html
74. Goodcorps aims to help business meet social goals, http://www.nytimes.com/2011/05/13/business/media/13adco.html?_r=2&
75. "Does sustainability factor into consumers' purchasing decisions? You bet!", http://sustainability.fleishmanhillard.com/tag/cone-2010-cause-evolution-study/
76. Save the ta-tas, http://www.savethetatas.org/
77. TOMS shoes, http://www.toms.com/
78. "Now is the time for cause marketing", http://www.selfishgiving.com/cause-marketing-news/cone-study-local-nonprofits-now-time-for-cause-marketing

79. "Grow with cause marketing", http://www.inc.com/marla-tabaka/grow-with-cause-marketing.html

CHAPTER 10: What's Happening to Our Children?

1. *It Takes a Village*, 10th Anniversary Edition, by Hillary Rodham Clinton, 12 December 2006, Simon and Schuster
2. United Nations Demographic Yearbook 2007, Economic and Social Affairs, http://unstats.un.org/unsd/demographic/products/dyb/dybsets/2007%20DYB.pdf
3. UN Population Division Policy Brief No. 2009/1 March 2009, United Nations Department of Economic and Social Affairs, http://www.un.org/esa/population/ publications/ UNPD_policybriefs/UNPD_policy_brief1.pdf
4. "World Mortality 2007," United Nations Department of Economic and Social Affairs, Population Division, http://www.un.org/esa/population/publications/worldmortality/ WMR2007_wallchart.pdf
5. UNICEF, http://www.unicef.org/
6. United Nations Statistics Division, Social Indicators, June 2009, http://unstats.un.org/unsd/demographic/products/socind/child&elderly.htm
7. United Nations Statistics Division, Demographic and Social Statistics, Table 23. Divorces and crude divorce rates by urban/rural residence: 2003–2007, http://unstats.un.org/unsd/demographic/products/dyb/dyb2007/Table23.pdf
8. Divorces leave nature in tatters: Study, by Jianguo Liu and Eunice Yu, Michigan State University, 7 Dec 2007, http://www.rediff.com/news/2007/dec/07divorce.htm
9. Divorce statistics in the United States, http://divorce.lovetoknow.com/Divorce_Statistics
10. U.S. Divorce Statistics, Divorce Magazine.com, http://www.divorcemag.com/statistics/ statsUS.shtml
11. Children of Divorce and Separation Statistics, http://fathersforlife.org/divorce/ chldrndiv.htm
12. The Growing Backlash Against Overparenting, by Nancy Gibbs, Time.com , 20 November 2009, http://www.time.com/time/nation/article/0,8599,1940395,00.html
13. "Simplicity Parenting: Using the Extraordinary Power of Less to Raise Calmer, Happier, and More Secure Kids", by Kim John Payne and Lisa M. Ross, 31 August 2010, Ballantine Books
14. "Grown Up Digital: How the Net Generation is Changing Your World", by Don Tapscott, 3 October 2008, McGraw-Hill
15. Youth Culture and New Technologies, M/Clopedia of New Media, http://wiki.media-culture.org.au/index.php/Youth_Culture_and_New_Technologies
16. Computer and Video Games: Effects on Young Children and their Construction of Reality, M/Clopedia of New Media, http://wiki.media-culture.org.au/index.php/ Computer_Games_-_Children_and_Reality
17. "Most US kids play videogames", *The Straits Times*, 4 December 2009, http://www.straitstimes.com/BreakingNews/TechandScience/Story/STIStory_462415.html

18. "Gaming fanatics show hallmarks of drug addiction", by Alison Motluk, New Scientist, 16 Nov 2005, http://www.newscientist.com/article/dn8327
19. Pathological Video Game Use Among Youths: A Two-Year Longitudinal Study, Pediatrics online, 17 January 2011, http://pediatrics.aappublications.org/ cgi/content/full/127/2/e319
20. "20-year-old gets breathless, stops gasping and dies", *The Straits Times*, 1 June 2007, http://www.straitstimes.com/Free/Story/STIStory_124759.html
21. "Campaigning to make gaming safer", BBC, 19 June 2008, http://news.bbc.co.uk/cbbcnews/hi/newsid_7460000/newsid_7462000/7462030.stm
22. "Cyberbullies hit primary schools", BBC, 16 November 2009, http://news.bbc.co.uk/cbbcnews/hi/newsid_8360000/newsid_8361700/8361792.stm?ls
23. "Quarter of girls cyber bullied", BBC, 15 June 2009, http://news.bbc.co.uk/cbbcnews/hi/ newsid_8100000/newsid_8100000/8100077.stm
24. "Suicide bid girl 'bullied online'", BBC, 19 September 2006, http://news.bbc.co.uk/2/hi/uk_news/wales/north_west/5356078.stm
25. "Cyber bullying rises in South Korea", by Dan Simmons, BBC, 3 November 2006, http://news.bbc.co.uk/2/hi/programmes/click_online/6112754.stm
26. "Teens fall prey to online predators", Singapore Sex News Blogspot, 31 December 2008, http://sgsexnews.blogspot.com/2008/12/teens-fall-prey-to-online-predators.html
27. Child-porn owning student an MOE scholar, Channel NewsAsia, 19 November 2010, http://news.xin.msn.com/en/singapore/article.aspx?cp-documentid=4469405
28. "Fast Food Creates Fat Kids", by Jeanie Lerche Davis, WebMD Health News, 5 January 2004, http://www.webmd.com/parenting/news/20040105/fast-food-creates-fat-kids
29. "Fast food branding makes children prefer happy meals", by Roxanne Khamsi, *New Scientist*, 6 August 2007
30. Study: Fast-Food Ad Ban Could Curb Childhood Obesity, by Patrick Sauer, Health.com , 26 November 2008, http://news.health.com/2008/11/26/fast-food-ad-ban-child-obesity/?pkw=righthealth-sm&topic=Fast food
31. "Diabetes in Children", *Reader's Digest*, November 2008, http://www.rdasia.com/diabetes_in_children_3920
32. America's Children: Key National Indicators of Well-Being, 2009, http://childstats.gov/americaschildren/index3.asp
33. "Type 2 Diabetes in Kids", by Heather M. Ross, About.com , 24 November 2008, http://diabetes.about.com/lw/Health-Medicine/Conditions-and-diseases/Type-2-Diabetes-in-Children-and-Adolescents.htm
34. "Diabetes, obesity on rise for children in China", Reuters, 15 November 2007, http://www.reuters.com/article/2007/11/15/us-china-diabetes-idUSPEK14346920071115?feedType=RSS&feedName=healthNews
35. "Malignant Tumors a Major Killer of Children in China", *The Epoch Times*, 15 February 2009, http://www.theepochtimes.com/n2/content/view/12059/
36. Cancer in Children, Centers for Disease Control and Prevention, http://www.cdc.gov/ Features/dsCancerInChildren/

37. UN urges protection of children against marketing of unhealthy food, http://www.uicc.org/general-news/un-urges-protection-children-against-marketing-unhealthy-food
38. "Coping with School Stress", by Katherine Kam, WebMD, http://www.webmd.com/parenting/features/coping-school-stress
39. School Violence, Wikipedia.org, http://en.wikipedia.org/wiki/School_violence
40. "School Violence: How Prevalent is it?", by Melissa Kelly, About.com Guide, http://712educators.about.com/cs/schoolviolence/a/schoolviolence.htm
41. American Foundation for Suicide Prevention, Facts and Figures: International Statistics, http://www.afsp.org/index.cfm?fuseaction=home.viewpage&page_id=0512CA68-B182-FBB3-2E4CB905983C0AB8
42. "60% of [Taiwan's] youth think suicide", *The Straits Times*, 19 November 2009, http://www.straitstimes.com/BreakingNews/Asia/Story/STIStory_456475.html
43. "More underage sex offenders [in Taiwan]", *The Straits Times*, 19 November 2009, http://www.straitstimes.com/BreakingNews/Asia/Story/STIStory_456464.html
44. "Suffer the little children", *Bangkok Post*, 26 October 2008, http://www.bangkokpost.com/261008_News/26Oct2008_news08.php
45. Sexual Risk Behaviors, National Center for Chronic Disease Prevention and Health Promotion, http://www.cdc.gov/HealthyYouth/sexualbehaviors/
46. Promoting and safeguarding the sexual and reproductive health of adolescents, World Health Organization, Department of Reproductive Health and Research Policy Brief 4, 2006, http://whqlibdoc.who.int/hq/2006/RHR_policybrief4_eng.pdf
47. Map: Child Soldiers Fighting Around the World, Wide Angle, 29 July 2008, http://www.pbs.org/wnet/wideangle/episodes/lords-children/map-child-soldiers-fighting-around-the-world/2097/
48. The State of the World's Children – Special Edition: Celebrating 20 Years of the Convention on the Rights of the Child, UNICEF, 20 November 2009, http://www.unicef.org/rightsite/sowc/pdfs/SOWC_SpecEd_CRC_Executive Summary_EN_091009.pdf
49. Analysis of the Youth Global Consultation 2005 – 07, International Federation of Red Cross Societies, http://www.ifrc.org/Docs/Youth/IFRC_Youth_Consultation_ Report_FINAL_EN.pdf
50. Severn Cullis-Suzuki, Environmental Activist, http://news.bbc.co.uk/2/hi/south_asia/8299780.stm
51. Malawi windmill boy with big fans, BBC, 1 October 2009, http://news.bbc.co.uk/2/hi/africa/8257153.stm
52. "At 16 years old, Babar Ali must be the youngest headmaster", BBC, 12 October 2009, http://news.bbc.co.uk/2/hi/south_asia/8299780.stm
53. "Major Religions of the Word Ranked by Number of Adherents", http://www.adherents.com/Religions_By_Adherents.html
54. "Children in Islam: Their Care, Development and Protection", Al-Azhar University, International Islamic Center for Population Studies and Research, UNICEF, 2005, http://www.unicef.org/egypt/Egy-homepage-Childreninislamengsum(1).pdf

55. Hinduism and Children: Hindu view in Children and Parenting, http://www.experiencefestival.com/a/Hinduism_Children_and_Parenting/id/54154
56. What Buddhists Believe, Chapter 12—Buddhist Views on Marriage, Dr. K. Sri Dhammananda, http://www.sinc.sunysb.edu/Clubs/buddhism/dhammananda/237.htm
57. "A Happy Married Life—A Buddhist View", Dr. K. Sri Dhammananda, http://www.urbandharma.org/udharma7/marriedlife.html
58. "How Children are Valued in the Jewish Tradition", Rabbi David Rosen, http://rabbidavidrosen.net/articles.htm
59. List of Generations, Wikipedia, http://en.wikipedia.org/wiki/List_of_generations#List_of_generations
60. "What Do Gen Xers Want?", by Anne Fisher, Fortune, 20 January 2006, http://money.cnn.com/2006/01/17/news/companies/bestcos_genx/index.htm
61. Tips to improve interaction among the generations, http://honolulu.hawaii.edu/intranet/committees/FacDevCom/guidebk/teachtip/intergencomm.htm
62. Tips for green parenting, http://www.pbs.org/parents/special/article-earthday-greenparenting.html
63. American Psychological Association, http://www.apa.org/monitor/2012/10/parenting.aspx
64. Green Parenting, http://ecochildsplay.com/2011/02/08/what-is-green-parenting-its-all-about-health-and-happiness/
65. Education and activities, http://www.ehow.com/ehow-mom/education-and-activities/blog/
66. Conscious parenting, http://www.metaphysical-mom.com/conscious-parent-creed.html
67. Conscious parents, http://consciousparents.org
68. "Climate change: What price will future generations pay?", BBC, 23 November 2009, http://news.bbc.co.uk/2/hi/science/nature/8374965.stm

Chapter 11: What's Happening to Our Lives?

1. Dollar a day nations, http://econ.worldbank.org/external/default/main?pagePK=64165259&piPK=64165421&theSitePK=469372&menuPK=64216926&entityID=000158349_20080902095754
2. The world hunger problem; facts, figures and statistics, http://library.thinkquest.org/C002291/high/present/stats.htm
3. Highly Indebted Poor Countries, http://web.worldbank.org/WBSITE/EXTERNAL/TOPICS/EXTDEBTDEPT/0,,contentMDK:20260411~menuPK:64166739~pagePK:64166689~piPK:64166646~theSitePK:469043,00.html
4. The world's billionaires, http://www.forbes.com/2009/03/11/worlds-richest-people-billionaires-2009-billionaires_land.html
5. Ireland's banks are rescued, http://www.telegraph.co.uk/finance/financetopics/financialcrisis/3111122/Financial-crisis-Irelands-banks-are-rescued.html
6. Suicide rates on the rise in Asia, http://in.reuters.com/article/lifestyleMolt/idINTRE51O02P20090225
7. The overworked American, http://users.ipfw.edu/ruflethe/american.html

8. Psychotrends: sexuality, familty and relationships, http://www.psychologytoday.com/articles/199401/psychotrends
9. Cohabitation and marriage, http://www.vifamily.ca/library/cft/cohabitation.html
10. Relationship trends, http://news-for-two.cloudworth.com/relationship-trends.php
11. Divorce statistics collection, http://www.divorcereform.org/rates.html
12. The internet—on the verge of exploding, *Discovery Channel Magazine*, July Issue, 2009
13. Internet pornography statistics, http://internet-filter-review.toptenreviews.com/internet-pornography-statistics.html#anchor4
14. Pornography': impact on the family an marriage, http://www.heritage.org/research/family/tst111405a.cfm
15. DFC intelligence 2005 forecast, http://www.thefreelibrary.com/DFC+Intelligence+Forecasts+Video+Game+Industry+to+Rival+Size+of...-a0138466909
16. Impact of computers of children, http://www.pamf.org/preteen/parents/videogames.html
17. Computer infantalises the brain, http://www.telegraph.co.uk/technology/news/5316735/Computers-could-be-fuelling-obesity-crisis-says-Baroness-Susan-Greenfield.html
18. The effect of computer on child outcomes, www.columbia.edu/~cp2124/papers/computer.pdf
19. Technology overuse could be affecting development, http://www.bonnersprings.com/news/2008/nov/19/technology-overuse-could-be-affecting-development/
20. Relationship of internet use to depression and isolation, http://findarticles.com/p/articles/mi_m2248/is_138_35/ai_66171001/
21. American Sociological Review (2006), http://www.dukenews.duke.edu/2006/06socialisolation.html
22. *From Science to God* (2003), by Peter Russell, New World Library
23. Consumerism and consumption, http://www.globalissues.org/issue/235/consumption-and-consumerism
24. U.N. Human Development Reports 2007/2008, http://hdr.undp.org/en/reports/global/hdr2007-2008/
25. Global Problems and the Culture of Capitalism, http://faculty.plattsburgh.edu/richard.robbins/legacy/
26. Victor Lebow, http://whatdoino-steve.blogspot.com/2008/01/victor-lebow-bio.html
27. Planned Obsolescence, http://www.adbusters.org/category/tags/obsolescence
28. Perceived obsolescence, http://www.greenlivingtips.com/articles/188/1/Perceived-obsolescence.html
29. Conspicuous consumption, http://www.theatlantic.com/doc/200807/consumption?ca=2l7oDJW7kruafuw%2BAmt%2FgGV%2BRts3U%2F5faeR%2BTMotW8A%3D
30. Creating the consumer; psychology of shoppers, http://www.globalissues.org/article/236/creating-the-consumer
31. *Ecological Intelligence* (2009), by Daniel Goleman, Broadway Books
32. GoodGuide: ratings of natural, green and healthy products, http://www.goodguide.com/
33. 1998 U.N. Human development report. http://hdr.undp.org/en/reports/global/hdr1998/

34. International crime victim survey 1980-2000, http://search.icpsr.umich.edu/NACJD/query.html?col=abstract&op0=%2B&tx0=international+crime+victimization+survey+(icvs)+series&ty0=p&fl0=series%3A&op1=tx1=restricted&ty1=w&fl1=availability%3A&op2=%2B&tx2=NACJD&ty2=w&fl2=archive%3A&nh=50&rf=3
35. Rapes statistics from U.N. Survey of Crime Trends (21995-1997), http://www.unodc.org/unodc/en/data-and-analysis/Sixth-United-Nations-Survey-on-Crime-Trends-and-the-Operations-of-Criminal-Justice-Systems.html
36. Rapes statistics from U.N. Survey of Crime Trends (2001-2002), http://www.unodc.org/unodc/en/data-and-analysis/Eighth-United-Nations-Survey-on-Crime-Trends-and-the-Operations-of-Criminal-Justice-Systems.html
37. Fatal blasts hit Jakarta hotels, http://news.bbc.co.uk/2/hi/asia-pacific/8155084.stm
38. Terrorism statistics 2000-2006, http://www.nationmaster.com/graph/ter_ter_act_200_inc-terrorist-acts-2000-2006-incidences
39. Terrorism guide, http://uk.oneworld.net/guides/terrorism?gclid=CLKPIPCJyJs CFcEtpAod-U9fLg#Global%20Jihad
40. World becoming safer for the U.S., http://www.americanprogress.org/issues/2008/08 /terrorism_index.html
41. National counterterrorism centre (U.S.) 2007 report. http://wits.nctc.gov/reports/crot2007 nctcannexfinal.pdf
42. Deadliest terrorist strikes, worldwide, http://www.johnstonsarchive.net/terrorism/wrjp255i.html
43. Diversifying counterterrorism efforts, http://www.thewashingtonnote.com/archives/2008/07/terrorism_salon_11/
44. A society frightened by crime, http://openlearn.open.ac.uk/mod/resource/view.php?id=184684
45. The 2007 Shift Report, the Institute of Noetic Sciences, http://www.shiftreport.org/
46. Kaiser Family Foundation (prescription drugs), http://www.kff.org/rxdrugs/index.cfm
47. Eggs and nest symbols, http://www.bellaonline.com/articles/art20069.asp
48. Epigenetic: methyl groups influences on genes, http://researchnews.wsu.edu/health/97.html
49. Easter Island mystery revealed using mathematical model, http://news.mongabay. com/2005/0901-easter_island.html
50. The Flat Earth, http://en.wikipedia.org/wiki/Flat_Earth
51. Claudius Ptolemy, http://library.thinkquest.org/23830/ptolemy.htm
52. Nicholas Copernicus, http://www.blupete.com/Literature/Biographies/Science/Copernicus.htm
53. Galileo Galilei, major role in scientific revolution, http://www.crystalinks.com/ galileo.html
54. *The Structure of Scientific Revolutions* (1962), by Thomas Kuhn, University of Chicago Press
55. Isaac Newton, http://www.newton.ac.uk/newtlife.html
56. A brief history of scientific revolution, http://openlearn.open.ac.uk/mod/resource/view.php?id=189036

57. The paradigm shift in consciousness, http://www.learntovisualize.com/Articles/How%20 Change%20Occurs%20I.htm
58. The 2008 Shift Report, the Institute of Noetic Sciences, http://www.shiftreport.org/
59. "The Moral Psychology of Capitalism", http://goinside.com/01/4/capital.html
60. GDP definition, http://en.wikipedia.org/wiki/Gross_domestic_product
61. GDP poor gauge of well-being, http://ipsnews.net/news.asp?idnews=40180
62. Beyond GDP International Conference website, http://www.beyond=gdp.eu/links.html
63. Interesting facts and quotes on GDP and measuring progress, http://www.beyond-gdp.edu/download/interesting_facts.pdf
64. Genuine progress indicator—redefining progress, http://www.rprogress.org/ sustainability_indicators/genuine_progress_indicator.htm
65. Shattered lives: arms report; Oxfam, http://www.oxfam.org.uk/download/?download =http://www.oxfam.org.uk/what_we_do/issues/conflict_disasters/downloads/shattered _eng_summ.pdf
66. War and conflict, http://www.worldrevolution.org/projects/globalissuesoverview/ overview2/peacenew.htm
67. Arms trade and world military spending, http://www.globalissues.org/issue/73/arms-trade-a-major-cause-of-suffering
68. Duane Elgin: "Collective Consciousness and Cultural Healing, http://www.awakeningearth.org/articles
69. "The Technological Adolescent Age", http://www.iar-onicet.gov.ar/SETI/Technological_ Adolesc.pdf
70. Humanity in adolescence, http://www.spiritofpeacesf.org/Sermons/06_humanity%20in% 20adolescence.htm
71. Link between income and happiness, http://www.princeton.edu/main/news/archive/ S15/15/09S18/index.xml?section=topstories
72. *Affluenza: the All-consuming Epidemic*, 2005, by John de Graaf, David Wann and Thomas Naylor, Berrett-Koehler Publishers, Inc.
73. "The Story of Stuff", http://www.youtube.com/watch?v=gLBE5QAYXp8 , by Annie Leonard
74. "Journal of Happiness Studies", http://www.springerlink.com/content/1389-4978
75. World Database of Happiness, http://worlddatabaseofhappiness.eur.nl/hap_cor/cor_fp.htm
76. *Cosmos* (2008), by Ervin Laszlo and Jude Currivan, Hay House Inc.
77. What is simple living?, http://www.simpleliving.net/main/
78. What imitation tells us about social cognition, http://www.ncbi.nlm.nih.gov/pmc/articles/ instance/1351349/
79. *Social Intelligence: The New Science of Human Relationship* (2006), by Daniel Goleman, Bantam
80. "Is empathic emotion a source of altruistic motivation?", Journal of personality and social psychology (1981), by Daniel Batson

Chapter 12: What Does It All Mean?

1. *Cataclysm!* (1997), by D.S. Allan and J. B. Delair, Bear & Company
2. "Deep ice cores tell long climate story", http://news.bbc.co.uk/2/hi/science/nature/ 5314592.stm

3. 420,000 years of atmospheric history revealed, http://www.cnrs.fr/cw/en/pres/compress/mist030699.html
4. IPCC synthesis report confirms global warming, http://www.grist.org/article/ipcc
5. Intergovernmental Panel on Climate Change, http://www.ipcc.ch/
6. Kyoto Protocol: the next generation, http://edition.cnn.com/2007/TECH/science/05/08/kyoto.protocol/index.html
7. U.N. Climate Change Conference, http://en.cop15.dk/news/view+news?newsid=876
8. "Meat strike" by French VIPs during Copenhagen summit, http://suprememastertv.com/save-our-planet/Meat-strike-by-French-VIPs-during-Copenhagen-summit.html
9. United Nation Foundation; managing the unavoidable, http://www.unfoundation.org/global-issues/climate-and-energy/sigma-xi.html
10. *Blessed Unrest* (2007), by Paul Hawken, New York Times Bestseller, Penguin Books
11. "The Weather Makers", http://www.theweathermakers.org/weathermakers/
12. Emissions of Greenhouse Gases in the U.S. 2008, ftp://ftp.eia.doe.gov/pub/oiaf/1605/cdrom/pdf/ggrpt/057308.pdf
13. The building sector; the hidden culprit, http://www.architecture2030.org/current_situation/building_sector.html
14. "Saving energy starts at home", *National Geographic Magazine*, March 2009 Issue
15. "Copenhagan may have failed but we can't afford to", http://www.thesundayleader.lk/2009/12/27/copenhagen-may-have-failed-but-we-cant-afford-to/
16. "How close are we to runaway climate change?" http://www.guardian.co.uk/environment/2006/oct/18/bookextracts.books
17. "Earth in crisis, warns NASA's top climate scientist", http://afp.google.com/article/ALeqM5g2Wkbo6PcynAVeJzSPWDQZaWAI8g
18. IPCC targets lead to 54 percent odds of catastrophic climate change, http://www.securegreenfuture.org/content/ipcc-targets-lead-least-54-odds-catastrophic-climate-change
19. Duane Elgin Ph.D., global consciousness change, http://www.awakeningearth.org/PDF/global_consciousness.pdf
20. "Gaia Hypothesis", http://www.experiencefestival.com/gaia_hypothesis
21. James Lovelock's significant scientific contributions, http://www.jameslovelock.org/page3.html
22. Sri Aurobindo; Supramental existence, http://www.kheper.net/topics/Aurobindo/Supramentalisation.htm
23. "An Integral Theory of Consciousness", http://www.imprint.co.uk/Wilber.htm
24. An overview of the work of Jean Gebser, http://www.gaiamind.org/Gebser.html
25. The integral vision; second-tier integral consciousness, http://www.fudomouth.net/thinktank/now_integralvision.htm
26. Cybernetics and human knowing; planetary consciousness, http://imprint.co.uk/C&HK/vol4/v4-4laszlo.htm
27. "The Planetary Bargain", http://www.earthscan.co.uk/?tabid=435

28. Environmental activism, http://uk.oneworld.net/guides/environmentalactivism
29. Evidence for the Akashic field from moderm consciousness research, Stanislav Grof, M.D. http://www.stanislavgrof.com/pdf/Akashic%20Field%20Evidence.PDF
30. The Global Consciousness Project, http://noosphere.princeton.edu/
31. "Dawn of a New Consciousness", http://www.iamall-thewayhome.co.za/2012_Nov_Get_It.pdf
32. "The ages of man", http://www.maicar.com/GML/AgesOfMan.html#golden
33. The five ages of man, http://ancienthistory.about.com/cs/grecoromanmyth1/a/hesiodagesofman.htm
34. *Waking Up in Time* (1998), by Peter Russell, Origin Press Inc.
35. *The End of Certainty* (1997) – the study of chaotic systems, by Ilya Prigogine, The Free Press
36. The story of the oneness project; worldshift 2012, by Barbara Marx Hubbard, http://oneness.posterous.com/
37. *Macroshift - Breakthrough or Breakdown*, by Ervin Lazslo, Berrett-Koehler Publishers
38. *Translucent Revolution* (2005), by Arjuna Ardagh, New World Library

Chapter 13: What Is Stopping Us from Evolving?

1. Civilisational Analysis: "A paradigm in the making", http://www.eolss.net/ebooks/Sample%20Chapters/C04/E6-97-01-00.pdf
2. "The Axial Age", http://spiralpathways.files.wordpress.com/2009/10/the-axial-age.ppt
3. Einstein and Buddha (2002), by Thomas J. McFarlane, Ulysses Press
4. Scientism, http://www.newworldencyclopedia.org/entry/Scientism#Scientific _imperialism
5. Scientific spirituality, http://www.akhandjyoti.org/?Akhand-Jyoti/2003/Jan-Feb/ScientificSpirituality/
6. Awakening Earth, Duane Elgin, http://www.awakeningearth.org/articles
7. Quantum Shift in the Global *Brain* (2008), by Ervin Laszlo, Inner Traditions
8. Four Myth Perceptions of the Apocalypse, http://stanford.wellsphere.com/raw-food-article/4-myth-perceptions-of-the-apocalypse/803162
9. *The Divine Matrix* (2007), by Gregg Braden, Hay House Inc.
10. Institute of Noetic Sciences, the Bleep Study Guide, http://www.noetic.org/research/files/Bleep_Study_Guide.pdf
11. The scientific revolution, http://www.wsu.edu/~dee/ENLIGHT/SCIREV.HTM
12. American Institute of Physics; Sir Joseph J. Thomson, http://www.aip.org/history/electron/jjhome.htm
13. Encyclopedia Britannica; Max Planck, http://www.britannica.com/EBchecked/topic/ 462888/Max-Planck
14. Nature's holism, http://www.ecotao.com/holism/
15. New physics theory; by James A. Putnam, http://newphysicstheory.com/Table_of_Contents.htm
16. Nobel prize foundation; Albert Einstein, Arthur Compton, Niels Bohr, Werner Heisenberg etc., http://search.nobelprize.org/search/nobel/

17. Stanford Encyclopedia of Philosophy; determinism, http://plato.stanford.edu/entries/determinism-causal/
18. The complete works of Charles Darwin, http://darwin-online.org.uk/content/frameset?itemID=F373&viewtype=text&pageseq=1
19. "The Other Darwin", *National Geographic Magazine*, Dec 2008 Issue
20. Strange science; Charles Lyell, http://www.strangescience.net/lyell.htm
21. "Principle of Population", by Robert Malthus, http://www.faculty.rsu.edu/~felwell/Theorists/Malthus/essay2.htm
22. *Spontaneous Evolution* (2009), by Bruce Lipton Ph.D., and Steve Bhaeman, Hay House Inc.
23. "Is there a purpose in nature?", By Tim Lenton, http://www.cts.cuni.cz/conf98/lenton.htm#24
24. The quotation page, http://www.quotationspage.com/quotes/Albert_Einstein/
25. Brainy quote, http://www.brainyquote.com/quotes/authors/d/dalai_lama.html
26. Worldcat identities; Richard Dawkins, http://worldcat.org/identities/lccn-n81-74298
27. "Thoughts worth thinking", http://www.rjgeib.com/thoughts/nature/hobbes-bio.html
28. "The Wealth of Nations and Theory of Moral Sentiments", by Adam Smith, http://metalibri.wikidot.com/authors:adam-smith
29. John Stuart Mill; utilitarianism, http://www.earlymoderntexts.com/millu.html
30. "Empathy on the Brain", Christian Keysers, http://www.bu.edu/sjmag/scimag2005/features/mirrorneurons.htm
31. *Social Intelligence* (2006), by Daniel Goleman, Bantam Books
32. "The compassionate instinct", by Dacher Keltner, http://integral-options.blogspot.com/2010/02/dacher-keltner-compassionate-instinct.html
33. "Born to Be Good", *The New York Times,* http://www.nytimes.com/2009/01/19/books/chapters/chapter-born-to-be-good.html,
34. "Those little rats!", Darlene Francis and Michael Meaney, http://www.flatrock.org.nz/topics/animals/maternal_care.htm
35. Paul Zak: Oxytocin, trust and greed, http://www.hugthemonkey.com/2007/03/paul_zak_oxytoc.html
36. MIT Press Journals; "A Theory of Fairness, Competition and Cooperation", http://www.mitpressjournals.org/doi/abs/10.1162/003355399556151
37. "Toward Homo Noeticus", http://www.kingsleydennis.com/John%20White%20-%20Toward%20Homo%20Noeticus.pdf
38. "The wolf of hate", *Psychology Today*, http://www.psychologytoday.com/blog/your-wise-brain/201002/the-wolf-hate
39. *Power vs. Force* (1995), by Dr. David Hawkins, Veritas Publishing
40. *Transcending the Levels of Consciousness* (2006), Dr. David Hawkins, Veritas Publishing
41. The Heartmath Institute; Bioelectromagnetic communication, http://www.heartmath.org/research/rp-energetic-heart-bioelectromagnetic-communication-within-and-between-people.html
42. *The Power of Intention* (2004), by Dr. Wayne Dyer, Hay House Inc.

43. The Kryon website; Lee Carroll, http://www.kryon.com/Leebio.html
44. *The Parables of Kryon* (1996), by Lee Carroll, Hay House Inc.
45. Richard Feynman quotes, http://en.wikiquote.org/wiki/Richard_Feynman
46. Buddha quotes, http://en.wikiquote.org/wiki/Buddha
47. Albert Einstein quotes, http://en.wikiquote.org/wiki/Albert_einstein
48. Sri Aurobindo quotes, http://en.wikiquote.org/wiki/Sri_Aurobindo

Chapter 14: What New Knowledge Can Help Us?

1. Lao Tzu quotes, http://www.brainyquote.com/quotes/quotes/l/laotzu133381.html
2. *Writings on Physics and Philosophy* (1994), by Wolfgang Pauli, Berlin: Springer-Verlag
3. *The Tao of Physics* (1975), by Fritjof Capra, Shambhala Publications
4. Convergence of science and spirituality, http://www.bibliotecapleyades.net/ciencia/ciencia_sciencespirituality.htm
5. Spiritual Science, Rudolf Steiner, http://en.wikipedia.org/wiki/Anthroposophy
6. *Introducing Quantum Theory* (2004), by J.P. McEvoy & Oscar Zarate, Icon Books Ltd.
7. Thomas Young's double slit experiment, http://physics.about.com/od/lightoptics/a/doubleslit.htm
8. Quantum theory demonstrated: observation affects reality, http://www.sciencedaily.com/releases/1998/02/980227055013.htm
9. *Einstein and Buddha* (2002), by Thomas J. McFarlane, Ulysses Press
10. Bell's Theorem, http://www.upscale.utoronto.ca/PVB/Harrison/BellsTheorem/BellsTheorem.html
11. Alain Aspect: Shedding new light on light and atoms, http://www2.cnrs.fr/en/447.htm
12. Quantum entanglement answered, http://www.youtube.com/watch?v=RgI_PTs2X7Q
13. How holograms work, http://www.howstuffworks.com/hologram.htm
14. Holographic universe may herald new era in fundamental physics, http://www.scienceblog.com/cms/holographic-universe-may-herald-new-era-fundamental-physics-18469.html
15. *Measuring The Immeasurable* (2008), by Sounds True
16. *The Intention Experiment* (2007), by Lynne McTaggart, Free Press
17. The scientific and spiritual implications of psychic abilities, http://www.espresearch.com/espgeneral/doc-AT.shtml
18. Glen Rein and Rollin McCraty, Coherent heart states affects DNA, http://www.laskow.net/articles/ModDNAbyCohHeartFreq.pdf
19. Modulation of DNA conformation by heart-focused intention, http://www.vitality-living.com/resources/Modulation_of_DNA.pdf
20. NASA, Astrophysics, http://science.nasa.gov/astrophysics/focus-areas/what-is-dark-energy/
21. "It's confirmed: Matter is merely vacuum fluctuations", http://www.newscientist.com/article/dn16095-its-confirmed-matter-is-merely-vacuum-fluctuations.html
22. *The Divine Matrix* (2007), by Gregg Braden, Hay House Inc.
23. *Science and The Akashic Field* (2004), by Ervin Laszlo, Inner Traditions

24. "The Real and the Apparent Man" (Akasha & Prana), by Swami Vivekananda, http://www.hinduism.co.za/thereal.htm
25. Harmonic resonance, http://www.imune.org/index/pdf/Harmonic%20Resonance.pdf
26. Global Oneness Project, by Dean Radin, Ph.D., http://dotsub.com/view/6178f577-f20e-497a-8539-6c4a30e0abb0
27. Foundation of Physic Letter; Correlations of continuous random data with major world events, by Nelson, Radin, Shoup and Bancel, http://www.boundaryinstitute.org/bi/articles/FoPL_nelson-pp.pdf
28. Random Number Generators and Geosynchronous Operational Environmental Satellites; September 11, 2001, http://www.glcoherence.org/monitoring-system/about-system.html
29. Global heart coherence can affect the Earth's field, http://www.glcoherence.org/about-us/about.html
30. *The Genie In Your Genes* (2007), by Dawson Church, Elite Books
31. *The Biology Of Belief* (2005), by Bruce Lipton, Elite Books
32. The central dogma of molecular biology, http://www.euchromatin.org/Crick01.htm
33. Howard M. Temin, Genetic data flowed backwards, http://www.nndb.com/people/368/000130975/
34. "Adaptive Mutation: Has the Unicorn Landed" (John Cairns research), by Patricia L. Foster, www.ncbi.nlm.nih.gov/pmc/articles/PMC1460081/pdf/9560365.pdf
35. "Borrowed Genes", *Discovery Channel Magazine*, May 2010 Issue
36. Randy Jirtle, nutritional effects on epigenetic gene regulation, http://sciencewatch.com/ana/st/epigen/09augEpiJirt/
37. Michael Skinner, "Transgenerational actions of endocrine disruptors", http://sciencewatch.com/ana/st/epigen/09marEpiSkin/
38. Michael Skinner, "Sins of the Grandfathers", Newsweek, http://www.newsweek.com/2010/10/30/how-your-experiences-change-your-sperm-and-eggs.html
39. *Spontaneous Evolution* (2009), by Bruce Lipton and Steve Bhaerman, Hay House Inc.
40. *The God Code* (2004), by Gregg Braden, Hay House Inc.
41. The Power of God Is with You, Bhagavad Gita, Chapter 18, http://www.gita4free.com/english_completegita18.html
42. Healing Research; Bernard Grad, McGill University, http://www.reiki.org/reikinews/reikin24.html
43. The ciba-geigy effect, http://www.arguewitheveryone.com/science-technology/54446-ciba-geigy-effect.html
44. Pulsed Magnetic Stimulation, http://www.earthpulse.net/MS.htm
45. Harmonic resonance, http://www.imune.org/index/pdf/Harmonic%20Resonance.pdf
46. "The electrical patterns of life", Dr Harold S. Burr, http://www.wrf.org/men-women-medicine/dr-harold-s-burr.php
47. "Highly diluted antigen increases coronary flow of isolated heart from immunized guinea-pigs", by Jacque Benveniste, http://www.homeopathy.org/research/basic/Silva.pdf

48. Human Proteome Organization, "Challenges in deriving high-confidence protein identification", **http://www.hupo.org/communications/publications/naturebiotech _2006.pdf**
49. "The placebo effect: redefining the role of the mind", by Christine Kaminski, http://serendip.brynmawr.edu/bb/neuro/neuro03/web1/ckaminski.html
50. "The depressing news about antidepressants", research by Irving Kirsch and Guy Sapirstein, http://www.newsweek.com/id/232781
51. "Influences of suggestions on airways reactivity in asthmatic subjects", Luparello et. al. http://www.psychosomaticmedicine.org/cgi/content/abstract/30/6/819
52. Alterations in brain and immune function produce by mindfulness meditation, http://www.psychosomaticmedicine.org/cgi/content/full/65/4/564
53. The effect of meditation on the brain activity in Tibetan meditators, http://www.andrew newberg.com/qna.asp
54. Breakthrough study of EEG of meditation, http://www.quantumconsciousness.org/ EEGmeditation.htm
55. *Measuring The Immeasurable* (2008), Compilations, Sounds True
56. *Journal of the American Medical Association,* "Incidence of adverse drug reactions in hospitalized patients", http://jama.ama-assn.org/cgi/content/abstract/279/15/1200
57. General systems theory, by David S. Walonick, http://www.survey-software-solutions. com/walonick/systems-theory.htm
58. "Bertalanffy's General Systems Theory", by Greg Mitchell, http://www.trans4mind. com/mind-development/systems.html
59. *Introducing Chaos* (2004), by Ziauddin Sardar and Iwona Abrams, Icon Books Ltd.
60. *The Essence of Chaos* (1993), by Edward Lorenz, University of Washington Press
61. Strange attractors, http://www.stsci.edu/~lbradley/seminar/attractors.html
62. Fractals, http://www.stsci.edu/~lbradley/seminar/fractals.html
63. Ontogeny and phylogeny, http://evolution.berkeley.edu/evosite/evo101/ZIIIC6a Ontogeny.shtml
64. Bifurcation, http://www.exploratorium.edu/complexity/CompLexicon/bifurcation.html
65. Ilya Prigogine, chaos and dissipative structures, http://www.osti.gov/accomplishments/ prigogine.html
66. *The Chaos Point* (2006), by Ervin Laszlo, Hampton Roads Publishing Company, Inc.
67. "Influences of Suprasystems on Systemic Change", http://www.springerlink.com/content/ d707234432676w10/
68. Kolam, http://en.wikipedia.org/wiki/Kolam
69. The 8 Trigrams (Ba Gua), http://www.kheper.net/topics/I_Ching/trigrams.htm
70. *Introducing Chaos* (2004), by Ziauddin Sardar and Iwona Abrams, Icon Books Ltd.
71. "The Five Eyes", by Dr. Shen Chia Theng, http://www.baus.org/baus/library/shen.html
72. Five Eyes, http://www.bhaisajyaguru.com/buddhist-ayurveda-encylopedia/five_eyes_pancha-chaksus_panca-cakkhuni_wu-yan_chaksu.htm
73. "A five minute introduction to Buddhism", http://www.buddhanet.net/e-learning/ 5minbud.htm
74. Buddhism A to Z, http://online.sfsu.edu/~rone/Buddhism/BuddhistDict/BDF.html

75. Introduction to M-theory, http://mkaku.org/home/?page_id=262
76. Imagining the Tenth Dimension, by Rob Bryanton, http://www.youtube.com/watch?v=JkxieS-6WuA
77. *Sayings of Buddha* (1998), by E. E. Ho, Ph.D. and W. L. Rathje, Ph.D., Asiapac Publication

Chapter 15: What New Presuppositions Must We Uphold to Evolve?

1. *The Structure of Scientific Revolutions* (1962), by Thomas S. Kuhn, The University of Chicago Press
2. *Encyclopedia of Systemic Neuro-Linguistic Programming* (2000), by Robert Dilts and Judith Delozier, NLP University Press
3. Articulating science and theology: presuppositions and implications of science, http://www.unav.es/cryf/articulatingsciencieandtheology.html
4. Newton's errors, http://www.therealskeptic.com/column06-06.html
5. Einstein as a philosopher, http://www.pantaneto.co.uk/issue17/weinert
6. "John S. Bell: The part-timer who proved Einstein wrong", http://www.freshbrainz.com/2008/05/ john-stewart-bell-part-timer-who-proved.html
7. John S. Bell and the most profound discovery of science, http://physicsworld.com/cws/article/ print/1332
8. Paradigm shifts, http://en.wikipedia.org/wiki/Paradigm_shift
9. Cognitive science, http://plato.stanford.edu/entries/cognitive-science/#Con
10. The Keynesian revolution, http://homes.chass.utoronto.ca/~reak/eco100/100_14.htm
11. Monetarism, http://en.wikipedia.org/wiki/Monetarism
12. Ecological principles, http://www.ecoliteracy.org/nature-our-teacher/ecological-principles
13. David W. Orr, Ecoliteracy, http://www.davidworr.com/
14. "Strange Attractors of Meaning", by Vladimir Dimitrov, http://www.zulenet.com/VladimirDimitrov/pages/SAM.html
15. Philosophical definition of datum, http://www.answers.com/topic/datum
16. "A Fractal Universe?", http://www3.amherst.edu/~rloldershaw/NOF.HTM
17. Holism, http://www.ecotao.com/holism/
18. *Science and the Akashic Field* (2004), by Ervin Laszlo, Inner Traditions
19. Macrocosm and microcosm, http://en.academic.ru/dic.nsf/enwiki/154800
20. The origin of "as above so below," http://radiantwoman.wordpress.com/2006/12/25/ origin-of-as-above-so-below/
21. *Spontaneous Evolution* (2009), by Bruce Lipton and Steve Bhaerman, Hay House Inc.
22. Bertalanffy's general systems theory, by Greg Mitchell, http://www.trans4mind.com/mind-development/systems.html
23. General Systems Theory, http://en.wikipedia.org/wiki/User:Mdd/General_systems_theory
24. "Society, spirit and ritual: morphic resonance and the collective unconscious", by Rupert Sheldrake, http://www.alice.id.tue.nl/references/sheldrake-1987part2.pdf
25. "It's confirmed: matter is merely vacuum fluctuations", http://www.newscientist.com/article/dn16095-its-confirmed-matter-is-merely-vacuum-fluctuations.html

26. Global oneness project, by Dean Radin, Ph.D., http://dotsub.com/view/6178f577-f20e-497a-8539-6c4a30e0abb0
27. Global heart coherence can affect the Earth's field, http://www.glcoherence.org/about-us/about.html
28. Man of the millennium, http://news.bbc.co.uk/hi/english/static/events/millennium/ default.stm
29. Karma Yoga: an NLP approach to using your life's work as a spiritual path by Dr. Richard Bolstad
30. *The Words of Gandhi* (1982), New Market Press
31. *Power vs. Force* (1995), by Dr. David Hawkins, Veritas Publishing
32. Values change the world, http://www.worldvaluessurvey.org/wvs/articles/older_published/article_base_110
33. Share the world's resources (STWR), economic sharing: a shift in global values, http://www.stwr.org/economic-sharing-alternatives/economic-sharing-a-shift-in-global-values.html
34. "Choosing Simplicity" (2000), by Linda B. Pierce, http://www.gallagherpress.com/pierce/book-cs.htm
35. "The High Price of Materialism", by Tim Kasser, http://www.human-nature.com/nibbs/03/kasser.html
36. Shifting values in response to climate change, www.worldwatch.org/files/pdf/SOW09_CC_values.pdf
37. *Voluntary simplicity* (1981), by Duane Elgin, Harper-Collins Publishers
38. Nietzsche's contribution to our understanding of the relationship between morality and religion (values), http://www.helium.com/items/1104068-nietzsche-relationship-between-morality-and-religion
39. Abundance and happiness; the law of resonance, http://www.abundance-and-happiness.com/law-of-resonance.html
40. Spiritual abundance—finding your spiritual sense of purpose, http://www.psitek.net/ pages/PsiTek-how-to-fill-your-life-with-unlimited-abundance-6.html
41. Alchemy, http://en.wikipedia.org/wiki/Alchemy
42. "Isaac Newton and the Philosopher's Stone" (alchemy), *Discover Magazine*, July/August 2010
43. Phoenix mythology, new world encyclopedia, http://www.newworldencyclopedia.org/entry/ Phoenix_(mythology)
44. The dark night of the soul, http://www.themystic.org/dark-night/
45. St John of the Cross, http://www.ccel.org/ccel/john_cross/dark_night.html
46. Resonance, http://en.wikipedia.org/wiki/Resonance
47. "The Power of Resonance", by Joel Kotarski and Christopher Galtenberg, http://www.scribd.com/doc/2200541/ The-Power-of-Resonance
48. Donald O. Hebb, Hebbian Theory, http://www.scholarpedia.org/article/Donald_ Olding_Hebb
49. Morphic resonance and morphic fields, http://www.sheldrake.org/Articles&Papers/papers/morphic/morphic_intro.html
50. Definition of choice, http://dictionary.reference.com/browse/choice
51. National Centre for Biotechnology Information; the lunar cycle; effects on human and animal behavior and physiology, http://www.ncbi.nlm.nih.gov/pubmed/16407788

52. Wholeness, Interconnectedness and Co-creativity, http://www.co-intelligence.org/I-whole_interconn_cocreatv.html
53. The chrysalis age, http://www.thechrysalisage.com/excerptsintro02.html
54. Coevolution, http://biomed.brown.edu/Courses/BIO48/27.Coevolution.HTML
55. Creation, evolution and co-creation: the conscious nature of God and humans, http://www.stonyhill .com/articles/creation.htm
56. A summary of "The tipping point," http://www.wikisummaries.org/The_Tipping_Point
57. The butterfly effect, http://en.wikipedia.org/wiki/Butterfly_effect
58. John Casti's "Mood Matters", http://www.moodmatters.net/
59. Global coherence initiative's monitoring system, http://www.glcoherence.org/monitoring-system/about-system.html

CHAPTER 16: What Are the Signs of Our Next Evolution?

1. Discover presents Origins: Human Evolution, Summer 2010
2. Hominid Species Timeline, http://www.wsu.edu:8001/vwsu/gened/learn-modules/top_longfor/timeline/timeline.html
3. "Neanderthal genes survive in us," BBC, 6 May 2010, http://news.bbc.co.uk/2/hi/science/nature/8660940.stm
4. *The Future of the Body: Explorations into the Further Evolution of Human Nature*, by Michael Murphy, 1993, Tarcher
5. World Values Survey, http://www.worldvaluessurvey.org/
6. "Human Beliefs and Values," by Ronald Inglehart, 2004, Siglo XXI
7. "Climato-economic Roots of Survival versus Self-expression Cultures," by Evert Van de Vliert,
8. Social and Organizational Psychology, University of Groningen, *Journal of Cross-Cultural Psychology*, Vol. 38, No. 2, 156-172, 2007, http://jcc.sagepub.com/cgi/content/ abstract/38/2/156
9. World Database of Happiness, Trend in Nations, by R. Veenhoven, Erasmus University Rotterdam, www.worlddatabaseofhappiness.eur.nl/trendnat/framepage.htm
10. *Spiral Dynamics*, by Don Edward Beck and Christopher C. Cowan, 1996, 2006, Blackwell Publishing Ltd.
11. "The Integral Vision at the Millennium," by Ken Wilber, excerpted from "Introduction to the Collected Works of Ken Wilber, Vol. 7," 2000, Shambhala Publications, http://www.fudomouth.net/thinktank/now_integralvision.htm
12. "In Our Own Words 2000 Research Program", http://www.inourownwords.org/index.html
13. *The Cultural Creatives: How 50 Million People Are Changing The World*, by Paul H. Ray Ph.D. and Sherry Ruth Anderson, 2001, Three Rivers Press
14. Cultural Creatives, The Global Oneness Commitment, http://www.experiencefestival. com/cultural_creatives
15. "Cultural Creatives and LOHAS," by Nancy Nachman-Hunt, *Natural Business LOHAS Journal* March/April 2000, The Global Oneness Commitment, http://www.experiencefestival.com/a/Cultural_Creatives__LOHAS/id/9542

16. *The Evolution of Consciousness*, by Robert Ornstein, 1991, Touchstone, Simon and Schuster
17. *The Consciousness Revolution*, by Ervin Laszlo, Stanislav Grof and Peter Russell, 2003, Elf Rock Productions
18. Institute of Noetic Sciences, http://www.noetic.org/
19. *Discovery of the Presence of God*, by David R. Hawkins, M.D., Ph.D., 2006, Veritas Publishing
20. Transcending the Levels of Consciousness, by David R. Hawkins, M.D., Ph.D., 2006, Veritas Publishing
21. "A Research-Based Model of Consciousness Transformations," Cassandra Vieten, Ph.D., Shift No.23, Summer 2009, Institute of Noetic Sciences
22. *Living Deeply—The Art and Science of Transformation in Everyday Life*, by Marilyn Schlitz, Ph.D., Cassandra Vieten, Ph.D., Tina Amorok, Psy.D., 2007, New Harbinger Publications
23. *The Wisdom of the Enneagram*, by Don Riso and Russ Hudson, 1999, Bantam Books
24. "The Enneagram Development Guide", by Ginger Lapid-Bogda, Ph.D., TheEnneagramInBusiness.com
25. Vladimir Vernadsky, Biosphere and Noosphere, http://espg.sr.unh.edu/preceptorial/Summaries_2004/Vernadsky_Pap_ITru.html
26. Sri Aurobindo, http://en.wikipedia.org/wiki/Sri_Aurobindo
27. Teilhard de Chardin and Transhumanism, http://jetpress.org/v20/steinhart.htm
28. Alberto Villoldo, PhD., Homo Luminous, http://www.thefourwinds.com/resources-articles.php
29. *"I": Reality and Subjectivity* (2003), by Dr. David Hawkins, Veritas Publishing
30. Conscious evolution, http://ervinlaszlo.com/forum/2010/07/30/conscious-evolution-as-a-context-for-the-integration-of-science-and-spirituality/
31. *Encyclopedia of systemic Neuro-Linguistic Programming* (2000), by Robert Dilts and Judith Delozier, NLP University Press
32. *Einstein and Buddha* (2002), edited by Thomas J. McFarlane, Ulysses Press
33. "Rummaging for a Final Theory", *Scientific American Magazine*, September 2010
34. *We Are the Ones We Have Been Waiting for*, 2006, by Alice Walker, The New Press
35. The spirit of ma'at, message from Hopi Elders, http://www.spiritofmaat.com/messages/oct28/hopi.htm

CHAPTER 17: Are We Ready to Be Part of the Solution?

1. "Avatar a reality for Indian tribe fighting mining company," CNN, 9 February 2010, http://edition.cnn.com/2010/WORLD/asiapcf/02/09/india.avatar.tribe/index.html
2. "Falluja doctors report rise in birth defects," BBC, 4 March 2010, http://news.bbc.co.uk/2/hi/middle_east/8548707.stm
3. "Schools urge action on legal drug mephedrone," BBC, 17 Mar 2010, http://news.bbc.co.uk/2/hi/uk_news/8571599.stm
4. "Peru glacier collapses, 50 hurt," *The Straits Times*, 12 Apr 2010, http://www.straitstimes.com/BreakingNews/World/Story/STIStory_513504.html
5. "Web hit by hi-tech crime wave," BBC, 20 April 2010, http://news.bbc.co.uk/2/hi/technology/8630160.stm

6. "U.N. report: Eco-systems at 'tipping point'," CNN, 10 May 2010, http://edition.cnn.com /2010/WORLD/americas/05/10/biodiversity.loss.report/index.html?hpt=Sbin
7. "Pesticide link to ADHD in kids," *The Straits Times*, 18 May 2010, http://www.straitstimes.com/BreakingNews/TechandScience/Story/STIStory_ 528058.html
8. "More US kids on prescriptions," *The Straits Times*, 20 May 2010, http://www.straitstimes.com/BreakingNews/TechandScience/Story/STIStory_ 529068.html
9. "Artificial life breakthrough announced by scientists," BBC, 20 May 2010, http://news.bbc.co.uk/2/hi/science_and_environment/10134341.stm
10. "The Gulf's silent environmental crisis," CNN, 28 May 2010, http://edition.cnn.com/ 2010/US/05/28/gulf.oil.environment.disaster/index.html
11. "Costing the Earth", BBC Radio 4, http://www.bbc.co.uk/programmes/b006r4wn
12. "Material World", BBC Radio 4, http://www.bbc.co.uk/programmes/b006qyyb
13. Seawater Greenhouses, www.seawatergreenhouse.com
14. "Saltwater Saviour," by Anne Casselman, *Discovery Channel Magazine*, March 2010
15. "A skyscraper designed to make a rotten river run clean," CNN, 6 May 2010, http://edition.cnn.com/2010/WORLD/asiapcf/05/05/jakarta.skyscraper.ciliwung/ index.html
16. "Going Green", CNN complete coverage on Environment, http://edition.cnn.com/ SPECIALS/2009/environment/
17. "Can we feed the world and sustain the planet?", *Scientific American Magazine*, November 2011
18. The Story of Stuff Project, http://storyofstuff.org
19. *The Story of Stuff*, by Annie Leonard, 2010, Free Press
20. *Diet for a Hot Planet*, by Anna Lappe, 2010, Bloomsbury USA
21. Muhammad Yunus and Grameen Bank, Grameen Foundation website, http://www.grameenfoundation.org/who-we-are/awards-and-recognition
22. *Three Cups of Tea*, by Greg Mortenson and David Oliver Relin, 2007, Penguin Books
23. Tumbleweed Tiny House Company, http://www.tumbleweedhouses.com/
24. *Blessed Unrest*, by Paul Hawken, 2007, Penguin Books
25. WiserEarth, online community space connecting people, nonprofits and businesses working toward a just and sustainable world, http://www.wiserearth.org/
26. Earth Hour, http://www.earthhour.org/Homepage.aspx?intro=no
27. Millions worldwide turn off their lights for Earth Hour 2010, CNN, 28 March 2010, http://edition.cnn.com/video/#/video/world/2010/03/28/nat.earth.hour.cnn
28. Avatar Movie, by James Cameron, 2009, http://www.avatarmovie.com/
29. Avatar, Box Office Mojo, http://boxofficemojo.com/movies/?id=avatar.htm
30. "Avatar breaks US DVD sales record," CNN, 26 April 2010, http://news.bbc.co.uk/2/hi/ entertainment/8643539.stm

31. "Audiences experience 'Avatar' blues," CNN, 11 January 2010, http://edition.cnn.com/2010/SHOWBIZ/Movies/01/11/avatar.movie.blues/index.html
32. Earth Day Santa Barbara 2010 James Cameron Avatar Honored by CEC SB Breaking News, 18 April 2010, http://www.youtube.com/watch?v=pwkESMbH1tc
33. Earth Day 2010, Earthday network, http://ww2.earthday.net/earthday2010
34. "James Cameron joins real-life 'Avatar' battle," CNN, 21 April 2010, http://edition.cnn.com/2010/SHOWBIZ/04/20/james.cameron.rain.forest/index.html
35. *The Power of Apology*, by Beverly Engel, 2001, John Wiley & Sons, Inc.
36. John Denver, his career, political and humanitarian work, http://en.wikipedia.org/wiki/John_Denver

CHAPTER 18: Be the Solution in Our World—Today!

1. I to WE: "The role of consciousness transformation in compassion and altruism", http://onlinelibrary.wiley.com/doi/10.1111/j.1467-9744.2006.00788.x/abstract
2. Developing a "WE" consciousness in 2010, http://www.huffingtonpost.com/natasha-dern/developing-a-we-conscious_b_406230.html
3. One world: environmental activism guide, http://uk.oneworld.net/guides/environmentalactivism
4. Concepts and issues of sustainability in countries in transition, ftp://ftp.fao.org/seur/ceesa/concept.htm
5. Ecology, http://en.wikipedia.org/wiki/Ecology
6. *Encyclopedia of Systemic Neuro-Linguistic Programming and the NLP New Coding*, 2000, by Robert. B. Dilts, NLP University Press
7. Our crisis of meaning, http://globaldialoguecenter.blogs.com/meaning/2007/12/our-crisis-of-m.html
8. Global depression statistics, http://www.sciencedaily.com/releases/2011/07/110725202240.htm
9. "Simplifying your life", http://fatherhood.about.com/od/workingfathers/a/simplify_2.htm
10. 10 most important things to simplify in life, http://www.becomingminimalist.com/the-10-most-important-things-to-simplify-in-your-life/
11. Reframing, http://www.successmeasures.com/reframing.htm
12. "The power of mind and the promise of placebo", http://www.wrf.org/alternative-therapies/power-of-mind-placebo.php
13. Wants vs. needs, http://frugalliving.about.com/od/frugalliving101/qt/ Wants_vs_Needs.htm
14. Difference between a want and a need, http://www.differencebetween.net/language/difference-between-a-want-and-a-need/
15. Congruence, http://en.wikipedia.org/wiki/Congruence
16. Embodied cognition, http://en.wikipedia.org/wiki/Embodied_cognition
17. Embodying the change you want to see in the world, http://thebrightarmy.com/you-must-be-the-change-you-want-to-see-in-the-world/
18. Resonance, http://en.wikipedia.org/wiki/Resonance

19. Morphic resonance and morphic fields, http://www.sheldrake.org/Articles&Papers/papers/morphic/morphic_intro.html
20. "What is intention?" http://life.gaiam.com/article/what-intention
21. *Living deeply*, 2007, by Marilyn M. Schlitz Ph.D., Cassandra Vieten Ph.D., Tina Amorok, Psy.D., New Harbinger Publications
22. *Entangled minds*, 2006, by Dean Radin Ph.D., Paraview Pocket Books
23. The Higgs Boson 'God Particle' discovery explained in the context of conscious cosmology, http://www.naturalnews.com/036428_consciousness_intention_Higgs _boson.html
24. Standard Model—Particle Physics, http://en.wikipedia.org/wiki/Standard_Model
25. "What is Self Transcendence?", http://www.selftranscendence.org/self_transcendence/what_is_self_transcendence/
26. "Spirituality at Work", by Seth Wax, https://docs.google.com/viewer?a=v&q=cache:UupVDiuLB3QJ:pzweb.harvard.edu/ebookstore/pdfs/goodwork41.pdf+seth+wax+spirituality+at+work
27. Finding God in our work, http://kensmessage.blogspot/2011/09/finding-god-in-our-work2011sep11.html
28. The mirror of relationships, http://www.chopra.com/relationship
29. What are our mirror reflections trying to teach us?, http://healing.about.com/od/selfpower/a/mirror-images.htm
30. U.N.: "Meat consumption must be cut to reduce greenhouse gases", http://www.enn.com/top_stories/article/44270
31. Cut down on meat to save planet (IPCC), http://www.thenational.ae/news/uae-news/environment/cut-down-on-meat-to-save-planet
32. Happiest countries in the world 2012 (Happy Planet Index), http://www.nowpublic.com/world/happiest-countries-world-2012-happy-nations-complete-list-2946693.html
33. Singapore is the richest country per capita, http://www.asianewsnet.net/news-35053.html
34. Singapore household expenditure survey (HES) 2007/2008, http://www.singstat.gov.sg/pubn/hhld.html#hes
35. U.N.: "Global resource consumption to triple by 2050", http://phys.org/news/2011-05-global-resource-consumption-triple.html
36. W.W.F.: Radical clean energy shift save 4 trillion euros, http://phys.org/news/2011-02-radical-energy-shift-tn-euros.html
37. How to reduce energy consumption, http://www.nrdc.org/air/energy/genergy.asp
38. Reduce consumption, waste and pollution, http://www.startupnation.com/steps/93/9024/2/1/reduce-consumption-waste-pollution.htm
39. Hybrid cars and alternate fuels, http://alternativefuels.about.com/od/2008ngvavailable/a/2008CNGvehicles.htm
40. Humanitarianism, http://en.wikipedia.org/wiki/Humanitarianism
41. The humanitarian forum, http://www.humanitarianforum.org/
42. "Animals are conscious and should be treated as such", http://www.newscientist.com/article/mg21528836.200-animals-are-conscious-and-should-be-treated-as-such.html

43. Examples of humanitarian acts, http://examples.yourdictionary.com/examples/examples-of-humanitarian-acts.html
44. Take action to help the environment/correct social injustice, http://www.globalshift now.com/practices_actions_pgs/take_action.html
45. Connecting health and environment, http://www.guardian.co.uk/sustainable-business/ connecting-health-and-environment
46. The healing power of a walk in the woods, http://eartheasy.com/blog/2011/07/the-healing-power-of-a-walk-in-the-woods/
47. Guerrilla gardening, http://www.guerrillagardening.org/
48. Teaching as a vocation, http://tanabok.hubpages.com/hub/teaching-as-a-vocation
49. "Mind: The many faces of happiness", *Scientific American*, September /October 2011
50. Income influence on happiness, http://wws.princeton.edu/news/Income_Happiness/
51. *Global Shift (Finding your unique purpose)*, 2009, by Edmund Bourne, Ph.D., New Harbinger Publications
52. International ranking of Singapore, http://en.wikipedia.org/wiki/International_rankings_ of_Singapore
53. Global footprint network, http://www.footprintnetwork.org/en/index.php/GFN/page/ world_footprint/
54. High-rise Farming, *Discovery Channel Magazine*, Volume 1, 2008
55. High-rise urban farms of the future, http://www.greenbiz.com/news/2009/11/24/high-rise-urban-farms-future
56. Vertical farming, http://en.wikipedia.org/wiki/Vertical_farming
57. Vertical farming in Singapore, http://skygreens.appsfly.com/Media
58. Happiness economics, http://en.wikipedia.org/wiki/Happiness_economics
59. Will global happiness index ever beat out GDP?, http://business.time.com/2011/05/24/is-a-global-happiness-index-on-the-horizon/
60. "Happiness is the ultimate economic indicator", by Richard Heinberg, http://www.fastco exist.com/1679289/happiness-is-the-ultimate-economic-indicator
61. The Genuine Progress Indicator, http://www.sustainwellbeing.net/gpi.html
62. Genuine Progress Indicator, http://en.wikipedia.org/wiki/Genuine_progress_indicator
63. Measuring prosperity: Maryland's GPI, http://www.thesolutionsjournal.com/node/1070
64. "The 2008 Shift Report", by the Institute of Noetic Sciences
65. State of the world 2008, Innovations for a sustainable economy, http://www.worldwatch. org/files/pdf/SOW08_chapter_2.pdf
66. The only way out is in, so look deep within, http://articles.timesofindia.indiatimes.com/ 2010-07-13/holistic-living/28300421_1_joy-life-experience
67. *Jesus, Buddha, Krishna and Lao Tzu: The parallel sayings*, 2007, by Richard Hooper

About the Authors

Barney Wee is the founder of Mind Transformations, an established training, consulting, and coaching company in Southeast Asia, specialised in Neuro-Linguistic Programming, Enneagram, and Quantum Psychology. His expertise has been featured in Singapore's newspapers, radio and TV networks. On the international stage, his ideas on NLP were introduced in the NLP Now Telesummit and the Happiness at Work Telesummit.

Barney is also an Organisational Development Consultant and he has worked with several large organisations, including the Basic Military Training Centre of Singapore, National Healthcare Group Polyclinics, DHL, DBS, Elsevier, Atos Origin, BBC Worldwide Distribution, the National University of Singapore, and many more.

This book is the result of Barney's extensive research and application of the information into all aspects of his life; he and this book are the living proof that one does not need to be a politician, scientist, billionaire or an expert in ecology to educate oneself, take a stand, and make a positive difference in the world. Barney has set an example for those who want to make the world a better place. He is showing everyone that one doesn't have to wait for some external authority to take the lead and make a difference. Regardless of education and professional background, *anyone* can be life-supportive.

Barney is now part of the solution to our global issues by living a life-supportive existence and striving to educate and lead others into a new living paradigm. The motto that has been the driving force in his life is, "In the power to change oneself, is the same power to change the world."—Anwar Sadat.

Agnes Lau is a Director, Trainer, NLP & Enneagram Coach of Mind Transformations, and Barney's business partner. Her passion and expertise is in coaching, personality profiling and complementary health, where she focuses on increasing one's overall wellbeing holistically.

She has been featured in Singapore's magazines, radio and TV networks, and worked with many organisations including the Institute of Public Administration & Management, Armstrong Industrial Corporation Ltd, Standard Chartered Bank, Rio Tinto Minerals, Rohde & Schwarz Systems & Communications, Asia Business Forum and more.

Previously, Agnes enjoyed a career in research in the civil service and corporate sectors, with a Bachelor of Science (Hons) degree from University College London, U.K., and an M.B.A. from Monash University, Australia. In 1997, she left to find true meaning and purpose in life. Now she feels blessed to be able to share her body-mind-spirit modalities, awakening people to a higher consciousness and connecting with their multi-dimensional selves.

This book gives Agnes the avenue to reach a wider audience, to connect the current reality with multiple ways one can choose to be life-supportive, and the reasons why personal responsibility for our collective future is in everyone's hands. She believes that our innate desire for a better world does not have to be a wish; it can be expressed through intelligent and practical choices in our daily lives. This book is Agnes's way to share her beliefs and practices on how she strives to make the world more harmonious, sustainable and humane.

CPSIA information can be obtained at www.ICGtesting.com
Printed in the USA
LVOW07s1533111214

418356LV00003B/483/P

9 781938 459535